A TEXTBOOK OF
B.Sc. MATHEMATICS

For Third Semester
Gulbarga University
(As per New Syllabus)

AF661101

G.K. RANGANATH
Formerly Professor of Mathematics
AES National College
Gauribidanur-561 208

S. CHAND
PUBLISHING
empowering minds

S. CHAND & COMPANY PVT. LTD.
(AN ISO 9001 : 2008 COMPANY)
RAM NAGAR, NEW DELHI - 110 055

S. CHAND & COMPANY PVT. LTD.

S. CHAND PUBLISHING
empowering minds

(An ISO 9001 : 2008 Company)

Head Office: 7361, RAM NAGAR, NEW DELHI - 110 055
Phone: 23672080-81-82, 9899107446, 9911310888 Fax: 91-11-23677446
Shop at: **schandgroup.com**; e-mail: **info@schandgroup.com**

Branches :

AHMEDABAD : 1st Floor, Heritage, Near Gujarat Vidhyapeeth, Ashram Road, **Ahmedabad** - 380 014, Ph: 27541965, 27542369, ahmedabad@schandgroup.com

BENGALURU : No. 6, Ahuja Chambers, 1st Cross, Kumara Krupa Road, **Bengaluru** - 560 001, Ph: 22268048, 22354008, bangalore@schandgroup.com

BHOPAL : Bajaj Tower, Plot No. 2&3, Lala Lajpat Rai Colony, Raisen Road, **Bhopal** - 462 011, Ph: 4274723, 4209587. bhopal@schandgroup.com

CHANDIGARH : S.C.O. 2419-20, First Floor, Sector - 22-C (Near Aroma Hotel), **Chandigarh** -160 022, Ph: 2725443, 2725446, chandigarh@schandgroup.com

CHENNAI : No.1, Whites Road, Opposite Express Avenue, Royapettah, **Chennai** - 600014 Ph. 28410027, 28410058, chennai@schandgroup.com

COIMBATORE : 1790, Trichy Road, LGB Colony, Ramanathapuram, **Coimbatore** -6410045, Ph: 2323620, 4217136 coimbatore@schandgroup.com **(Marketing Office)**

CUTTACK : 1st Floor, Bhartia Tower, Badambadi, **Cuttack** - 753 009, Ph: 2332580; 2332581, cuttack@schandgroup.com

DEHRADUN : 1st Floor, 20, New Road, Near Dwarka Store, **Dehradun** - 248 001, Ph: 2711101, 2710861, dehradun@schandgroup.com

GUWAHATI : Dilip Commercial (Ist floor), M.N. Road, Pan Bazar, **Guwahati** - 781 001, Ph: 2738811, 2735640 guwahati@schandgroup.com

HALDWANI : Bhatt Colony, Talli Bamori, Mukhani, **Haldwani** -263139 **(Marketing Office)** Mob. 09452294584

HYDERABAD : Padma Plaza, H.No. 3-4-630, Opp. Ratna College, Narayanaguda, **Hyderabad** - 500 029, Ph: 27550194, 27550195, hyderabad@schandgroup.com

JAIPUR : 1st Floor, Nand Plaza, Hawa Sadak, Ajmer Road, **Jaipur** - 302 006, Ph: 2219175, 2219176, jaipur@schandgroup.com

JALANDHAR : Mai Hiran Gate, **Jalandhar** - 144 008, Ph: 2401630, 5000630, jalandhar@schandgroup.com

KOCHI : Kachapilly Square, Mullassery Canal Road, Ernakulam, **Kochi** - 682 011, Ph: 2378740, 2378207-08, cochin@schandgroup.com

KOLKATA : 285/J, Bipin Bihari Ganguli Street, **Kolkata** - 700 012, Ph: 22367459, 22373914, kolkata@schandgroup.com

LUCKNOW : 1st Floor, Arya Pratinidhi Sabha, 5-Mirabai Marg, Near Narhi Sabzi Mandi, Hazratganj, Lucknow - 226 001, Ph: 4065646, 4026791, 4027188, 4022948 lucknow@schandgroup.com

MUMBAI : Blackie House, IInd Floor, 103/5, Walchand Hirachand Marg, Opp. G.P.O., **Mumbai** - 400 001, Ph: 22690881, 22610885, mumbai@schandgroup.com

NAGPUR : Karnal Bagh, Near Model Mill Chowk, **Nagpur** - 440 032, Ph: 2720523, 2777666 nagpur@schandgroup.com

PATNA : 104, Citicentre Ashok, Mahima Palace , Govind Mitra Road, **Patna** - 800 004, Ph: 2300489, 2302100, patna@schandgroup.com

PUNE : Sadguru Enclave, Ground floor, Survey No. 114/3, Plot no. 8 Alandi Road , Vishrantwadi **Pune** – 411015 Ph: 64017298 pune@schandgroup.com

RAIPUR : Kailash Residency, Plot No. 4B, Bottle House Road, Shankar Nagar, **Raipur** - 492 007, Ph: 2443142,Mb. : 09981200834, raipur@schandgroup.com **(Marketing Office)**

RANCHI : Shanti Deep Tower, Opp.Hotel Maharaja, Radium Road, **Ranchi**-834001 Mob. 09430246440 ranchi@schandgroup.com

SILIGURI : 122, Raja Ram Mohan Roy Road, East Vivekanandapally, P.O., Siliguri, **Siliguri**-734001, Dist., Jalpaiguri, (W.B.) Ph. 0353-2520750 **(Marketing Office)** siliguri@schandgroup.com

VISAKHAPATNAM : No. 49-54-15/53/8, Plot No. 7, 1st Floor, Opp. Radhakrishna Towers, Seethammadhara North Extn., **Visakhapatnam** - 530 013, Ph-2782609 (M) 09440100555, visakhapatnam@schandgroup.com **(Marketing Office)**

First Edition 1990

Reprints 1992, 93, 94, 95 (Twice), 96, 98, 99, 2000, 2001, 2004, 2005, 2006, 2007, 2008, 2011,

Revised Edition 2012, Reprint 2013

Second Revised and Enlarged Edition 2015

ISBN : 97-89-3856-7600-0 **Code :** 1014B 556

PRINTED IN INDIA

By Vikas Publishing House Pvt. Ltd., Plot 20/4, Site-IV, Industrial Area Sahibabad, Ghaziabad-201010 and published by S. Chand & Company Pvt. Ltd., 7361, Ram Nagar, New Delhi -110 055.

PREFACE

Recently, the course in Mathematics for the twelfth standard of Karnataka state has been revised and therefore, it was necessary to restructure the syllabus for the three-year degree course of Gulbarga University. The revised syllabus has been introduced for the semester system of B.Sc. course, effective from 2014-15 onwards.

This present book, *A Textbook of B.Sc. Mathematics*, covers the syllabus for the third semester, which includes chapter on Group, Rings, Line and Multiple Integrals, Riemann Integration and Ordinary Differential Equations.

An attempt has been made to present the subject matter in as precise and concise a manner as possible. It is earnestly hoped that community of teachers and students will find the book useful to them.

I am highly thankful to M/s. S. Chand & Company Pvt. Ltd., New Delhi for publishing this book.

Suggestions for the improvement of the book are most welcome.

G.K. Ranganath

Syllabus

BSM–3.1

Groups, Rings, Line and Multiple Integral

Introduction to Groups

Binary operation, Group and Sub-Group. The necessary and sufficient condition for a sub-group. Properties of groups and examples.

Order of an element of a group. Properties of the order of an element

Cycle group – Definition and properties, Cosets definition and properties. Index of a sub group. Language's Theorem Consequences of Lagranger's Theorem Euler's phi function and Format's Theoram.

Normal sub groups of a group - Equivalent versions of the criterion for a sub group to be normal sub group - Factor groups – problems illustrating the concept and results. Homomorphism, Isomorphism of groups, Kernal, normality fundamental theorem of homomorphism.

Permutation groups of degree *n*. Cayley's theorem – problems illustrating concept and results

Normality of the Kernal – Fundamental Theorem of homomorphism. Iso morphism theorems – permutation group of degree '*n*' Cayleys theorem – problems illustrating the concept and results

(30 hrs)

Rings

Definition of Ring, Subrings definition and properties, integral domain, field and ideals of rings properties and examples

(10 hrs)

Line and multiple integral

Line integral, basic properties and examples

Double and Triple integrals, Evaluation of double integrals, and triple integrals (1) under given limits (2) in regions bounded by giver curves. Change of variables to polar, cylindrical and spherical polar co-ordinates examples thereon.

(24 hrs)

BSM–3.2

Riemann Integration and Ordinary Differential Equations

Riemann Integrability:

Real valued functions of single real variable – Integral primitive of function – Inequalities. Absolute values, lower and upper bounds – Bounded function (re-capitulation) – lower and upper sums – lower and upper Riemann integrals – necessary and sufficient conditions for Remann Intigrability – Reimann Integral Darboux's theorem (statement only) Integrability of (a) continuous function (b) monotonic function. Properties of Integrable functions – Integrability of sum, difference, Integrability of modules of an integrable function – Fundamental theorem of integral calculus - continuity and derivability of integrable function – First and second mean value theorem of integral calculus. Leibnitz's result to evaluate the examples of differentation under the integral sign. (24 hrs)

Ordinary Differential Equations

Differential equations of first order and higher degree. Equations solvable for p, x, y – Clairault's equations. General and singular solutions Linear Differential Equation of finding complementary function – Non-homogeneous differential equation – Methods of finding particular integrals- When RHS function is of the form $e^{ax} x^n$, $\sin ax$, $\cos ax$, $e^{ax}F$ where F is a function of x – Cauchy Euler Differential Equations of order two – Simultaneous differential equations – Linear differential equations of IInd order with variable coefficients – Solutions by the following methods

(a) When a part of CF is given

(b) Changing the independent variable

(c) Changing the dependent variable

(d) When first integral is given

(e) Variation of parameters (40 hrs)

Semester –III

Paper BSM 3.1 – Groups, Rings, Line and Multiple Integrals

Section – A

Answer Any Ten of the following. Marks $10 \times 2 = 20$

1–6 Groups

7–8 Rings

9-12 Line and Multiple Integrals

Section – B

Answer Any Five of the following. Marks $5 \times 5 = 25$

1–6 Groups

7–8 Rings

Section – C

Answer Any Three of the following. Marks $3 \times 5 = 15$

1–5 Line and Multiple Integral.

Semester – III

BSM 3.2 - Riemann Integration and Ordinary Differential Equations

Section–A

Answer Any Ten of the following: Marks $10 \times 2 = 20$

1–3 Riemann Integration

4-6 Ordinary Differential Equation (up to singular solutions)

7-10 Linear Differential Equations of second order with variable coefficients

11-12 Solutions by the different methods

Section–B

Answer Any Three of the following: Marks $3 \times 5 = 15$

1–5 Riemann Integration

Section–C

Answer Any Ten of the following: Marks $5 \times 5 = 25$

1–2 Ordinary Differential Equations (upto singular solutions)

3–6 Linear Differential Equations (upto simultaneous differential equations)

7–8 Solutions by the different methods

CONTENTS

BSM — 3.1

BSM – 3.2

BSM – 3.1

Chapter 1. Groups
Chapter 2. Rings, Integral Domain and Fields
Chapter 3. Line and Multiple Integral

1 Groups

1.1 Introduction

In this chapter we shall see an algebraic structure with one binary operation called Group and some of the properties associated with it.

We use the following notations to represent the set of numbers mentioned below.

R = **Set of all real numbers**

Q = **Set of all rational numbers**

R^* = **Set of all non zero real numbers**

Q^* = **Set of all non zero rational numbers**

Q' = **Set of all irrational numbers**

N = **Set of all natural numbers**

Z or I = **Set of all integers**

C = **Set of all complex numbers**

C^* = **Set of all non zero complex numbers.**

1.2 Binary Operations (Compositions)

Definition : Let S be a non empty set. The mapping, $* : S \times S \to S$, which associates each ordered pair (a, b) of the elements of S to a unique element of S, denoted by $a * b$, is called a binary operation or a binary composition on S.

From the above definition, it follows that an operation $*$ on a non empty set S is a binary operation or a binary composition if and only if

$$\forall a \in S, b \in S \Rightarrow a * b \in S$$

If $*$ is a binary operation defined on the set S, then we say **"S is closed under binary operation $*$"** or simply **"S is closed under $*$".**

Examples :

1. The usual addition (+) and multiplication (×) are binary operations on the sets of **(i)** natural numbers, **(ii)** Z of integers, **(iii)** Q of rationals, **(iv)** R of real numbers and **(v)** C of complex numbers.
2. The usual subtraction is not a binary operation on the set N of natural numbers, for $3 \in N, 5 \in N$, but $3 - 5 = -2 \notin N$. Whereas subtraction is a binary operation on the sets Z, Q and R.
3. The usual division is not a binary operation on the set R and Q. However, it is a binary operation on $R*$ and $Q*$.

4. The matrix addition is a binary operation on the set of all matrices of order $m \times n$, over the set R, N, Q or C.
5. The matrix multiplication is a binary operation on the set M_n of all $n \times n$ matrices.

1.3 Laws of Binary Operation

A Binary operation $*$ on a non–empty set S is said

(a) to be commutative, if $a * b = b * a \quad \forall\, a, b \in S$

(b) to be associative, if $a * (b * c) = (a * b) * c \quad \forall\, a, b, c \in S.$

(c) The set S is said to have an **identity element** w.r.t * if there exist an element $e \in S$, such that

$$e * a = a * e = e \qquad \forall\, a \in S$$

(d) If e is an identity element with respect to $*$, then an element $a \in S$ is said to have **an inverse,** if there exists an element $b \in S$ such that

$$a * b = b * a = e$$

and in this case we say ***b* is an inverse of *a*.**

Example 1. A binary operation $*$ is defined on the set Z of integers by $a * b = 1 + ab$, $\forall\, a, b \in Z$. Show that $*$ is commutative but not associative.

Solution : By definition, $a * b = 1 + ab$

Now, $b * a = 1 + ba = 1 + ab$ (usual multiplication is commutative)

$\therefore$ $b * a = a * b \Rightarrow *$ **is commutative.**

Let a, b, c be any three elements of Z.

Now, $a * (b * c) = a * (1 + bc) = 1 + a(1 + bc) = 1 + a + abc$

Now, $(a * b) * c = (1 + ab) * c = 1 + (1 + ab) . c = 1 + c + abc$

Clearly, $a * (b * c) \neq (a * b) * c$

Thus, $*$ is not associative.

Example 2. A binary operation $*$ is defined on the set Z of integers by $a * b = a + b - 1$. Show that 1 is an identity element and inverse of 2 is 0.

Solution : By data, $a * b = a + b - 1$, $\forall\, a, b \in Z$.

Let $c \in Z$. Now, consider, $c * 1 = c + 1 - 1 \quad \Rightarrow c * 1 = c$

Again, $1 * c = 1 + c - 1 \quad \Rightarrow 1 * c = c$

Thus, 1 acts as identity element in Z w.r.t. $*$.

Let, $a * 2 = 1 \quad \Rightarrow a + 2 - 1 = 1 \quad \Rightarrow a + 2 = 2 \quad \Rightarrow a = 0$

Again, $2 * a = 1 \quad \Rightarrow 2 + a - 1 = 1 \quad \Rightarrow a + 2 = 2 \quad \Rightarrow a = 0$

Thus, **the inverse of 2 is 0.**

Example 3. In the set $R*$ of non zero real numbers the binary operations $*$ is defined by $a * b = \dfrac{ab}{2}, \forall\, a, b \in R$. (i) Find the identity element, (ii) Find the inverse of the element 4.

Solution : By data, $a * b = \dfrac{ab}{2}$, $\forall\, a, b \in R$

(i) Let, $a * b = a \Rightarrow \dfrac{ab}{2} = a \Rightarrow b = 2 \qquad (\because a \neq 0)$

Again, $b * a = a \Rightarrow \dfrac{ba}{2} = a \Rightarrow b = 2 \qquad (\because a \neq 0)$

Thus, the element **2 acts as an identity element.**

(ii) Let x be the inverse of 4 under the binary operation *.

$$\Rightarrow \qquad x * 4 = 4 * x = 2 \Rightarrow \frac{4x}{2} = 2 \Rightarrow x = 1$$

Thus, **1 acts as a inverse of 4.**

Example 4. Find the identity in the set of rationals except 1 w.r.t. $*$ which is defined by

$$\boldsymbol{a * b = a + b - ab.}$$

Solution : By data, $a * b = a + b - ab, a \neq 1, b \neq 1$. Let e be the identity element. Then

$$\begin{aligned} a * e = e * a = a \quad &\Rightarrow \quad a + e - ae = a \\ &\Rightarrow \quad e(1 - a) = 0 \\ &\Rightarrow \quad e = 0 \text{ or } 1 - a = 0 \\ &\Rightarrow \quad e = 0 \text{ or } a = 1 \Rightarrow e = 0 \qquad (\because a \neq 1) \end{aligned}$$

Thus, **the identity element is 0.**

EXERCISES

1. On the set Q^+ of all positive rational numbers the binary operation $*$ is defined by $a * b = \frac{ab}{2}$. Show that $*$ is both commutative and associative. Find the identity element in Q^+.
2. A binary operation $*$ on the set R of real numbers except 1 is defined by $a * b = a + b - ab$, $\forall a, b \in R$. Show that the identity element is 0 and the inverse of 2 is 2.
3. A binary operation $*$ on the set $R*$ of non zero real numbers is defined by $a * b = \frac{ab}{3}$, $\forall a, b \in R*$. Show that the inverse of 9 is 1.
4. A binary operation $*$ on the set Z of all integers is defined by $a * b = 3a + 2b, \forall a, b \in Z$, show that $a * b = a * c \Rightarrow b = c$.
5. A binary operation $*$ on the set Z of all integers is defined by $a * b = a + b + 1$. Show that the identity element is -1 and the inverse of -2 is 0.
6. If in the set of real numbers other than -1, the operation $*$ is defined by $a * b = a + b + ab, \forall a, b \in R - \{-1\}$, show that the identity element is 0 and the inverse of 4 is $-\frac{4}{5}$.
7. If the binary operation $*$ on the set R of real numbers is defined by $a * b = a + b - 7$, $\forall a, b \in R$, show that $*$ is commutative, associative, the identity element is 7. Also find the inverse of 14.
8. Show that the binary operation $*$ on the set Q of rationals by $a * b = \frac{ab}{7}$, is both commutative and associative.

1.4 Algebraic Structures

Definition : A non empty set G together with one or more binary compositions on G is called an algebraic structure or an algebraic system.

For example, each of the following is an algebraic structure.

(i) $(N, +)$ **(ii)** $(Q, +, \times)$ **(iii)** $(R, +)$ **(iv)** $(Z, +)$

We shall see the algebraic structure with one binary composition called a **group** in this book.

Group

Definition : A system $(G, *)$, where G is a non empty set and $*$ is a binary operation on G, is called a group, if it satisfies the following axioms.

1. Associative law: $a * (b * c) = (a * b) * c \quad \forall\ a, b, c \in G$

2. Existence of identity: There exists an element $e \in G$, such that

$$a * e = e * a = a \qquad \forall\, a \in G$$

e is called an identity element.

3. Existence of inverse: For every element $a \in G$, there exists an element denoted by $a^{-1} \in G$ such that $a * a^{-1} = a^{-1} * a = e$, a^{-1} is called inverse element of a.

Note : The element a^{-1} does not mean $\frac{1}{a}$.

Definition : A group $(G, *)$ is called an Abelian group or a commutative group, if

$$a * b = b * a,\ \forall\ a, b \in G$$

This is called commutative law of a group.

If the system $(G, *)$ is a group, we say that G is a group under binary operation *.

Definition : If the number of elements in the group $(G, *)$ is finite then it is called finite group; otherwise it is called an infinite group.

Definition (Semi group) : A system $(G, *)$ where G is a non empty set and $*$ is a binary operation on G, is called a semi group if it satisfies the following axiom.

(1) Associative Law : $a * (b * c) = (a * b) * c,\ \forall\ a, b, c \in G$

Definition (Order of a group) : The number of elements in a group G is called the order of the group and denoted by $O(G)$.

1.5 Properties of a Group

In this section we shall see few elementary properties of a Group in the form theorems.

Theorem 1 : The identity element of a group is unique.

Proof : Let $(G, *)$ be a group. If possible let e and e' be two identity elements.

Since e is an identity element and $e' \in G$ we have,

$$e * e' = e' \quad \text{and} \quad e' * e = e' \qquad \ldots(1)$$

Again e' is an identity element and $e \in G$ we have,

$$e' * e = e \quad \text{and} \quad e * e' = e \qquad \ldots(2)$$

From (1) and (2) it follows $e = e'$.

Hence, **the identity element in a group is unique.**

Theorem 2 : The inverse of an element of a group is unique.

Proof : Let e be the identity element and a be an element of the group $(G, *)$.

If possible let b and c be two inverses of a in G.

Since b is an inverse of a, we have, $a * b = b * a = e$ ...(1)

Again as c is an inverse of a, we have, $a * c = c * a = e$ ...(2)

Now, $b = b * e$ ($\because$ e is the identity element)

$\Rightarrow$ $b = b * (a * c)$ (by (2))

$\Rightarrow$ $b = (b * a) * c$ (by associative law)

$\Rightarrow$ $b = e * c$ (by (1))

$\Rightarrow$ $b = c$ ($\because$ e is the identity element)

Hence, **the inverse of an element is unique.**

Theorem 3 : If a is an element of the group $(G, *)$ then $(a^{-1})^{-1} = a$.

Proof : Let $x = a^{-1}$. We shall show that $x^{-1} = a$.

To show that $x^{-1} = a,$ we have to establish $a * x = x * a = e.$

Now consider, $a * x = a * a^{-1} = e$ ...(1)

Again, $x * a = a^{-1} * a = e$ ...(2)

(1) and (2) $\Rightarrow$ $x^{-1} = a \Rightarrow (a^{-1})^{-1} = a$ $(\because x = a^{-1})$

Theorem 4 : Let $(G, *)$ be a group and $a, b \in G$. Then, $(a * b)^{-1} = b^{-1} * a^{-1}$.

Proof : We have to show that $b^{-1} * a^{-1}$ is the inverse of $a * b$.

Let, $x = b^{-1} * a^{-1}$ and $y = a * b$

Consider, $x * y = (b^{-1} * a^{-1}) * (a * b)$

$\Rightarrow$ $x * y = b^{-1} * (a^{-1} * a) * b$ (by associative law)

$\Rightarrow$ $x * y = (b^{-1} * e) * b$ $(\because a^{-1} * a = e)$

$\Rightarrow$ $x * y = b^{-1} * b$ $(\because$ e is the identity)

$\Rightarrow$ $x * y = e$ ...(1)

Again consider, $y * x = (a * b) * (b^{-1} * a^{-1})$

$\Rightarrow$ $y * x = a * (b * b^{-1}) * a^{-1}$ (by associative law)

$\Rightarrow$ $y * x = (a * e) * a^{-1}$ $(\because b^{-1} * a = e)$

$\Rightarrow$ $y * x = a * a^{-1}$ $(\because$ c is the identity)

$\Rightarrow$ $y * x = e$ ...(2)

(1) and (2) $\Rightarrow$ x is the inverse of y

$\Rightarrow$ $x = y^{-1} \Rightarrow b^{-1} * a^{-1} = (a * b)^{-1}$ $\quad \therefore$ $(a * b)^{-1} = b^{-1} * a^{-1}$

This result is also called **"reversal law of inverse"**.

Note: 1. The above result can be extended to any finite number of elements $(G, *)$.

$$(a * b * c * .....)^{-1} = (..... c^{-1} * b^{-1} * a^{-1}), \text{ where } a, b, c, ... \in G$$

2. If $(G, *)$ is an Abelian group, then we have,

$$a * b = b * a, \forall a, b \in G$$

As $a^{-1}, b^{-1} \in G$, we have, $a^{-1} * b^{-1} = b^{-1} * a^{-1}$

$\Rightarrow$ $(a * b)^{-1} = b^{-1} * a^{-1} = a^{-1} * b^{-1}$

Theorem 5 : In a group $(G, *)$ for all $a, x, y \in G$.

(i) $a * x = a * y \Rightarrow x = y$ **(left cancellation law)**

(ii) $x * a = y * a \Rightarrow x = y$ **(right cancellation law)**

Proof : **(i)** By data, $a * x = a * y$

$\Rightarrow$ $a^{-1} * (a * x) = a^{-1} * (a * y)$ (operating a^{-1} on both sides)

$\Rightarrow$ $(a^{-1} * a) * x = (a^{-1} * a) * y$ (by associative law)

$\Rightarrow$ $e * x = e * y$ $(\because a^{-1} * a = e)$

$\Rightarrow$ $x = y$ $(\because$ e is the identity)

Hence, $\boldsymbol{a * x = a * y \Rightarrow x = y}$

(ii) By data, $x * a = y * a$

$\Rightarrow$ $(x * a) * a^{-1} = (y * a) * a^{-1}$ (operating a^{-1} on both sides)

$\Rightarrow$ $x * (a * a^{-1}) = y * (a * a^{-1})$ (by associative law)

$\Rightarrow \quad x * e = y * e \quad (\because a^{-1} * a = e)$

$\Rightarrow \quad x = y \quad (\because e \text{ is the identity})$

Hence, $\quad x * a = y * a \Rightarrow x = y$

Theorem 6: Given two elements a and b of the group $(G, *)$ there exists unique solution for the equation $a * x = b$ and $y * a = b$.

Proof : Consider the equation $a * x = b$

$a^{-1} * (a * x) = a^{-1} * b \Rightarrow (a^{-1} * a) * x = a^{-1} * b \Rightarrow e * x = a^{-1} * b \Rightarrow x = a^{-1} * b$

Thus, a solution of $a * x = b$ is $\mathbf{x = a^{-1} * b}$.

To establish the uniqueness of the solution, we shall assume that there are two solutions say x_1 and x_2.

i.e., $\quad a * x_1 = b$ and $a * x_2 = b \Rightarrow a * x_1 = a * x_2 \Rightarrow x_1 = x_2 \quad$ (by left cancellation law)

Hence, the solution is unique.

Again consider the equation $y * a = b$

$(y * a) * a^{-1} = b * a^{-1} \Rightarrow y * (a * a^{-1}) = b * a^{-1} \Rightarrow y * e = b * a^{-1} \Rightarrow y = b * a^{-1}$

Thus, a solution of $y * a = b$ is $\mathbf{y = b * a^{-1}}$.

To show that this solution is unique, we shall assume that there are two solutions say y_1 and y_2.

i.e., $\quad y_1 * a = b$ and $y_2 * a = b \Rightarrow y_1 * a = y_2 * a \Rightarrow y_1 = y_2 \quad$ (by right cancellation law)

Hence, the solution is unique.

1.6 Composition Table for Binary Operation on Finite Set

Let G be a non empty finite set and $*$ be a composition defined on it. While proving that $(G, *)$ is a group we make use of a table called **composition table** or **Cayley table** which we construct in the following way. Let $G = \{a_1, a_2, a_3, a_4\}$.

$*$	a_1	a_2	a_3	a_4
a_1	$a_1 * a_1$	$a_1 * a_2$	$a_1 * a_3$	$a_1 * a_4$
a_2	$a_2 * a_1$	$a_2 * a_2$	$a_2 * a_3$	$a_2 * a_4$
a_3	$a_3 * a_1$	$a_3 * a_2$	$a_3 * a_3$	$a_3 * a_4$
a_4	$a_4 * a_1$	$a_4 * a_2$	$a_4 * a_3$	$a_4 * a_4$

We write the elements of G in a horizontal row as well as in a vertical column. Consider any two elements of G say a_2 and a_3. We write the element $a_2 * a_3$ obtained by combining a_2 with a_3 at the intersection of the row headed by a_2 and the column headed by a_3. Like this we fill up all the entries of the table, called the composition table (see the table above).

Example 1. Show that the set Z of all integers is an abelian group under usual addition.

Solution : We know that the usual addition on the set of integers is a binary operation.

i.e., $\quad \forall\, a, b \in Z$, we have, $a + b \in \mathbf{Z}$

Also the usual addition + is associative in Z.

i.e., $\quad a + (b + c) = (a + b) + c \ \forall\, a, b, c \in \mathbf{Z}$.

Existence of identity: The integer $0 \in \mathbf{Z}$ acts as the identity element.

i.e., $\quad \forall\, a \in Z$, we have, $a + 0 = 0 + a = a$

Existence of inverse: $\forall\, a \in Z$, the integer $-a \in \mathbf{Z}$ acts as the inverse of a.

i.e., $\quad \forall\, a \in \mathbf{Z}$, we have, $a + (-a) = (-a) + a = 0$

Further addition on **Z**, is commutative

i.e., $\forall\, a, b \in \mathbf{Z}$, we have, $a + b = b + a$

Hence, **(Z, +) is an abelian group.**

The students are required to prove in the above way, that the following sets together with the binary operation mentioned thereby are groups.

1. The set Q of all rational numbers is a group under usual addition.

2. The set R of all real numbers is a group under usual addition.

Example 2. The set C of all complex numbers is a group under usual addition of complex numbers.

Solution : We know that the set **C** of complex numbers is closed under addition defined by

$$(a + ib) + (c + id) = (a + c) + i\,(b + d)$$

for any two complex numbers $a + ib$ and $c + id$.

Associative law: The addition of complex numbers is associative.

i.e., $\forall\, z_1, z_2, z_3 \in \mathbf{C}$, $z_1 + (z_2 + z_3) = (z_1 + z_2) + z_3$

Existence of identity: The complex number $0 = 0 + i0$, acts as the identity element under the addition

i.e., $\forall\, z \in \mathbf{C}$, $z + 0 = 0 + z = z$

Existence of inverse: $\forall\, z \in \mathbf{C}$, the complex number $-z \in \mathbf{C}$ acts as the inverse of z.

i.e., $\forall\, z \in \mathbf{C}$, $z + (-z) = (-z) + z = 0$

Commutative law: The addition of complex numbers is commutative.

i.e., $\forall\, z_1, z_2 \in C$, $z_1 + z_2 = z_2 + z_1$

Hence, **(C, +) is an abelian group.**

Example 3. Show that $(N, \times)$, where N is the set of natural numbers is a semi group with identity element.

Solution : We know that the set **N** of natural numbers is closed under multiplication. That is the product of any two natural numbers is again a natural number.

Associative law: $\forall\, a, b, c \in \mathbf{N}$, we have, $a \times (b \times c) = (a \times b) \times c$

Hence $(N, \times)$ is a semi group.

The integer $1 \in \mathbf{N}$, acts as the identity element in **N** – for $\forall\, a \in \mathbf{N}$, $a \times 1 = 1 \times a = a$

Hence $(N, \times)$ **is a semi group with identity element.**

Example 4. Show that the set N is not a group under multiplication.

Solution : From the example 3, we have seen that $(N, \times)$ is a semi group with identity element as 1. Now, $(N, \times)$ will be a group if every element of **N** has inverse in **N**.

Let $a \in N$ and $a \cdot b = 1$ for some $b \in N \Rightarrow b = \frac{1}{a}$, but $a \in N$ need not imply $\frac{1}{a} \in N$.

For example, $3 \in N$ but $\frac{1}{3} \notin N$. Hence a has no inverse under multiplication on **N**.

Hence, **$(N, \times)$ is not a group.**

Example 5. Show that $(Q, \times)$, where Q is the set of rationals, is not a group.

Solution : Observe all the axioms of the group are satisfied for the elements of Q under multiplication, except the existence of inverse element – for the element $0 \in Q$ has no inverse in Q. That is, there exists no element $a \in Q$, such that $0 \times a = a \times 0 = 1$. Hence, $(Q, \times)$ is **not a group.**

Similarly, the system $(R, \times)$ *is* not a group for the same reason that 0 **has no inverse under multiplication.**

However, the systems $(Q^*, \times)$ and $(R^*, \times)$ where Q^* and R^* are respectively the set of non zero rationals and the set of non zero real numbers, are abelian groups.

Example 6. Show that the square roots of unity form a group under usual multiplication.

Solution: The square roots of unity are 1 and –1.

Let, $G = \{1, -1\}$. We shall write the multiplication table for G.

From the table it follows that G is closed under multiplication.

$\times$	1	–1
1	1	–1
–1	–1	1

(i) Associative law: The usual multiplication is associative.

(ii) Existence of identity: The element 1 acts as the identity element – this can be observed from the table.

(iii) Existence of inverse: Again from the table it follows inverse of 1 is 1 and that of – 1 is –1.

(iv) Commutative law: Usual multiplication is commutative.

Hence $(G, \times)$ is an abelian group.

Example 7. Show that the cube roots of unity form a group under multiplication.

Solution : The cube roots of unity are 1, ω and ω^2.

Let, $G = \{1, \omega, \omega^2\}$. The multiplication table is as shown in the figure.

All the entries in the table are same as the elements in G. Thus, G is closed under usual multiplication.

$\times$	1	ω	ω^2
1	1	ω	ω^2
ω	ω	ω^2	1
ω^2	ω^2	1	ω

(i) Associative law: Consider

$$1 \cdot (\omega \cdot \omega^2) = 1 \cdot \omega^3 = 1 \cdot 1 = 1$$
$$(1 \cdot \omega) \cdot \omega^2 = \omega \cdot \omega^2 = \omega^3 = 1$$
$$\Rightarrow \quad 1 \cdot (\omega \cdot \omega^2) = (1 \cdot \omega) \cdot \omega^2$$

Similarly, with any other three elements of G.

Thus, associative law is satisfied. (In fact, usual multiplication is associative).

(ii) Existence of identity: From the table it follows that 1 acts as the identity element.

(iii) Existence of inverse: The inverse of 1 is 1. Inverse of ω is ω^2 and inverse of ω^2 is ω.

(iv) Commutative law: The table is symmetric about the principal diagonal, therefore, commutative law holds.

Hence, $(G, \times)$ **is an abelian group.**

Example 8. Show that the set of all fourth roots of unity form a group under multiplication.

Solution : The fourth roots of unity are $1, -1, i, -i$.

Let, $G = \{1, -1, i, -i\}$. The multiplication table is as shown below.

$\times$	1	–1	i	$-i$
1	1	–1	i	$-i$
–1	–1	1	$-i$	i
i	i	$-i$	–1	1
$-i$	$-i$	i	1	–1

From the table it is clear that G is closed under multiplication.

(i) Associative law: Consider,

$$1 \times (-1 \times i) = 1 \times -i = -i; \quad (1 \times -1) \times i = -1 \times i = -i$$

Thus, $1 \times (-1 \times i) = (1 \times -1) \times i$

Similarly, with any other three elements. Hence associative law holds.

(ii) Existence of identity: From the table it follows that the element 1 acts as the identity element.

(iii) Existence of inverse: The inverse of 1 is 1, inverse of –1 is –1, inverse of i is $-i$ and inverse of $-i$ is i.

(iv) Commutative law: Usual multiplication is commutative.

Hence, **(G, ×) is an abelian group.**

Example 9. Show that (Q^+, *), where Q^+ is the set of all positive rational numbers, is a group; where * is defined by $a * b = \dfrac{ab}{5}, \forall\, a, b \in Q^+$.

Solution : By data, $a * b = \dfrac{ab}{5}$, $\forall\, a, b \in Q^+$. If $a, b \in Q^+$, then a and b are positive rationals and so is $a \cdot b$.

Thus, $\forall\, a, b \in Q^+$, $\dfrac{ab}{5} \in Q^+ \Rightarrow Q^+$ is closed under *.

(i) Associative law: $\forall\, a, b, c \in Q^+$, we have,

Now, $$a * (b * c) - a * \frac{bc}{5} = \frac{a \cdot \left(\frac{bc}{5}\right)}{5} = \frac{abc}{25}$$ (Note)

Now, $$(a * b) * c = \frac{ab}{5} * c = \frac{\left(\frac{ab}{5}\right) \cdot c}{5} = \frac{abc}{25}$$

$\therefore$ $a * (b * c) = (a * b) * c \Rightarrow$ * is associative

(ii) Existence of identity: Let $a \in Q^+$ and $e \in Q^+$ such that

Now, $a * e = a \Rightarrow \dfrac{ae}{5} = a \Rightarrow e = 5$ $(\because a \neq 0)$

Also, $e * a = a \Rightarrow \dfrac{ea}{5} = a \Rightarrow e = 5$

Thus, **e = 5 acts as identity element.**

(iii) Existence of inverse: Let a $\in Q^+$ and $b \in Q^+$ such that

$a * b = 5 \Rightarrow \dfrac{ab}{5} = 5 \Rightarrow ab = 25 \Rightarrow b = \dfrac{25}{a} \in Q^+$

Also, $b * a = 5 \Rightarrow b = \dfrac{25}{a} \in Q^+$

Thus, the element $\dfrac{25}{a} \in Q^+$, acts as the inverse element of a. Hence, (Q^+, *) is a group.

Further, $\forall\, a, b \in Q^+$, $a * b = \dfrac{ab}{5} = \dfrac{ba}{5} = b * a$. Thus, the commutative law holds good.

Hence, **(Q^+, *) is an abelian group.**

Example 10. Let G be the set of rationals except –1 and * be the binary operation on G defined by $a * b = a + b + ab$. Show that (G, *) is a group. Is it an abelian group?

Solution: Let $a, b \in G$. Thus, $a \neq -1$ and $b \neq -1$. We have, $a * b = a + b + ab$

To show that $a * b \in G$, we have to show that $a + b + ab \neq -1$.

Supposing $a + b + ab = -1$.

$\Rightarrow \quad a + 1 + b(a + 1) = 0 \quad \Rightarrow \quad (a + 1)(1 + b) = 0$

$\Rightarrow \quad a + 1 = 0$ or $1 + b = 0$

$\Rightarrow \quad a = -1$ or $b = -1$

But by data, $a \neq -1$ and $b \neq -1$. Thus, $a + b + ab \neq -1 \Rightarrow \forall\, a, b \in G, a * b \in G$

Thus, G is closed under the binary operation *.

1. Associative law: Let $a, b, c \in G$, we have

Now, $a * (b * c) = a * (b + c + bc)$

$= a + (b + c + bc) + a(b + c + bc) = a + b + c + bc + ab + ac + abc$

Again, $(a * b) * c = (a + b + ab) * c$

$= (a + b + ab) + c + (a + b + ab) \cdot c = a + b + c + bc + ab + ac + abc$

$\Rightarrow \quad a * (b * c) = (a * b) * c \Rightarrow$ * is associative

2. Existence of identity: Let $a \in G$, and $e \in G$, such that

$a * e = a \Rightarrow a + e + ae = a$

$\Rightarrow e(1 + a) = 0$

$\Rightarrow e = 0 \qquad (\because 1 + a \neq 0 \text{ as } a \neq -1)$

Similarly, $e * a = a \Rightarrow e = 0$.

Thus, $0 \in G$ acts as the identity element.

3. Existence of inverse: Let $a \in G$, and $b \in G$ such that

$a * b = 0 \Rightarrow a + b + ab = 0$

$\Rightarrow b(1 + a) = -a$

$\Rightarrow b = \dfrac{-a}{1+a} \qquad (\because 1 + a \neq 0 \text{ as } a \neq -1)$

Similarly, $b * a = 0 \Rightarrow b = \dfrac{-a}{1+a}$

Thus, he element $\dfrac{-a}{1+a} \in G$ acts as the inverse of the element $a \in G$. Hence, $(G, *)$ is a group.

Consider $a, b \in G$, then $a * b = a + b + ab$ and $b * a = b + a + ba \Rightarrow a * b = b * a$

Thus, the binary operation * is commutative.

Hence, **$(G, *)$ is an abelian group.**

Example 11. Show that the set of all 2 × 2 real non singular matrices form a group under matrix multiplication.

Solution: Let G be the set of all 2×2 real non singular matrices.

i.e., $$G = \left\{ \begin{pmatrix} a & b \\ c & d \end{pmatrix} \middle|\, a, b, c, d \in R, ad - bc \neq 0 \right\}$$

Clearly, the product of any two 2×2 non singular matrices is again 2×2 non singular matrix. Thus, G is closed under matrix multiplication.

(i) **Associative law:** Matrix multiplication is associative.

(ii) **Existence of identity:** The unit matrix I_2 of order 2×2 acts as the identity element i.e., $\forall A \in G, A \cdot I = I \cdot A = A$

(iii) **Existence of inverse:** Let $A \in G$. Thus, A is non singular 2×2 matrix and therefore A^{-1} exists.

That is, $A \cdot A^{-1} = A \cdot A = I$

Hence, **(G, *) is a group.**

Note that G is **not an abelian group,** as the matrix multiplication is not commutative.

Example 12. Show that the set $G = \{(\cos\theta + i\sin\theta) \mid \theta \text{ is real}\}$ is an abelian group under multiplication.

Solution : Let $z_1, z_2 \in G$ and $z_1 = \cos\theta_1 + i\sin\theta_1$, $z_2 = \cos\theta_2 + i\sin\theta_2$

Consider, $z_1 \cdot z_2 = (\cos\theta_1 + i\sin\theta_1)(\cos\theta_2 + i\sin\theta_2) = \cos(\theta_1 + \theta_2) + i\sin(\theta_1 + \theta_2) \in G$

Thus, G is closed under multiplication.

(i) Associative law: Let $z_3 = \cos\theta_3 + i\sin\theta_3$

Consider, $z_1 \cdot (z_2 \cdot z_3) = (\cos\theta_1 + i\sin\theta_1)[(\cos(\theta_2 + \theta_3) + i\sin(\theta_2 + \theta_3)]$

$= \cos(\theta_1 + \theta_2 + \theta_3) + i\sin(\theta_1 + \theta_2 + \theta_3)$

Again, $(z_1 \cdot z_2) \cdot z_3 = [\cos(\theta_1 + \theta_2) + i\sin(\theta_1 + \theta_2)](\cos\theta_3 + i\sin(\theta_3)$

$= \cos(\theta_1 + \theta_2 + \theta_3) + i\sin(\theta_1 + \theta_2 + \theta_3)$

Thus, $z_1 \cdot (z_2 \cdot z_3) = (z_1 \cdot z_2) \cdot z_3 \Rightarrow$ multiplication is associative.

(ii) Existence of identity: The element $1 = \cos 0 + i\sin 0 \in G$ acts as the identity element in G.

Let $z = \cos\theta + i\sin\theta$, then

$$z \cdot 1 = (\cos\theta + i\sin\theta)(\cos 0 + i\sin 0) = \cos(\theta + 0) + i\sin(\theta + 0) = \cos\theta + i\sin\theta = z$$

Similarly, $1 \cdot z = z$

(iii) Existence of inverse: Let $z = \cos\theta + i\sin\theta \in G$. Consider,

$$(\cos\theta + i\sin\theta)(\cos(-\theta) + i\sin(-\theta)) = \cos(\theta - \theta) + i\sin(\theta - \theta) = \cos 0 + i\sin 0 = 1$$

Thus, the inverse of $(\cos\theta + i\sin\theta)$ is $\cos(-\theta) + i\sin(-\theta)$.

(iv) Commutative law: Consider,

$z_1 \cdot z_2 = \cos(\theta_1 + \theta_2) + i\sin(\theta_1 + \theta_2)$ and $z_2 \cdot z_1 = \cos(\theta_2 + \theta_1) + i\sin(\theta_2 + \theta_1)$

Clearly, $z_1 \cdot z_2 = z_2 \cdot z_1$

Hence, **G is an abelian group under multiplication.**

Example 13. In the group R^* of non zero real numbers the binary operation * is defined by $a * b = \frac{ab}{5}$. Solve $2 * x * 5 = 10$ in R^*.

Solution : $2 * x * 5 = 10 \Rightarrow 2 * (x * 5) = 10$

$$\Rightarrow \quad 2 * \frac{5x}{5} = 10$$

$$\Rightarrow \quad 2 * x = 10 \Rightarrow \frac{2x}{5} = 10 \Rightarrow \mathbf{x = 25}$$

Example 14. In the group G of rational numbers except –1, the binary operation * is defined by $a * b = a + b + ab$. Solve $3 * 4 * x = 0$.

Solution : $3 * 4 * x = 0 \Rightarrow (3 * 4) * x = 0$

$\Rightarrow (3 + 4 + 12) * x = 0$

$\Rightarrow \quad 19 * x = 0$

$\Rightarrow 19 + x + 19x = 0 \Rightarrow 20x = -19 \Rightarrow \mathbf{x = \frac{-19}{20}}$

Example 15. If every element of a group G is its own inverse, show that G is abelian.

Solution : By data, $\forall\, a \in G, a^{-1} = a$. We have to show that,

$$x \cdot y = y \cdot x, \ \forall\, x, y \in G$$

[Here $x \cdot y$ is the composition of two elements x and y of G, it should not be mistaken for usual multiplication of the elements].

Consider, $(x \cdot y)^{-1} = y^{-1} \cdot x^{-1}$

$\Rightarrow$ $(x \cdot y)^{-1} = y \cdot x$ $\qquad (\because y^{-1} = y, x^{-1} = x)$

Now, $x, y \in G \Rightarrow x \cdot y \in G \Rightarrow (x \cdot y)^{-1} = y \cdot x$ (by data)

Thus, $(x \cdot y)^{-1} = y \cdot x \Rightarrow x \cdot y = y \cdot x$

Hence, G is abelian.

Example 16. If G is a group of even order, then show that there exists an element $a \in G$, (where $a \neq e$) such that $a^2 = e$ [i.e., $a^{-1} = a$].

Solution : A group G is said to be of even order if G has even number of elements. Also, $a^2 = a \cdot a$.

We know that each element of G has unique inverse and e is its own inverse.

When e is eliminated, then G will have odd number of elements.

Thus there exists an element $a\ (\neq e) \in G$ such that $a^{-1} = a$.

Now, $a = a^{-1} \Rightarrow a \cdot a = a^{-1} \cdot a \Rightarrow a^2 = e$

Example 17. Show that a group G is abelian if and only if $(ab)^2 = a^2b^2$, $\forall\ a, b \in G$

Solution : Let G be abelian, we shall prove $(ab)^2 = a^2\, b^2$

Now, $(ab)^2 = (ab)\,(ab)$

$\Rightarrow$ $(ab)^2 = a\,(ba)\,b$ (associative law)

$\Rightarrow$ $(ab)^2 = a\,(ab)\,b$ (G is abelian)

$\Rightarrow$ $(ab)^2 = (aa)\,(bb)$ (associative law)

$\Rightarrow$ $\mathbf{(ab)^2 = a^2 \cdot b^2}$

Now, let $(ab)^2 = a^2 \cdot b^2 \quad \forall\, a, b \in G$

$\Rightarrow$ $(ab)\,(ab) = (aa)\,(bb)$

$\Rightarrow$ $a\,(ba)\,b = a\,(ab)\,b$ (associative law)

$\Rightarrow$ $(ba)\,b = (ab)\,b$ (by left cancellation law)

$\Rightarrow$ $\mathbf{ba = ab}$ (by right cancellation law)

$\Rightarrow$ **G is abelian.**

Example 18. If in a group $(G, *)$, $a * a = a$, $\forall\ a \in G$, show that $a = e$, where e is the identity element.

Solution : By data, $a * a = a \forall\, a \in G$

$\Rightarrow$ $a^{-1} * (a * a) = a^{-1} * a$

$\Rightarrow$ $(a^{-1} * a) * a = e$ (associative law)

$\Rightarrow$ $e * a = e \Rightarrow \mathbf{a = e}$

Example 19. Show that, all groups of order 4 or less are abelian.

Solution : **1.** Let G be a group of order 1. That is G contains only one element say e.

Clearly, $e * e = e$. **Therefore, G is abelian.**

2. Let G be a group of order 2. That is G contains two elements one of which acts as the identity element e. Let $G = \{e, a\}$. Clearly, the inverse of a is itself (because the inverse of e is e). Further $e \cdot a = a \cdot e$ and $a \cdot a = e$. **Thus, G is abelian.**

3. Let G be a group of order 3 and $G = \{e, a, b\}$, where e is the identity element.

Here two cases arise.

Case (i): $a^{-1} = a$ and $b^{-1} = b$. That is inverse of each element is itself. Hence G is abelian.

Case (ii): $a^{-1} \neq a$ and $b^{-1} \neq b$. Thus, $a^{-1} = b$ and $b^{-1} = a$. $(\because e^{-1} = e)$

Now, $a \cdot b = e$ and $b \cdot a = e \Rightarrow a \cdot b = b \cdot a$

Hence, **G is abelian.**

Let, $G = \{e, a, b, c\}$ be a group of order 4, where e is the identity.

Case (i): If inverse of each element is itself, then G is abelian.

Case (ii): Now, G has four elements. Thus there exists an element say a $(\neq e)$ such that $a^{-1} = a$. That is $a \cdot a = e$.

Now, if $b = b^{-1}$ and $c = c^{-1}$ then clearly G is abelian. (by case (i))

Let, $b^{-1} = c$ and $c^{-1} = b$. That is $b \cdot c = e$, $c \cdot b = e$. Thus, $b \cdot c = c \cdot b$.

Now
$$\left.\begin{array}{l} a \cdot b = b \Rightarrow a = e, \text{ a controdiction} \\ a \cdot b = a \Rightarrow b = e, \text{ a controdiction} \\ a \cdot b = e \Rightarrow b^{-1} = a, \text{ a controdiction} \end{array}\right] \Rightarrow a \cdot b = c$$

Again
$$\left.\begin{array}{l} a \cdot b = b \Rightarrow a = e, \text{ a controdiction} \\ a \cdot b = a \Rightarrow b = e, \text{ a controdiction} \\ a \cdot b = e \Rightarrow b^{-1} = a, \text{ a controdiction} \end{array}\right] \Rightarrow b \cdot a = c$$

Thus, $\boldsymbol{a \cdot b = b \cdot a}$.

Now, $a \cdot b = c \Rightarrow a \cdot (a \cdot b) = a \cdot c \Rightarrow a^2 \cdot b = a \cdot c \Rightarrow a \cdot c = b$ $(\because a^2 = e)$

Again, $b \cdot a = c \Rightarrow (b \cdot a) \cdot a = c \cdot a \Rightarrow b \cdot a^2 = c \cdot c \Rightarrow c \cdot a = b$ $(\because a^2 = e)$

Thus, $a \cdot c = c \cdot a$. Hence **G is abelian.**

1.7 Semi Group

Definition: A set G with a binary composition * is said to be a semi group, if the binary composition * is associative.

i.e., $\forall\, a, b, c \in G,\ a * (b * c) = (a * b) * c$

Examples: (i) $\{Z, \cdot\}$ is a semi–group, **(ii)** $(Q, \cdot\}$ is a semi-group.

Note: In fact every group is a semi-group. The converse of this statement is not true. However, we have the following theorem:

Theorem: A finite semi-group with both cancellation laws is a group.

Proof: Let $\{G, \cdot\}$ be a finite semi-group in which both cancellation law holds.

Let, $G = \{a_1, a_2, a_3, \ldots\ldots, a_n\}$, where a_i's are the n distinct elements of the semi-group.

Let, $a_l \in G$ be arbitrary.

Consider the set, $S_1 = \{a_1 \cdot a_l, a_2 \cdot a_l, a_3 \cdot a_l, \ldots\ldots a_n \cdot a_l\}$ ...(1)

The elements of S_1, are the compositions of the elements of G with a_l.

Since $(\cdot)$ is a binary composition, the elements of S_1 are again in G.

Clearly the elements of S_1 are distinct, for if $a_i \cdot a_l = a_j \cdot a_l$ for $i \neq j$.

$\Rightarrow a_i = a_j$ (by right cancellation law) $\Rightarrow$ The elements of G are not distinct

This is a contradiction.

Thus, $\forall\, a_i \in G, 1 \leq i \leq n, \exists\, a_j \in G$ such that $a_i = a_j \cdot a_l$

In particular, $\exists\, a_k \in G$, such that $a_l = a_k \cdot a_l$

Let, $a_i \in G$ be arbitrary. Consider,

$$\Rightarrow \quad a_i \cdot a_l = a_i \cdot (a_k \cdot a_l) \qquad (\because a_l = a_k \cdot a_l)$$

$$\Rightarrow \quad a_i \cdot a_l = (a_i \cdot a_k) \cdot a_l \qquad \text{(associative law)}$$

$$a_i = a_i \cdot a_k \qquad \text{(by right cancellation law)}$$

Since a_i is arbitrary, the above identity is true for all i, $1 \le i \le n$. Thus, a_k acts as right identity in G.

Similarly by considering the set

$$S_2 = \{a_l \cdot a_1, a_l \cdot a_2, \ldots\ldots, a_l \cdot a_n\}$$

and using left cancellation law, we can show that there exists $a_m \in G$ such that,

$$a_m \cdot a_i = a_i \; \forall \; i. \; 1 \le i \le n$$

Thus, a_m acts as left identity in G.

Now we shall show that these two identities are equal.

Now, $\quad a_k \in G$ and a_m is left identity $\Rightarrow a_k = a_m \cdot a_k$ ...(2)

Again, $\quad a_m \in G$ and a_k is right identity $\Rightarrow a_m = a_m \cdot a_k$ ...(3)

Thus, $\quad$ (2) and (3) $\Rightarrow a_k = a_m$

Let us denote $a_k = a_m$ by 'e'. Thus G has the identity e.

As, $e \in G, \exists \; a_s \in G$ s.t. $e = a_s \cdot a_l$ (by (1)) $\Rightarrow a_s$ is the left inverse of a_l.

Similarly, $\exists \; a_u \in G$ s.t. $e = a_l \cdot a_u$ (by (2)) $\Rightarrow a_u$ is the right inverse of a_l.

We shall show that these two inverse are equal. To this end, consider,

$$a_u = e \,.\, a_u = (a_s \,.\, a_l) \,.\, a_u$$
$$= a_s \,(a_l \,.\, a_u)$$
$$= a_s \,.\, e = a_s$$

Thus, $a_u = a_s$ and we shall denote this by a_l^{-1}.

Since a_l is arbitrary, every element of G has inverse in G.

Thus, G has identity and every element has unique inverse in G. Hence **G is a group.**

EXERCISES

1. Is the set of all odd integers is a group under addition?
2. **(a)** Is (**Z**, *), with $a * b = a \cdot b, \forall \, a, b \in$ **Z** a group?
   (b) Is (**Z**, *), with $a * b = a - b, \forall \, a, b \in$ **Z** a group?
   (c) Is (*Q*, *), with $a * b = a \cdot b, \forall \, a, b \in$ **Q** a group?
3. Show that ($\mathbf{Z^+}$, +), where Z^+ is the set of all positive integers including 0 is not a group.
4. Give an example of a semi group which is not a group.
5. In the set Q^* of non zero rational numbers together with * defined by $a * b = a^b$, is $(Q^*, *)$ a group?
6. In the group $(Q^+, *)$, where Q^+ is the set of all positive rationals and $a * b = \dfrac{ab}{2}$, find the solution of the equation **(a)** $6 * x = 3$ **(b)** $y * 5 = 7$.
7. In the set of all rationals except 1, the binary operation * is defined by $a * b = a + b - ab$. Solve the equations
   (a) $6 * x = 4$ **(b)** $y * (3 * 2) = -5$ **(c)** $(4 * z) * 3 = -7$
8. In the set Z of all integers the binary operation * is defined by $a * b = a + b + 1$. Solve
   (a) $(4 * 3) * x = -2$ **(b)** $y * (7 * -3) = 7$.

II.

1. Show that the set R^* of all non zero real numbers is an abelian group under usual multiplication.
2. Show that the set C^* of all non zero complex numbers is a group under usual multiplication.
3. Show that the set M of all $m \times n$ matrices over real numbers is an abelian group under matrix addition.
4. Show that the set $G = \{2^n \mid n \text{ is an integer}\}$ is an abelian group under multiplication.
5. Show that the set $G = \{2n \mid n \in Z\}$ is a abelian group under addition.
6. Let Q^+ be the set of all positive rational numbers $*$ be defined and $a * b = \dfrac{ab}{2}$, $\forall\, a, b \in Q^*$. Show the $(Q^+, *)$ is an abelian group.
7. Show that $(\mathbf{Z}, *)$, where $\mathbf{Z}$ is the set of all integers and $*$ be defined by $a * b = a + b + 1$, $\forall\, a, b \in \mathbf{Z}$ is an abelian group.
8. In the set G of all rational numbers $\neq 1$, define $*$ by $a * b = a + b - ab$. Show that $(G, *)$ is an abelian group.
9. Show that the set of all matrices of the form $\begin{pmatrix} \cos\alpha & -\sin\alpha \\ \sin\alpha & \cos\alpha \end{pmatrix}$, where α is real, is a group under matrix multiplication.
10. Show that the set $G = \{a + b\sqrt{2} \mid \forall\, a, b \in \mathbf{Z}, a \neq 0\}$ is a group under multiplication.
11. Show that the set $G = \{A, B, C, D\}$ where $A = \begin{pmatrix} 1 & 0 \\ 0 & 1 \end{pmatrix}$, $B = \begin{pmatrix} -1 & 0 \\ 0 & 1 \end{pmatrix}$, $C = \begin{pmatrix} 1 & 0 \\ 0 & -1 \end{pmatrix}$, $D = \begin{pmatrix} -1 & 0 \\ 0 & -1 \end{pmatrix}$ is an abelian group under matrix multiplication.
12. Show that $(\mathbf{Z}, *)$ where $\mathbf{Z}$ is the set of integers and $a * b = a + b - 1$, $\forall\, a, b \in \mathbf{Z}$ is an abelian group.
13. Show that the set $G = \{....., a^{-2}, a^{-1}, a^0, a^1, a^2, a^3, ...\}$ where a is a non zero integer is an abelian group under multiplication.
14. Show that the set $M = \left\{ \begin{pmatrix} x & x \\ x & x \end{pmatrix} \middle| \begin{matrix} x \in R \\ x \neq 0 \end{matrix} \right\}$ is an abelian group under multiplication.
15. Show that the set $G = \left\{ \begin{pmatrix} x & x \\ 0 & 0 \end{pmatrix} \middle| \begin{matrix} x \in R \\ x \neq 0 \end{matrix} \right\}$ is an abelian group under matrix multiplication.
16. Show that the set $G = \{A, B, C, D\}$ where $A = \begin{pmatrix} 1 & 0 \\ 0 & 1 \end{pmatrix}$, $B = \begin{pmatrix} 0 & 1 \\ -1 & 0 \end{pmatrix}$, $C = \begin{pmatrix} -1 & 0 \\ 0 & -1 \end{pmatrix}$, $D = \begin{pmatrix} 0 & -1 \\ 1 & 0 \end{pmatrix}$ is an abelian group under matrix multiplication.
17. Show that the set $G = \{I, A, B, C, D, E\}$ where $I = \begin{pmatrix} 1 & 0 \\ 0 & 1 \end{pmatrix}$, $A = \begin{pmatrix} \omega & 0 \\ 0 & \omega^2 \end{pmatrix}$, $B = \begin{pmatrix} \omega^2 & 0 \\ 0 & \omega \end{pmatrix}$, $C = \begin{pmatrix} 0 & 1 \\ 1 & 0 \end{pmatrix}$, $D = \begin{pmatrix} 0 & \omega^2 \\ \omega & 0 \end{pmatrix}$, $E = \begin{pmatrix} 0 & \omega \\ \omega^2 & 0 \end{pmatrix}$ where ω is a complex cube root of unity is an abelian group under matrix multiplication.

1.8 Modular Systems

In this section we shall introduce two special types of binary operations on the set of integers. These two binary operations are called **addition modulo m** and **multiplication modulo m,** where m is a fixed positive integer. These binary operations are denoted by $+_m$ and $\times_m$.

1. Addition modulo *m*

Let m be a fixed positive integer and a and b be any two integers. The binary composition called **addition modulo *m*,** denoted and defined as

$a +_m b$ = least non negative remainder obtained by dividing the usual sum $a + b$ by *m*.

For example, under addition modulo 7, we have $4 +_7 6 = 3$ – since the usual sum $4 + 6 = 10$ leaves 3 as remainder when divided by 7.

Similarly, $4 +_4 5 = 1$, $3 +_2 7 = 0$, $2 +_5 7 = 4$ etc.

2. Multiplication modulo *m*

Let m be a fixed positive integer and a and b be any two integers. The binary composition called **multiplication modulo *m*,** denoted and defined as

$a \times_m b$ = least non negative remainder obtained by dividing the usual product $a \cdot b$ by *m*

For example, under multiplication modulo 6, we have $4 \times_6 6 = 0$ – since the usual product $4 \cdot 6 = 24$ leaves 0 as remainder when divided by 6.

Similarly, $4 \times_3 5 = 2$, $3 \times_3 7 = 0$, $2 \times_5 7 = 4$ etc.

Note : The addition modulo m and multiplication modulo m are both commutative and associative. The commutativity of $+_m$ and $\times_m$ can be established directly.

Since $a + b$ and $b + a$ both leave same remainder when divided by m, we have

$$a +_m b = b +_m a$$

Similarly, $a \cdot b$ and $b \cdot a$ both leave same remainder when divided by m, therefore, we have

$$a \times_m b = b \times_m a$$

However, to establish the associativity, the argument is not that direct as it is in the case of commutativity. It requires more of discussion. Further as it is beyond the scope of this book, we omit it and just assume the associativity of both addition and multiplication modulo m.

While working out a problem, we just establish the associative law by taking some three element of the set and say same will hold good with any other three elements of the set.

If one says that $a + (b + c)$ and $(a + b) + c$ leaves the same remainder when divided by m and therefore $a +_m (b +_m c) = (a +_m b) +_m c$, then it is not a proof of associativity or the justification for associativity of $+_m$.

The algebraic systems under the binary operations addition modulo m or multiplication modulo m are called **modular systems.**

Example 1. Form the table of multiplication mod 10 for the set {1, 3, 7, 9}.

Solution : The required multiplication mod 10 table is given below.

$\times_{10}$	1	3	7	9
1	1	3	7	9
3	3	9	1	7
7	7	1	9	3
9	9	7	3	1

Here, for example, $3 \times_{10} 7 = 1$, because 1 is the remainder obtained when $3 \cdot 7 = 21$ is divided by 10. Similarly other entries.

Example 2. Form the addition modulo 6 table for the set {0, 1, 2, 3, 4, 5}.

Solution : The required table is as shown at the next page.

$+_6$	0	1	2	3	4	5
0	0	1	2	3	4	5
1	1	2	3	4	5	0
2	2	3	4	5	0	1
3	3	4	5	0	1	2
4	4	5	0	1	2	3
5	5	0	1	2	3	4

Let m be any positive integer. Consider all possible non negative remainders that is obtained when any integer is divided by m. These are 0, 1, 2, 3, 4, ..., $m - 1$

Let the set of these integers be denoted by Z_m. Then

$$Z_m = \{0, 1, 2, 3, 4, ..., m - 1\}$$

For example, $Z_5 = (0, 1, 2, 3, 4\}$, $Z_3 = \{0, 1, 2\}$, $Z_2 = \{0, 1\}$ etc.

Example 3. Show that $(Z_6, +_6)$, where $Z_6 = \{0, 1, 2, 3, 4, 5\}$, is a group.

$+_6$	0	1	2	3	4	5
0	0	1	2	3	4	5
1	1	2	3	4	5	0
2	2	3	4	5	0	1
3	3	4	5	0	1	2
4	4	5	0	1	2	3
5	5	0	1	2	3	4

Solution : The $+_6$ table is as shown in the figure. From the table it follows that the Z_6 is closed under the operation $+_6$.

1. **Associative law:**

   Consider, $3 +_6 (4 +_6 2) = 3 +_6 0 = 3$. Also, $(3 +_6 4) +_6 2 = 1 +_6 2 = 3$

   $\Rightarrow \quad 3 +_6 (4 +_6 2) = (3 +_6 4) +_6 2$

   Similarly, with any other three elements of Z_6. Thus, $+_6$ is associative in Z_6.
2. **Existence of identity:** From the addition table it follows that 0 acts as the identity element.
3. **Existence of inverse:** Inverse of 0 is 0, inverse of 1 is 5, inverse of 2 is 4, inverse of 3 is 3, inverse of 4 is 2 and inverse of 5 is 1.
4. **Commutative law:** For any two elements $a, b \in Z_6$, $a + b$ and $b + a$ leaves the same remainder when divided by 6. Thus, $a +_6 b = b +_6 a$. Thus, $+_6$ is commutative.

   Hence, **$(Z_6, +_6)$ is an abelian group.**

Example 4. Show that $G = \{1, 2, 3, 4\}$ is an abelian group under multiplication modulo 5.

Solution : The multiplication modulo 5 table is given below.

From the table it follows that the G is closed under multiplication modulo 5.

$\times_5$	1	2	3	4
1	1	2	3	4
2	2	4	1	3
3	3	1	4	2
4	4	3	2	1

1. **Associative law:** Consider,

   $3 \times_5 (4 \times_5 2) = 3 \times_5 3 = 4$

   $(3 \times_5 4) \times_5 2 = 2 \times_5 2 = 4$

   $\Rightarrow \quad 3 \times_5 (4 \times_5 2) = (3 \times_5 4) \times_5 2$

   Similarly for any other three elements of G. Thus, $\times_5$ is associative.
2. **Existence of identity:** From the table it follows that 1 acts as the identity element.

3. **Existence of inverse:** The inverse of 1 is 1, inverse of 2 is 3, inverse of 3 is 2, inverse of 4 is 4.
4. **Commutative law:** For any two elements $a, b \in G, a \cdot b$ and $b \cdot a$ leaves the same remainder when divided by 5. Thus, $a \times_5 b = b \times_5 a \Rightarrow \times_5$ is commutative.

Hence, **$(G, \times_5)$ is an abelian group.**

Example 5. Assuming the associative law holds, show that the set $G = \{2, 4, 6, 8\}$ is a group under multiplication modulo 10.

Solution : The multiplication modulo 10 table is shown below. Clearly from the table it follows that G is closed under $\times_{10}$.

$\times_{10}$	2	4	6	8
2	4	8	2	6
4	8	6	4	2
6	2	4	6	8
8	6	2	8	4

Associative law: By data it is true.

Existence of identity: From the table it follows that 6 is the identity element.

Existence of inverse: Inverse of 2 is 8, inverse of 4 is 4, inverse of 6 is 6, inverse of 8 is 2. **$(G, \times_{10})$ is a group.**

Example 6. Write the multiplication modulo 4 table for the set $G = \{0, 1, 2, 3\}$. Is $(G, \times_4)$ a group?

Solution : The multiplication modulo 4 table is shown in the adjacent table.

$\times_4$	0	1	2	3
0	0	0	0	0
1	0	1	2	3
2	0	2	0	2
3	0	3	2	1

From the table it follows that 1 is the identity element.

$(G, \times_6)$ is not a group – for the element 0 has no inverse under $\times_4$.

Example 7. Is $G = \{1, 2, 3, 4, 5\}$ a group under multiplication modulo 6?

Solution : The multiplication modulo 6 table is shown in the figure.

$\times_6$	1	2	3	4	5
1	1	2	3	4	5
2	2	4	0	2	4
3	3	0	3	0	3
4	4	2	0	4	2
5	5	4	3	2	1

From the table we observe that $2 \times_6 3 = 0 \notin G$.

Also $3 \times_6 4 = 0 \notin G$. Thus, G is not closed under multiplication modulo 6. Therefore G is not a group under $\times_6$.

Example 8. In the group $(Z_6, +_6)$, find the element $(3 +_6 5^{-1})^{-1}$.

Solution : We have, $5^{-1} = 1$. ($\because$ 5 + 1 = 6 leave 0 as remainder when divided by 6)

Now, $3 +_6 5^{-1} = 3 +_6 1 = 4$; $(3 +_6 5^{-1})^{-1} = 4^{-1} = 2$ $(\because 4 + 2 = 6)$

Example 9. In the group $G = \{1, 2, 3, 4, 5, 6\}$ under multiplication modulo 7, find $4^{-1} \times_7 5$.

Solution : Clearly, $4^{-1} = 2$ $(\because 4 \times 2 = 8 = 7 \times 1 + 1)$

Now, $4^{-1} \times_7 5 = 2 \times_7 5 = 3$ $(\because 2 \times 5 = 10 = 7 \times 1 + 3)$

EXERCISES

I.

1. Write the multiplication modulo 12 table of the set {1, 5, 7, 11}.
2. Write the addition modulo 6 table for the set {0, 2, 4}.
3. Write the multiplication modulo 7 for the set (0, 1, 2, 3, 4, 5, 6}.
4. Find the inverse of 5 and 7 in the set (1, 5, 7, 11} under $\times_{12}$.
5. Find the inverse of 3 and 9 in the set {1, 3, 7, 9} under $\times_{10}$.
6. Find the inverse of the elements 4 and 3 in the set {0, 1, 2, 3, 4} under $+_5$.

II.

1. Find the identity element and inverse of 4 in the set {2, 4, 6, 8} w.r.t. multiplication modulo 10.
2. Write the multiplication modulo 2 table for the set $Z_2 = \{0, 1\}$. Is $(Z_2, \times_2)$ a group?
3. Write the multiplication modulo 4 table for the set $Z_4 = \{0, 1, 2, 3\}$. Is $(Z_4, \times_4)$ a group?
4. Show that the set $G = \{1, 2, 3\}$ is not a group under $\times_4$.
5. In the group (Z_6, +mod 6) find $2 +_6 4^{-1} +_6 3^{-1}$.

III.

1. Show that $(Z_4, +_4)$, where $Z_4 = \{0, 1, 2, 3\}$ is an abelian group.
2. Show that $(Z_5, +_5)$, where $Z_5 = \{0, 1, 2, 3, 4\}$ is an abelian group.
3. Show that $(G, \times_3)$, where $G = \{1, 2\}$ is an abelian group.
4. Show that $(G, \times_7)$, where $G = \{0, 1, 2, 3, 4, 5, 6\}$ is an abelian group.
5. Show that $G = \{1, 3, 7, 9\}$ is an abelian group under multiplication modulo 10.
6. Show that $G = \{1, 5, 7, 11\}$ is an abelian group under multiplication modulo 12.
7. Show that the set $G = \{3, 6, 9, 12\}$ forms an abelian group under multiplication modulo 15.
8. Show that $G = \{1, 3, 4, 5, 9\}$ is a group under multiplication modulo 11.
9. Show that the set $G = \{1, 3, 5, 7\}$ is a group under multiplication modulo 8.

1.9 Subgroup

Let $(G, *)$ be a group. This group G can have a subset S, which is a group by itselt under the same binary operation *. Such subsets of a group G are called **subgroups** of the group G. A formal definition of a subgroup is given below.

Definition : A subset *H* of a group (*G*, *) is said to be a subgroup of *G*, if *H* is a group by itself under the binary operation induced by * in *H*.

Note : 1. If H is a subgroup of a group G, then necessarily $H \neq \phi$.

2. If $(G, *)$ is a group, then it will have at least two subgroups one is G itself and the other is the subset $\{e\}$ consisting of the identity element e of G. These two subgroups are called **trivial** subgroups of the group G. These are also called improper subgroups* of G. The other subgroups (i.e., other than G and $\{e\}$) are called **non trivial subgroups** or **proper subgroups of *G*.**

Examples :

1. The group $(Q, +)$ of rational numbers under usual addition, is a subgroup of the group $(R, +)$ of real numbers, under usual addition.
2. The group $(Z, +)$ of integers is a subgroup of the group $(Q, +)$ and hence that of $(R, +)$.
3. The group $(R, +)$ of real numbers is a subgroup of the group $(C, +)$ of complex numbers.
4. The group $(Q^*, \times)$ of nonzero rationals is a subgroup of the group $(R^*, \times)$ of nonzero real numbers.

Now, we shall prove two theorems, which will give us the necessary and sufficient conditions for a subset H of a group G to be a subgroup of group G.

Theorem 1 : A non empty subset *H* of a group (*G*, *) is a subgroup of *G* if and only if

(i) $a, b \in H \Rightarrow a * b \in H$ (ii) $a \in H \Rightarrow a^{-1} \in H$

Proof : **(a)** Let H be a sub group of G. We shall show that condition (i) and (ii) are true.

* The word improper used here is not in the sense used to mention improper subset of a set.

By data H is a sub group of $(G, *)$. Thus H is a group by itself under the binary operation*. Therefore, H is closed under the binary operation *.

i.e., $\forall\, a, b \in H, a * b \in H$

Again H is a group under *, thus, by the existence of inverse axiom, we have,

$$a \in H \Rightarrow a^{-1} \text{ exists and } a^{-1} \in H$$

Hence, **(i) and (ii) are true.**

(b) Let H be a subset of G such that (i) and (ii) are true. We shall show that H is a group by itself under the binary operation *.

By data, $(a, b \in H \Rightarrow a * b \in H) \Rightarrow H$ is closed under the binary operation *.

Since the associative law is true in G, it is also true in $H \subset G$.

Again by data, $a \in H \Rightarrow a^{-1} \in H$ (by (ii))

Now, $a \in H, a^{-1} \in H \Rightarrow a * a^{-1} \in H$ (by (i)

$\Rightarrow e \in H$

Thus, the identity element of $e \in H$.

Again by data $a \in H \Rightarrow a^{-1} \in H$. Thus, the inverse of a exists in H.

Thus, H is a group by itself and hence, **it is *a* subgroup of G.**

Theorem 2 : A non empty subset *H* of a group (*G*, *) is a subgroup of *G* if and only if, $\forall\, a, b \in H, a * b^{-1} \in H$.

Proof : **(a)** Let H be a sub group of $(G, *)$. We shall show that

i.e., $\forall\, a, b \in H, \quad a * b^{-1} \in H$ ...(1)

By data H is a sub group of G and therefore H is a group by itself, Thus

$$b \in H \Rightarrow b^{-1} \in H \text{ and } a, b \in H, a * b \in H$$

Now, $a \in H, b \in H \Rightarrow a \in H, b^{-1} \in H \Rightarrow a * b^{-1} \in H$

Hence, (1) is true.

(b) Let H be a non empty subset of G such that

$$\forall\, a, b \in H \Rightarrow a * b^{-1} \in H$$

We shall show that H is a group under *.

Now let, $a \in H \Rightarrow a * a^{-1} \in H \Rightarrow e \in H$ (by (1))

Thus H contains the identity element.

Now, $e \in H, a \in H \Rightarrow e * a^{-1} \in H \Rightarrow a^{-1} \in H$ (by (1))

Thus, inverse of a exists in H.

Now, $\forall\, a, b \in H \Rightarrow a, b^{-1} \in H \Rightarrow a * (b^{-1})^{-1} = a * b \in H \Rightarrow H$ is closed under *.

The associative law holds in H, as it holds in G.

Thus, H is a group by itself and hence, **it is a subgroup of the group $(G, *)$.**

Note : The above condition in the additive notation is given by

$$\forall\, a, b \in H \Rightarrow a + (-b) \in H$$

Example 1. Show that the set of even integers including 0 is a subgroup of additive group of the set *Z* of integers.

Solution : Let H be the set of even integers including 0. Let $a, b \in H$.

Thus, $a = 2n$ and $b = 2m$ for some integers n and m.

Consider, $a + (-b) = 2n + (-2m)$ ($-b$ is the inverse of b in Z)

$= 2(n-m)$

$= 2q =$ even integer (where $q = n - m$)

Thus, $\forall\, a, b \in H,\ a + (-b) \in H$ Hence, **H is a sub group of Z.**

Example 2. Show that the set of all integral multiples of 5 is a subgroup of additive group of the set Z of integers.

Solution : Let H be the set of all integral multiples of 5. That is

$$H = \{\ldots. -15, -10, -5, 0, 5, 10, 15, \ldots..\}$$

Clearly, $H \subset Z$.

Let, $a, b \in H \Rightarrow a = 5n$ and $b = 5m$ for some $n, m \in Z$

Now, $a + (-b) = 5n + (-5m) = 5(n - m) = 5q =$ multiple of 5 (where $q = n - m$)

Thus, $\forall\, a, b \in H,\ a + (-b) \in H$. Hence, **$H$ is a subgroup of Z.**

Example 3. Show that the set of square roots of unity is a subgroup of the group of fourth roots of unity under multiplication.

Solution : Let $H = \{1, -1\}$ the set of square roots of unity. Let $G = \{1, -1, i, -i\}$. Clearly $H \subset G$.

Now,
$$1 \in H, 1 \in H \Rightarrow 1 \cdot 1^{-1} = 1 \cdot 1 = 1 \in H$$
$$-1 \in H, -1 \in H \Rightarrow -1 \cdot (-1)^{-1} = (-1) \cdot (-1) = 1 \in H$$
$$1 \in H, -1 \in H \Rightarrow 1 \cdot (-1)^{-1} = 1 \cdot (-1) = -1 \in H$$
$$-1 \in H, 1 \in H \Rightarrow (-1) \cdot 1^{\ 1} = (-1) \cdot 1 - -1 \in H$$

Thus, $\forall\, a, b \in H,\ a \cdot b^{-1} \in H$. Thus, **$H$ is a subgroup of G.**

Aliter : Show independently H is a group by itself. As H is a subset of a group G, H will be a subgroup of G.

Example 4. Show that $H = \{0, 2, 4\}$ is a subgroup of the group $(G, +_6)$ where $G = \{0, 1, 2, 3, 4, 5\}$.

Solution : Clearly $H \subset G$. We shall show that H is a group by itself under addition modulo 6.

$+_6$	0	2	4
0	0	2	4
2	2	4	0
4	4	0	2

Consider the addition modulo 6 table for the set H.

Clearly H is closed under addition modulo 6.

1. **Associative law:** $H \subset G$ and the associative law holds good in G. Thus it 2 is also true in H.
2. **Existence of identity:** The element 0 acts as the identity element.
3. **Existence of inverse:** The inverse of 0 is 0, inverse of 2 is 4 and inverse of 4 is 2.

Thus, $(H, +_6)$ is a group by itself and hence a subgroup of G.

Example 5. Show that the intersection of two subgroups of a group G is again a subgroup.

Solution : Let H and K be any two subgroups of G. We shall show that $H \cap K$ is a subgroup of G.

Now,
$$x, y \in H \cap K \Rightarrow x, y \in H \text{ and } x, y \in K$$
$$\Rightarrow x \cdot y^{-1} \in H \text{ and } x \cdot y^{-1} \in K$$

(because H and K are subgroups of G)

$$\Rightarrow x \cdot y^{-1} \in H \cap K$$

Thus, $x, y \in H \cap K$ $\Rightarrow x \cdot y^{-1} \in H \cap K$. Hence, **$H \cap K$ is a subgroup of G.**

Example 6. Show that the union of two subgroups of a group need not be a subgroup of G.

Solution : Consider the group $(Z_6, *)$, where $Z_6 = \{0, 1, 2, 3, 4, 5\}$.

Let, $H = \{0, 3\}$ and $K = \{0, 2, 4\}$. Clearly, H and K are subgroups of Z_6.

Consider $H \cup K = \{0, 2, 3, 4\}$. This is not a subgroup of Z_6. Because $3 +_6 4 = 1 \notin H \cup K$. That is, $H \cup K$ is not closed under addition.

Few more useful results.

1. **The set** $M = \left\{\begin{pmatrix} a & 0 \\ 0 & b \end{pmatrix} : \begin{matrix} a, b \in R \\ a \neq 0, b \neq 0 \end{matrix}\right\}$ **is an abelian group under multiplication.**

   Identity element is $\begin{pmatrix} 1 & 0 \\ 0 & 1 \end{pmatrix}$ **and inverse of** $\begin{pmatrix} a & 0 \\ 0 & b \end{pmatrix}$ **is** $\frac{1}{ab}\begin{pmatrix} b & 0 \\ 0 & a \end{pmatrix}$ **i.e.,** $\begin{pmatrix} \frac{1}{a} & 0 \\ 0 & \frac{1}{b} \end{pmatrix}$.

2. **The set** $M = \left\{\begin{pmatrix} x & -x \\ -x & x \end{pmatrix} : x \in R,\ x \neq 0\right\}$ **is an abelian group under matrix multiplication.**

   Identity element $= \begin{pmatrix} \frac{1}{2} & -\frac{1}{2} \\ -\frac{1}{2} & \frac{1}{2} \end{pmatrix}$; **Inverse of** $\begin{pmatrix} a & -a \\ -a & a \end{pmatrix}$ **is** $\begin{pmatrix} \frac{1}{4a} & -\frac{1}{4a} \\ -\frac{1}{4a} & \frac{1}{4a} \end{pmatrix}$.

3. **The set** $G = \begin{pmatrix} a & b \\ -b & a \end{pmatrix} : a, b \in R,\ a^2 - b^2 \neq 0\}$ **is an abelian group under matrix multiplication. Identity element** $= \begin{pmatrix} 1 & 0 \\ 0 & 1 \end{pmatrix}$ **and inverse of** $\begin{pmatrix} a & b \\ -b & a \end{pmatrix}$ *is* $\frac{1}{a^2+b^2}\begin{pmatrix} a & -b \\ b & a \end{pmatrix}$.

EXERCISES

1. Show that the set $H = \{1, 9\}$ *is a* subgroup of $(G, \times_{10})$, where $G = \{1, 3, 7, 9\}$.
2. Consider the group $(G, \times_{12})$, where $G = \{1, 5, 7, 11\}$. Show that
   (a) $H_1 = \{1, 5\}$ is a subgroup of G, **(b)** $H_2 = \{1, 7\}$ is a subgroup of G,
   (c) $H_3 = \{1, 11\}$ is a subgroup of G.
3. Show that the set $H = \{0, 3, 6, 9\}$ is a subgroup of $(Z_{12}, +_{12})$.
4. Show that the set of all multiples of 4 is a subgroup of the group Z of integers under addition.
5. Let G be the group of all non singular real matrices of order 2×2, under multiplication. Let $H = \left\{\begin{pmatrix} 1 & a \\ 0 & 1 \end{pmatrix} \middle| a \in R\right\}$. Show that H is a subgroup of G.

Some more examples of subgroups

1. The subset $H = \{1, -1\}$ is a subgroup of the multiplicative group $G = \{1, -1, i, -i\}$ of fourth root of unity.
2. The group $(Z, +)$ of integers is a subgroup of the group $(R, +)$ of reals and the group $(Q, +)$ of rationals.
3. For any integer N, the subset $nZ = \{n\, z \mid z \in Z\}$ of the group $(Z, +)$ is a subgroup of $(Z, +)$. Later we shall see that any subgroup of the group $(Z, +)$ is of the form $n\,Z$, for some integer n and conversely.
4. If H is a subgroup of a group G, and $g \in G$, then the set $\{g\, h\, g^{-1} \mid h \in H\}$ is a subgroup G.
5. If G is a group and $a \in G$, then the set $N(a) = \{x \in G \mid x\,a = a\,x\}$ is a subgroup of G. This subgroup $N(a)$ of G is called the **normalizer of a** in G. If $a = e$, then $N(e) = G$ and if G is abelian then $N(a) = G$ for every $a \in G$.
6. If G is a group then the set $C = \{x \in G \mid ax = xa \;\forall\; a \in G\}$ is a subgroup of G. This subgroup C is called the **centre of the group G.**
7. The intersection of two subgroups of a group is again a subgroup of the group. However the union of two subgroups is not always of a subgroup of G.

1.10 Result associated to subgroups

Now we shall see few more results related to subgroups ot a group, which are useful in the study of the theory of groups.

Let H and k be any two subsets of a group $(G, .)$. The "product" of H and K denoted by HK is a sub set of G, defined by

$$HK = \{h\,k \mid h \in H, k \in H\}$$

In other words the HK is the set of all possible compositions of the elements of H with the elements of K.

Let H be a subset of a group G. The set H^{-1} is defined by

$$H^{-1} = \{a^{-1} \mid a \in H\}$$

That is the set H^{-1} is the set of all inverses of the elements of H.

We shall see few results in the form of theorems.

Theorem 1. If H and K are any two subsets of G, then $(H\,K)^{-1} = K^{-1} . H^{-1}$

Proof : Let $x \in (H\,K)^{-1} \Rightarrow \exists\, h \in H$ and $k \in K$, such that $x = (h\,k)^{-1}$

Now, $x = (h\,k)^{-1} \Rightarrow x = k^{-1} . h^{-1} \in K^{-1} . H^{-1}$ [Since $k^{-1} \in K^{-1}, h^{-1} \in H^{-1}$]

Thus, $(H\,K)^{-1} \subseteq K^{-1} H^{-1}$ ... (1)

Again, let $y \in K^{-1} H^{-1} \Rightarrow \exists\, k \in K^{-1}$ and $h \in H^{-1}$, such that $y = k^{-1} . h^{-1}$

Now, $y = k^{-1} . h^{-1} \Rightarrow y = (h\,.\,k)^{-1} \Rightarrow y \in (H\,K)^{-1}$

Thus, $K^{-1} H^{-1} \subseteq (H\,K)^{-1}$ ... (2)

From (1) and (2) it follows,

$$(H\,K)^{-1} = K^{-1} H^{-1}$$

Theorem 2. If H is a subgroup of a group G, then $H^{-1} = H$.

Proof : Let $h \in H$ be arbitrary, $\Rightarrow h^{-1} \in H$, as H is a subgroup.

Now, $h^{-1} \in H \Rightarrow (h^{-1})^{-1} \in H^{-1} \Rightarrow h \in H^{-1} \Rightarrow H \subset H^{-1}$

Again H is a subgroup of G, it must contain the inverses of all its elements. Thus $H^{-1} \subset H$. Hence $\mathbf{H^{-1} = H}$

Note : The converse of the above theorem is not always true. Consider the group $G = \{-1, 1\}$ of square roots of unity and $H = \{-1\}$. Clearly $H^{-1} = H$. But H is not a subgroup of G.

The following theorem gives a useful criteria for a subset of a group to be a subgroup of it.

Theorem 3. A subset H of a group G is a subgroup of G if and only if $H . H^{-1} = H$

Proof : Let H be a subgroup of the group G.

Let $h . k^{-1} \in H . H^{-1}$ for some $h, k \in H$

Now, $h, k \in H \Rightarrow h . k^{-1} \in H$, as H is a subgroup of G

Thus, $h . k^{-1} \in H . H^{-1} \Rightarrow h . k^{-1} \in H \Rightarrow HH^{-1} \subseteq H$ ... (1)

Now, let $a \in H$ be arbitrary.

$\Rightarrow a . e^{-1} \in H . H^{-1}$ [Since $e \in H$]

$\Rightarrow a . e \in H . H^{-1}$ [Since $e^{-1} = e$]

$\Rightarrow a \in H . H^{-1} \quad \Rightarrow H \subseteq H H^{-1}$ ... (2)

From (1) and (2) it follows that $H . H^{-1} = H$.

Conversely, let H be a subset of G, such that $H . H^{-1} = H$. We shall show that H is a subgroup of G.

Let $h, k \in H$ be arbitrary. $\Rightarrow h, k^{-1} \in H . H^{-1} \Rightarrow h . k^{-1} \in H$ [Since $H . H^{-1} = H$]

Thus H is a subgroup of G.

Theorem 4. If H, K and L are any three subsets of the group G, then

$$(H K) L = H (K L)$$

Proof : Exercise.

The following theorem gives a criteria for the subset HK of the group G to be a subgroup of it, where H and K are any two subgroups of G.

Theorem 5. If H and K are any two subgroups of a group G, then HK is a subgroup of G if and only if $HK = KH$

Proof : Let HK be a subgroup of G. By data H and K are subgroups of G

$\Rightarrow H = H^{-1}$ and $K = K^{-1}$

Now, $(H K) = (H K)^{-1} \Rightarrow H K = K^{-1} H^{-1}$

$\Rightarrow H K = K H$ [Since $H = H^{-1}, K = K^{-1}$]

Conversely, let H and K be any two subgroup of G, such that $HK = KH$

We shall show that HK is a subgroup of G. To this end consider,

$$\begin{aligned}(H K) . (H K)^{-1} &= (H K) . (K^{-1} H^{-1}) \\ &= H (K K^{-1}) H^{-1} \\ &= (H K) H^{-1} && \text{[Since } K \text{ is a subgroup]} \\ &= (K H) H^{-1} && \text{[By data]} \\ &= K (H H^{-1}) \\ &= K H && \text{[Since } H \text{ is a subgroup]} \\ &= H K && \text{[By data]}\end{aligned}$$

Thus we have $(HK) . (HK)^{-1} = HK$. Hence HK is a subgroup of G.

The following theorem gives us the characterization of any subgroup H of the group $(Z, +)$ of integers.

Theorem 6. If H is a subgroup of $(Z, +)$, then either $H = \{0\}$ or there exists a positive integer p such that $H = \{kp \mid k \in Z\}$.

Proof : Let H be a subgroup of $(Z, +)$ and $H \neq \{0\}$

We shall show that there exists a positive integer p, such that $H = \{kp \mid k \in Z\}$.

Consider

$$S = \{j \mid j > 0, j \in H\}$$

That is the set of all positive integers that are in H. Clearly $S \subset H$.

By data $H \neq \{0\}$. Thus there exists at least one element $n \neq 0$ in Z, such that $n \in H$. Further H being a subgroup of Z, $-n \in H$. Hence there exists at least one positive integer n or $-n$ in H and therefore in S. This implies s is non empty.

Thus there exists a least element p in S*

$\Rightarrow$ p is the least positive integer that belongs to H.

Now if $k > 0$ and $k \in Z$, then

$$kp = p + p + \dots\dots\dots + p \ (k \text{ Times}) \in H \qquad \dots (1)$$

Again as $p \in H$, we have $-p \in H$. Thus if $k < 0$, then,

$$kp = (-p) + (-p) + \dots\dots\dots + (-p) \ (k \text{ Times}) \in H \qquad \dots (2)$$

Also $$0 . p = 0 \in H \qquad \dots (3)$$

Thus from (1), (2) and (3) it follows that

$$\{kp \mid k \in Z\} \subset H \qquad \dots (4)$$

Now let, $m \in H$ be arbitrary. By division algorithm, there exists integers q and r such that

$$m = qp + r, \quad 0 \le r < p \quad \Rightarrow \quad r = m - qp, \quad 0 \le r < p$$

By data H is a subgroup. Thus $m, qp \in H$

$\Rightarrow$ $m - qp = r \in H \quad \Rightarrow \quad r\,(< p) \in H$

But p is the least positive integer that belongs to S, $\Rightarrow r = 0 \Rightarrow m = qp$

Thus $\forall \ m \in H, \exists\, q \in Z$, such that $m = qp$

$\Rightarrow$ $$H \subset \{kp \mid k \in Z\} \qquad \dots (5)$$

From (4) and (5) it follows that

$$H = \{kp \mid k \in Z\}$$

Hence the proof.

1.11 Integral Powers of an Element of a Group.

Let a be an element of a group G. Then for every integer n, by the element a^n, we mean,

(i) $a^n = a . a . a \dots\dots\dots\dots\dots a$, n times, if n is a positive integer.

(ii) $(a^{-1})^n = (a^n)^{-1}$, i.e. a^{-n} is the inverse of a^n.

(iii) $a^0 = e$, the identity element of G.

Note : If the group operation is denoted by +, then we write $a + a + a + a \dots\dots\dots\dots + a$, by na instead of a^n (n, a positive integer). We define $(-na) = -(na)$ and $0 . a = 0$, the identity element of a group.

For example, in $(Z_4, +_4)$, $6(3) = 3 +_4 3 +_4 3 +_4 3 +_4 3 +_4 3 = 2$

Here $3 \in Z_4$ and 6 is the integer by which times the element 3 is composed with itself.

In $(Z^*_5, \times_5)$, where $Z^*_5 = \{1, 2, 3, 4\}$, $3^4 = 3 \times_5 3 \times_5 3 \times_5 3 = 2$

* This follows from the well ordering principle, i.e. every non empty subset of the set n of natural numbers, contains a least element.

The following index laws hold.

If a is any element of a group G, then,

(i) $a^m \cdot a^n = a^{m+n}$

(ii) $(a^m)^n = a^{mn}$, for any integers m and n.

(iii) If G is abelian, then $\forall\ a, b \in G, (a \cdot b)^n = a^n \cdot b^n$

Example 1. If G is a group in which $(a\,b)^n = a^n \cdot b^n$ for three consecutive integers n, for all $a, b \in G$, then show that G must be abelian.

Solution : By data, $\forall\ a, b \in G$

$$(a\,b)^n = a^n \cdot b^n \quad \text{... (1)}$$

$$(a\,b)^{n+1} = a^{n+1} \cdot b^{n+1} \quad \text{... (2)}$$

$$(a\,b)^{n+2} = a^{n+2} \cdot b^{n+2} \quad \text{... (3)}$$

We shall prove that $ab = ab$, $\forall\ a, b \in G$

$(a\,b)^{n+1} = a^{n+1} \cdot b^{n+1} \quad \Rightarrow \quad (ab)^n \cdot ab = (a^n \cdot a)(b^n \cdot b)$

$\Rightarrow (a^n b^n) \cdot (ab) = (a^n \cdot a)(b^n \cdot b)$ ... [from (1)]

$\Rightarrow a^n \cdot (b^n \cdot a)\, b = a^n (a \cdot b^n)\, b$

$\Rightarrow b^n \cdot a = a \cdot b^n$ ... (4)

[By cancellation law]

Again consider,

$(a\,b)^{n+2} = a^{n+2} \cdot b^{n+2} \quad \Rightarrow \quad (ab)^{n+1} \cdot ab = (a^{n+1} \cdot a)(b^{n+1} \cdot b)$

$\Rightarrow (a^{n+1} b^{n+1}) \cdot (ab) = a^{n+1} \cdot (a \cdot b^{n+1})\, b$ ... [from (2)]

$\Rightarrow a^{n+1} (b^{n+1} a)\, b = a^{n+1} (a\, b^{n+1})\, b$

$\Rightarrow b^{n+1} \cdot a = a \cdot b^{n+1}$ [By cancellation law]

$\Rightarrow b^n (b\,a) = (a\,b^n)\, b$

$\Rightarrow b^n (b\,a) = (b^n\, a)\, b$ ... [from (4)]

$\Rightarrow b^n (b\,a) = b^n (a\,b)$

$\Rightarrow b\,a = a\,b$ [By cancellation law]

Thus for, $\forall\ a, b \in G$, we have $ab = ba$. **Hence G is abelian.**

Example 2. If G is a finite group, show that there exists a positive integer n such that $a^n = e$ for all $a \in G$.

Solution : Let $a \in G$ be arbitrary. Then the elements,

$$a, a^2, a^3, \ldots\ldots\ldots\ldots$$

are again elements of G. (This is because G is closed under the binary operation). By data G is finite. Thus these elements cannot all be distinct. Therefore there exists integers r and s with $r > s > 0$, such that $a^r = a^s$.

$\Rightarrow \quad a^r \cdot a^{-s} = a^s \cdot a^{-s} \Rightarrow a^{r-s} = e$ with $r - s > 0$

Let S be the set such that

$$S = \{q \mid a^q = e, q, \text{a positive integer}\}$$

Clearly $S \neq 0$, because $r - s \in S$

Thus S is a non empty subset of natural numbers. Therefore by well ordering principle, S has a least element say n. Hence $a^n = e$.

Example 3. If a and x are any two elements of a group G, then show that $(x\,a\,x^{-1})^n = x\,a^n\,x^{-1}$, for any positive integer n.

Solution : We shall prove the statement by induction.

Let $$(x\,a\,x^{-1})^n = x\,a^n\,x^{-1} \quad \text{... (1)}$$

Clearly $$(x\,a\,x^{-1})^n = x\,a^n\,x^{-1}, \text{ for } n = 1.$$

Let us assume that the statement (1) is true for $n = k > 1$.

i.e. $$(x\,a\,x^{-1})^k = x\,a^k\,x^{-1} \quad \text{... (2)}$$

Now, $$\begin{aligned}(x\,a\,x^{-1})^{k+1} &= (x\,a\,x^{-1})^k \,.\, (x\,a\,x^{-1}) \\ &= (x\,a^k\,x^{-1}) \,.\, (x\,a\,x^{-1}) \quad \text{[By (2)]} \\ &= (x\,a^k)\,(x^{-1}\,x)\,(a\,x^{-1}) \\ &= (x\,a^k)\,(e\,a\,x^{-1}) = x\,(a^k\,a)\,x^{-1} = x\,a^{k+1}\,x^{-1}\end{aligned}$$

This shows that (1) is true for $n = k + 1$, whenever it is true for $n = k$.

Hence by induction (1) is true for all positive integer n.

Example 4. If G is a group and $a \in G$ is a fixed element of G, then the set $H = a^n \mid n \in z$, is a subgroup of G.

Solution : By definition, $a^0 = e \in H$. Thus $H \neq \phi$.

Let $x, y \in H$ be arbitrary elements. $\Rightarrow \exists\, m, n \in Z$, such that $x = a^m$ and $y = a^n$

Now, $x\,y^{-1} = a^m \,.\, (a^n)^{-1} \Rightarrow x\,y^{-1} = a^m \,.\, a^{-n} = a^{m-n}$

Now, $m, n \in \mathbf{Z} \Rightarrow m - n \in Z \Rightarrow a^{m-n} \in H$

Thus, $\forall\, x, y \in H, x\,y^{-1} \in H \Rightarrow$ **H is a subgroup of G.**

The subgroup H in the above example, is called the **cyclic subgroup** (or simply subgroup) **generated by the element a** of the group G.

1.12 Order of an Element

Definition : The order of an element a of a group G, is the least positive integer m such that $a^m = e$. If there exists no such integer for which $a^m = e$, we say the element a is of infinite order.

The order of the element a is denoted by $O\,(a)$, read as order of a.

Note :

1. For any group, the identity element is the only element of order 1.
2. If for $a \in G$, where G is a group, $a^n = e$, for some positive integer n, then $O\,(a) \leq n$.
3. If $O\,(a) \leq n$, then $a^m \neq e$ for $0 < m < n$.

Example 1. Find the orders of the elements of the multiplicative group $G = \{1, -1, i, -i\}$ of fourth root of unity.

Solution : The element 1 is the identity element.

Thus $O\,(1) = 1$. Now $(-1)^2 = 1$. Thus $O\,(-1) = 2$. Again, we have $i^4 = 1$ and for no positive integer $m < 4$, $i^m = 1$. Thus $O\,(i) = 4$. Again $(-i)^4 = 1$ and therefore $O\,(-i) = 4$.

Example 2. Find the orders of the elements of the multiplicative group $G = \{1, \omega, \omega^2\}$ of cube roots of unity.

Solution : The identity element of the group is 1. Thus $O\,(1) = 1$.

Now $$\omega^3 = 1 \Rightarrow O\,(\omega) = 3$$

Again, $$(\omega^2)^3 = (\omega^3)^2 = 1 \Rightarrow O\,(\omega^2) = 3$$

Example 3. Find the orders of the elements of the group $G = (Z_4, +_4)$.

Solution : The given set $Z_4 = \{0, 1, 2, 3\}$ and the binary operation is in the addition notation. The identity element is 0. Thus $O(0) = 1$.

Consider, $1 +_4 1 \neq 0, 1 +_4 1 +_4 1 \neq 0$, but $1 +_4 1 +_4 1 +_4 1 = 0 \Rightarrow O(1) = 4$

Again, $2 +_4 2 = 0 \Rightarrow O(2) = 2$

Also, $3 +_4 3 +_4 3 +_4 3 = 0 \Rightarrow O(3) = 4$

Note :

1. In the group $(Z, +)$ of integers, every element except identity 0, is of infinite order.
2. In the multiplicative group of positive rational numbers, every element except the identity element 1, is of the infinite order.

1.13 Properties Related to Order of an Element of a Group.

In this section we shall see few important properties related to order of an element of a group.

1. In a group G, $O(a) = O(a^{-1})$, for every $a \in G$.

Proof : We shall consider two cases.

Case (i) : Let $O(a)$ be infinite. In this case we shall show that $O(a^{-1})$ is also infinite.

Let us assume the contrary – i.e., let if possible $O(a^{-1})$ be finite, say $O(a^{-1}) = m$.

$\Rightarrow$ m is the least positive integer, such that $(a^{-1})^m = e$, where e is the identity element.

Now, $O(a^{-1})^m = e \quad \Rightarrow \quad a^{-m} = e$

$\Rightarrow \quad a^{-m} . a^m = e = e . a^m \Rightarrow e = a^m$

$\Rightarrow O(a)$ is finite and $\leq m$. But this contradicts our data. Hence $O(a^{-1})$ is also infinite.

Case (ii) : Let $O(a)$ be finite and $O(a) = n$. That is, n is the least positive integer, such that $a^n = e$.

Now, $a^n = e \Rightarrow a^n . a^{-n} = e . a^{-n}$

$\Rightarrow e = a^{-n}$

$\Rightarrow (a^{-1})^n = e \Rightarrow O(a^{-1})$ is also finite and $O(a^{-1}) \leq n$.

Let $O(a^{-1}) = m$. Thus $(a^{-1})^m = e$

Now, $(a^{-1})^m = e \Rightarrow a^{-m} . a^m = e . a^m$

$\Rightarrow e = a^m \Rightarrow O(a) \leq m$

Now, $\left.\begin{array}{l} O(a^{-1}) \leq n \Rightarrow m \leq n \\ O(a^{-1}) \leq m \Rightarrow n \leq m \end{array}\right] \Rightarrow n = m$

Hence, $O(a) = O(a^{-1})$

2. If a is any element of the group G, is of order n, then $a^m = e$, for any integer m, if and only if n divides m.

Proof : Let $a^m = e$. We shall prove that n divides m.

Now $a^m = e$ and $O(a) = n \Rightarrow n \leq m$ (Because n is the least positive integer such that $a^n = e$).

Let, $m = q\,n + r, q, r \in Z$ and $0 \leq r \leq n$

Now, $a^m = e \Rightarrow a^{(q\,n + r)} = e$

$\Rightarrow \quad a^{qn} . a^r = e$

$\Rightarrow \quad (a^n)^q . a^r = e$

$\Rightarrow \quad e^q . a^r = e$ [Since $a^n = e$]

$\Rightarrow \quad a^r = e$

By data $O(a) = n$ and therefore n is the least positive integer such that $a^n = e$

Thus $0 \leq r \leq n$ and $a^r = e$ is a contradiction.

Hence $r = 0$ and therefore $m = q\,r$. This implies n divides m.

Conversely, let n divide m. Then we shall show that $a^m = e$.

Now, $n \mid m \Rightarrow m = n\,q$, q being an integer.

$\Rightarrow a^m = a^{nq}$

$\Rightarrow a^m = (a^n)^q$

$\Rightarrow a^m = e^q$ [Since $a^n = e$]

$\Rightarrow a^m = e$

Hence the proof.

Note : It follows from the above property that if $a \in G$ is of finite order and $a^m = e$, then $O(a) \leq m$.

3. If a and x be any two elements of the group G then $O(a) = O(x\,a\,x^{-1})$

Proof : Let $O(a) = n$. Then n is the least positive integer such that $a^n = e$.

Now, $(x\,a\,x^{-1})^n = x\,a^n\,x^{-1}$ [See example 3 of section 1.4]

$\Rightarrow (x\,a\,x^{-1})^n = x\,e\,x^{-1}$ [Since $a^n = e$]

$\Rightarrow (x\,a\,x^{-1})^n = x\,x^{-1} = e \Rightarrow O(x\,a\,x^{-1})$ is finite and is $\leq n$ ... (1)

Let, $O(x\,a\,x^{-1}) = m$. Thus $m \leq n$

Now, $O(x\,a\,x^{-1}) = m \Rightarrow (x\,a\,x^{-1})^m = e$

$\Rightarrow x\,a^m\,x^{-1} = e$

$\Rightarrow x\,a^m\,x^{-1} = x\,e x^{-1}$

$\Rightarrow a^m = e$ [By cancellation law]

$\Rightarrow O(a) \leq m \Rightarrow n \leq m$ ... (2)

From (1) and (2), we have $m = n$. Hence $O(a) = O(x\,a\,x^{-1})$.

Corollary : $O(a\,b) = O(b\,a)$, $\forall\ a, b \in G$, where G is a group.

Proof : From the above property, we have

$O(a) = O(x\,a\,x^{-1}), \quad \forall x \in G$

$\Rightarrow \quad O(a\,b) = O(x\,a\,b\,x^{-1}), \quad \forall x \in G$

In particular,

$O(a\,b) = O[b\,(a\,b)\,b^{-1})] \Rightarrow O(a\,b) = O[(b\,a)\,(b\,b^{-1})]$

$\Rightarrow$ $\mathbf{O(a\,b) = O[(b\,a)]}$ [Since $b\,b^{-1} = e$]

4. In a group G, the order of any element a cannot exceed that of the group. That is $O(a) \leq O(G)$

Proof : Let G be a group and $a \in G$ be arbitrary. Let H be the subgroup of G generated by the element a.

Let, $O(a) = n$. Thus, $H = \{e, a, a^2, a^3, \ldots\ldots\ldots, a^{n-1}\}$

H contains n elements and therefore $O(H) = n$.

We shall consider two cases.

Case (i) : Let $H = G$

$\Rightarrow$ $O(H) = O(G) = n \Rightarrow O(G) = O(a)$

Case (ii) : Let $H \neq G$

$\Rightarrow$ $O(H) < O(G) \Rightarrow n < O(G) \Rightarrow O(a) < O(G)$

Combining both the cases we have, $\mathbf{O(a) \leq O(G)}$

5. The order of any power of an element of a group cannot exceed the order of the element.

Proof : Let G be a group and $a \in G$ be arbitrary.

Let, $O(a) = n$. Thus n is the least positive integer, such that $a^n = e$

Let, a^k be any power of a.

Now, $a^n = e \Rightarrow (a^n)^k = e \Rightarrow (a^k)^n = e \Rightarrow O(a^k)$ is finite and $\leq n$

Hence, $\mathbf{O(a^k) \leq O(a)}$

6. Let G be a group and $a \in G$. If $O(a) = n$ and $(m, n) = 1$, then $O(a^m) = n$.

Proof : Let $O(a^m) = k$. By the property 5, $k \leq n$

$O(a^m) = k \quad \Rightarrow (a^m)^k = e \Rightarrow a^{mk} = e \Rightarrow n \mid mk$ [By the Property 2]

Now, by data, we have $(m, n) = 1$

Thus, $n \mid mk$ and $(m, n) = 1 \Rightarrow n \mid k$

Further, $n \mid k$ and $k \leq n \Rightarrow k = n$. Hence, $O(a^m) = n = O(a)$.

7. If $O(a) = n$, where a is an element of the group G and $d = (n, m)$ then $O(a^m) = \dfrac{n}{d}$.

Proof : Let $O(a^m) = k$. Then k is the least positive integer such that $(a^m)^k = e$

By data $\quad d = (n, m) \quad \Rightarrow d \mid n$ and $d \mid m$

$\Rightarrow n = dq$ and $m = dp$

where p and q are positive integers with $(q, p) = 1$.

Now, $\quad (a^m)^k = e \quad \Rightarrow a^{mk} = e$

$\Rightarrow n \mid mk$ [By the property 2]

$\Rightarrow dq \mid mk$

$\Rightarrow dq \mid (dp)k$

$\Rightarrow q \mid pk$

$\Rightarrow q \mid k$ [Since $(q, p) = 1$]

Consider, $\quad (a^m)^q = a^{mq} \quad \Rightarrow (a^m)^q = a^{dpq}$

$\Rightarrow (a^m)^q = (a^{dq})^p$

$\Rightarrow (a^m)^q = (a^n)^p$ [Since $n = dq$]

$\Rightarrow (a^m)^q = e^p$ [Since $a^n = e$]

$\Rightarrow (a^m)^q = e$

Now, $O(a^m) = k$ and $(a^m)^q = e \Rightarrow k \mid q$ [By property 2]

Now, $\quad q \mid k$ and $k \mid q \Rightarrow q = k$

Thus, $\quad n = dk \Rightarrow k = \dfrac{n}{d} \Rightarrow O(a^m) = \dfrac{n}{d}$.

Example 1. Give an example of a group to show that $O(ab)$ need not be equal to $O(a) . O(b)$.

Solution : Consider the group $G = \{1, -1, i, -i\}$ of fourth root of unity under multiplication.

Now, $(-1)^2 = 1 \Rightarrow O(-1) = 2, \quad (-i)^4 = 1 \Rightarrow O(-i) = 4$

Now, $(-1)(-i) = i$ and $(i)^4 = 1 \Rightarrow 0[(-1)(-i)] = 4$

But, $O(-1) . O(-i) = 2 . 4 = 8$

Thus, $O(-1) . O(-i) \neq O[(-1) . (-i)]$

Example 2. Let a, b be two elements of a group G such that $ab = ba$. If $O(a)$ and $O(b)$ are relatively prime, show that $O(ab) = O(a) . O(b)$.

Solution : Let $O(a) = n$ and $O(b) = m$

Thus, $a^n = e$ and $b^m = e$. Also by data $(m, n) = 1$

Now, $(ab)^{mn} = a^{mn} . b^{mn}$ [Since $ab = ba$]

$= (a^n)^m . (b^m)^n = e^m . e^n = e$

$\Rightarrow$ $O(ab)$ is finite and $\leq mn$

Let $O(ab) = l$. Then $l \leq mn$ and $l \mid (mn)$ ... (1)

Now, $(ab)^l = e \Rightarrow a^l b^l = e$ [Since $ab = ba$]

$\Rightarrow a^l = b^{-l}$

Now, $a^{lm} = (a^l)^m = (b^{-l})^m = (b^m)^{-l} = e \Rightarrow n \mid lm$ [Since $O(a) = n$]

Now, $(m, n) = 1$ and $n \mid lm \Rightarrow n \mid l$

Similarly, $m \mid l$. Thus $mn \mid l$ ... (2)

From (1) and (2), we have $l = mn$. Thus, $O(ab) = O(a) . O(b)$.

1.14 Cyclic Groups

In this section, we shall consider a special type of group called the **cyclic group** and its properties.

We have seen in the example 4 of § 1.4, that if G is a group and $a \in G$, the subset $H = \{a^n \mid n \in Z\}$ is a subgroup of G and is known as the cyclic subgroup of G generated by the element a. In this subgroup for every element $b \in H$, there exists an integer $m \in Z$, such that $b = a^m$. That is every element of H can be expressed as an integral power of a. For this reason, H is called the **"subgroup generated by a"**. If this property of expressing every element as an integral power of a single element, say a hold good for the whole group, then such group is called a **cyclic group** generated by the element a. The formal definition is given below.

Definition : A group G is said to be a cyclic group generated by the element $a \in G$ if $G = \{a^n \mid n \in Z\}$ and is denoted by $G = \langle a \rangle$.

In other words, a group G is said to be cyclic, if there exists an element $a \in G$, such that for every $b \in G$, there exists $n \in Z$ with $b = a^n$.

Note :

1. If the binary operation for the cyclic group $G = \langle a \rangle$ is +, then every element of G is of the form na for some $n \in Z$.
2. A cyclic group may have more than one generator.
3. A cyclic group is said to be finite cyclic group, if the group is finite, otherwise it is said to be infinite cyclic group.

Example 1. Show that the multiplicative group of fourth roots of unity is a cyclic group.

Solution : The group is, $G = \{1, -1, i, -i\}$

Now, $(i)^1 = i$, $(i)^2 = -1$, $(i)^3 = -i$, $(i)^4 = 1$

Therefore i is a generator of G. Thus G is a cyclic group generated by the element i. The element $-i$ is an another generator of G.

Example 2. Show that the multiplicative group of cube roots of unity is a cyclic group.

Solution : The group is $G = \{1, \omega, \omega^2\}$

Consider, $(\omega)^1 = \omega$, $(\omega)^2 = \omega^2$, $(\omega)^3 = 1$

Therefore, $\omega \in G$, generates the elements of G.

Thus G is a cyclic group with ω as a generator.

Example 3. Show that the group $(Z_5, +_5)$ is a cyclic group and every non zero element of Z_5 is a generator.

Solution : Here $Z_5 = \{0, 1, 2, 3, 4\}$. The group operation is in additive notation.

Consider an element say $4 \in Z_5$

Now, $4 = 1 . 4$,

$$4 +_5 4 = 3 \Rightarrow 3 = 2 . 4$$
$$4 +_5 4 +_5 4 = 2 \Rightarrow 2 = 3 . 4$$
$$4 +_5 4 +_5 4 +_5 4 = 1 \Rightarrow 1 = 4 . 4$$
$$4 +_5 4 +_5 4 +_5 4 +_5 4 = 0 \Rightarrow 0 = 5 . 4$$

Thus every element of Z_5 can be expressed as some integral multiple of the element $4 \in Z_5$. Therefore Z_5 is a cyclic group with 4 as a generator. Similarly one can verify any other non zero element of Z_5 is also a generator.

Note :

1. $(Z, +)$ is a infinite cyclic group with 1 as a generator.
2. $(Z_n, +_n)$ is a finite cyclic group with 1 as a generator.
3. The group $(Q, +)$ of all rational numbers is not cyclic. This can be established as below. Supposing $(Q, +)$ is cyclic, with $q \in Q$ as a generator. Clearly $q \neq 0$.

Let $q = \dfrac{m}{n}$, $(m, n \in Z$ and $n \neq 0)$

Now, $\dfrac{1}{2n} \in Q \Rightarrow \dfrac{1}{2n} = r\,q$, for some, $r \in Z$,

$$\Rightarrow \frac{1}{2n} = \frac{rm}{n} \Rightarrow \frac{1}{2} = r\,m$$

This is impossible as the product of two integers is not equal to a fraction. Thus $(Q, +)$ cannot be a cyclic group.

1.15 Properties of Cyclic Group

In this section we shall see few properties of cyclic group in the form of theorems.

Theorem 1. Every cyclic group is abelian.

Proof : Let G be a cyclic group with `a' as a generator.

Thus, $G = \langle a \rangle = \{a^n \mid n \in Z\}$ and $x, y \in G$ be arbitrary.

$\Rightarrow \exists$ two integers say m and n, such that, $x = a^m$ and $x = a^n$

Now, $x . y = a^m . a^n \Rightarrow x . y = a^{m+n}$

$$\Rightarrow x . y = a^{n+m} \Rightarrow x . y = a^n . a^m \Rightarrow x . y = y . x$$

Thus, $\forall\ x, y \in G$, we have $x . y = y . x$. **Thus G is an abelian.**

Note : The converse of the above theorem is not true. That is every abelian group is not necessarily cyclic. For example consider the set, $G = \{v_1, v_2, v_3, v_4\}$

Consider in G, an associative binary composition $*$, given by the following table.

$*$	v_1	v_2	v_3	v_4
v_1	v_1	v_2	v_3	v_4
v_2	v_2	v_1	v_4	v_3
v_3	v_3	v_4	v_1	v_2
v_4	v_4	v_3	v_2	v_1

Clearly the element v_1 is the identity element. Further the inverses of v_1, v_2, v_3 and v_4 are respectively are themselves. Thus $(G, *)$ is an abelian group and known as **Klien's 4 group.**

Now, one can observe that $<v_1> = \{v_1\}, <v_2> = \{v_1, v_2\}, <v_3> = \{v_1, v_3\}, <v_4> = \{v_1, v_4\}$

Thus no element of G, generates the whole group G. Hence, **G is not cyclic.**

Theorem 2. If a is a generator of a cyclic group G, then a^{-1} is also a generator.

Proof : By data $G = <a> \Rightarrow \forall\ g \in G,\ \exists\ m \in Z,$ such that $g = a^m$

But g can be written as $g = (a^{-1})^{-m} = (a^{-1})^n$, where $n = -m$

Thus $\forall\ g \in G, \exists\ n \in Z$, such that $g = (a^{-1})^n \Rightarrow a^{-1}$ is also a generator of G.

Theorem 3. If a is a generator of a cyclic group G, then $O(a) = O(G)$.

Proof : Let $O(a) = n \Rightarrow n$ is the least positive integer such that $a^n = e$, where e is the identity element. Consider the subgroup H generated by the element a.

That is, $$H = \{a, a^2, a^3, \ldots\ldots\ldots, a^{n-1}, a^n = e\}$$

The elements of H are distinct. – for if $a^r = a^s$, where $0 < r < s \leq n$, then

$$a^r = a^s \Rightarrow a^{r-s} = e, \text{ for } 0 < r - s < n.$$

This contradicts the fact that n is the least positive integer, such that $a^n = e$. Thus the elements of H are distinct.

Let $g \in G$ be arbitrary. Then for some $m \in Z, g = a^m$.

By division algorithm, there exists two integers q and r, such that $m = q\,n + r, 0 \leq r < n$.

Now, $$g = a^m \Rightarrow g = a^{q\,n + r}$$
$$\Rightarrow g = (a^n)^q \,.\, a^r$$
$$\Rightarrow g = e \,.\, a^r \qquad [\text{Since } a^n = e]$$
$$\Rightarrow g = a^r$$

Since $r < n$, the element $a^r \in H \Rightarrow g \in H$.

Thus $\forall\ g \in G$, is also an element of $H \Rightarrow G \subseteq H$.

But $H \subseteq G$. Hence $G = H \Rightarrow O(G) = n = O(a)$.

Theorem 4. If G is a finite group of order n, containing an element of order n, then G is cyclic.

Proof : Let $a \in G$, be the element of order n. Let H be the cyclic subgroup of G generated by the element a.

By data $O(a) = n$, we have $O(H) = n = O(G)$. Further $H \subseteq G$. Thus $H = G$ and hence G is cyclic.

Theorem 5. Let G be a cyclic group of order k and a be a generator. If $a^m = a^n$ $(m \neq n)$, then $m \equiv n \pmod{k}$ and conversely.

Proof : By data $O(G) = k$ and $G = <a>$. Thus $O(a) = k$

Now, $$a^m = a^n \Leftrightarrow a^m \,.\, a^{-n} = a^n \,.\, a^{-n}$$
$$\Leftrightarrow a^{m-n} = e$$

$\Leftrightarrow \quad k \mid (m-n)$

$\Leftrightarrow \quad m \equiv n \pmod{k}$

Theorem 6. Let G be a cyclic group of order d and a be a generator. The element a^k $(k < d)$ is also a generator of G, if and only if $(k, d) = 1$.

Proof : By data $G = \langle a \rangle$ and $O(G) = d$. Thus $O(a) = d$. Let a^k $(k < d)$ be a generator of G.

$\Rightarrow \quad a = (a^k)^s$ for some integer s.

$\Rightarrow \quad a \,.\, a^{-1} = a^{ks} \,.\, a^{-1} \Rightarrow e = a^{ks-1}$

$\Rightarrow \quad d \mid ks - 1$ [Since $O(a) = d$]

$\Rightarrow \quad ks - 1 = qd$, for some integer q.

$\Rightarrow \quad ks - qd = 1$

$\Rightarrow \quad (k, d) = 1$

Conversely, let k be an integer such that $(k, d) = 1 \Rightarrow \exists$ two integers s and q such that $ks + qd = 1$

Now, $\quad a = a^1 \Rightarrow a = a^{ks+qd}$

$\Rightarrow \quad a = a^{ks} \,.\, (a^d)^q$

$\Rightarrow \quad a = a^{ks} \,.\, e$ [Since $a^d = e$]

$\Rightarrow \quad a = (a^k)^s$

Let, $g \in G$ be any element.

$\Rightarrow \quad g = a^n$ for some integer n as a is a generator. $\Rightarrow g = (a^{ks})^n \Rightarrow g = (a^k)^{sn}$

$\Rightarrow$ every element $g \in G$ can be expressed as some integral power of a^k.

$\Rightarrow$ **a^k is also a generator of G.**

Theorem 7. Every subgroup of a cyclic group is cyclic.

Proof : Let $G = \langle a \rangle$ be a cyclic group and H be a subgroup of G.

Let, $\quad S = \{m \in Z \mid a^m \in H\}$

H is a subgroup of G, thus $e \in H$, where e is the identity element of G.

$\Rightarrow \quad e = a^0 \in H \Rightarrow 0 \in S \Rightarrow S \neq \phi$

Let, $\quad m, n \in S$ be arbitrary $\Rightarrow a^m, a^n \in H$

By data H is a subgroup of G, thus $a^m \,.\, (a^n)^{-1} \in H$

Now, $\quad a^m \,.\, (a^n)^{-1} \in H \Rightarrow a^m \,.\, a^{-n} \in H \Rightarrow a^{m-n} \in H$

$\Rightarrow m - n \in S$, for all $m, n \in S$

Hence S is a subgroup of $(Z, +)$.

Therefore, $\exists$ a positive integer p such that, $S = \{kp \mid k \in Z\}$ or $S = \{0\}$

If $S = \{0\}$, then $H = \{e\} = \langle e \rangle$

If $S = \{kp \mid k \in Z\}$ then $H = \{a^m \mid m = kp, k \in Z\}$. Thus $H = \langle a^p \rangle$ therefore cyclic.

Definition : Let d be any positive integer. The number of positive integer less than d and prime to d, is denoted by $\varphi(d)$.

From the theory of numbers, we know that any integer n can be decomposed into product of prime factors, as

$$n = p_1^{\alpha_1} \cdot p_2^{\alpha_2} \cdot p_3^{\alpha_3} \ldots\ldots\ldots\ldots p_k^{\alpha_k}$$

where $p_1, p_2, \ldots\ldots\ldots\ldots p_k$ are prime factors.

Further we have

$$\varphi(n) = n\left(1 - \frac{1}{p_1}\right)\left(1 - \frac{1}{p_2}\right) \ldots\ldots\ldots \left(1 - \frac{1}{p_k}\right)$$

Theorem 8. A cyclic group of order d has $\varphi(d)$ generators.

Proof : Let G be a cyclic group, with generator a.

By data $O(G) = d$ and therefore $O(a) = d$.

The distinct elements of G are

$$G = \{a, a^2, a^3, \ldots\ldots\ldots\ldots, a^d = e\}$$

By theorem 6, we have that the generators of G are the elements of the form a^k, with $(k, d) = 1$. $\Rightarrow$ the number of generators of G is the number of positive integers less than d and prime to d. This is $\varphi(d)$.

Hence there are $\varphi(d)$ generators for the cyclic group G of order d.

Example 1. Find the number of generators of the cyclic group of order 60.

Solution : Let G be the cyclic group of order 60.

Now 60 can be written as $60 = 5 \,.\, 3 \,.\, 2^2$.

Now, $$\varphi(60) = 60\left(1-\frac{1}{5}\right)\left(1-\frac{1}{3}\right)\left(1-\frac{1}{2}\right) = 60\left(\frac{4}{5}\right)\left(\frac{2}{3}\right)\left(\frac{1}{2}\right) = 2\,.\,4\,.\,2\,.\,1 = 16$$

Thus **G has 16 generators.**

Example 2. Find the number of generators of the cyclic group G of order 8.

Solution : By data $O(G) = 8$. Also $8 = 2^3$

$$\varphi(8) = 8\left(1-\frac{1}{2}\right) = 8.\frac{1}{2} = 4$$

Hence **G has 4 generators.**

If a is one of the generator, then other generators are of the form a^k, with $(k, 8) = 1$.

The integers that are less than 8 and prime to 8 are 1, 3, 5, 7. Thus the four generators are a, a^3, a^5 and a^7.

Note : If the group operation is in addition form, and a is one of the generator, then other generators are of the form $k \,.\, a$ with $(k, n) = 1$, where $O(G) = n$.

Example 3. Find the number of generators of the cyclic group $(Z_{18}, +_{18})$. Write all the generators.

Solution : The order of $Z_{18} = 18$. Now $18 = 2 \,.\, 3^2$.

The number of generators is $\varphi(18) = 18\left(1-\frac{1}{2}\right)\left(1-\frac{1}{3}\right) = \mathbf{6}$.

We know that 1 is a generator. Thus the other generators of Z_{18} are of the form $k \,.\, 1$, with $(k, 18) = 1$. The integers that are less than 18 and prime to 18 are 1, 5, 7, 11, 13 and 17.

Thus the generators are **1, 5, 7, 11, 13 and 17.**

Example 4. How many generators are there for the cyclic group of order 10. If a is a generator, what are the other generators?

Solution : Let G be a cyclic group of order 10.

Now, $$10 = 5 \,.\, 2 \Rightarrow \varphi(10) = 10\left(1-\frac{1}{5}\right)\left(1-\frac{1}{2}\right) = 4.$$

Thus **there are 4 generators.**

By data a is a generator. The other generators are of the form a^k, with $(k, 10) = 1$.

Hence $k = 1, 3, 7, 9$. Thus the generators are $\boldsymbol{a, a^3, a^7}$ **and** $\boldsymbol{a^9}$.

1.16 Coset Decomposition of a Group

Let $(G, *)$ be a group and H be a subgroup of G. We shall define a relation called **"Congruence modulo H" on G as below.**

For $a, b \in G$ **"a is said to be congruent to b modulo H"** or aRb, if and only if $\boldsymbol{a.b^{-1} \in H}$.

We shall now show this relation is an equivalence relation.

Theorem 1. The relation R on a group G defined by aRb if and only if $a\,.\,b^{-1} \in H$, where H is a subgroup is an equivalence relation.

Proof : 1. For every $a \in G, a\,.\,a^{-1} = e \in H$ [Since H is a subgroup of G].

$\Rightarrow$ $aRa, \forall\ a \in G$. **Hence R is reflexive.**

2. Let $a, b \in G$, such that aRB.

$\Rightarrow a\,.\,b^{-1} \in H \Rightarrow (a\,.\,b^{-1})^{-1} \in H$ [Since H is a subgroup]

$\Rightarrow (b^{-1})^{-1}.\,a^{-1} \in H \Rightarrow b\,.\,a^{-1} \in H \Rightarrow bRa$. **Hence R is symmetric.**

3. Let $a, b, c \in G$, such that aRb. and $bRc \Rightarrow a\,.\,b^{-1} \in H$ and $b\,.\,c^{-1} \in H$

Since H is a subgroup $(a\,.\,b^{-1})(b\,.\,c^{-1}) \in H \Rightarrow a\,(b^{-1}b)\,c^{-1} \in H \Rightarrow a\,.\,c^{-1} \in H \Rightarrow aRc$

Hence R is transitive. Thus, R is an equivalence relation.

If a is congruent to b mod H, then we write $a \equiv b$ (mod H)

The relation $a \equiv b$ (mod H) is an equivalence relation. Hence this relation gives a partition of G, into disjoint classes. Each member in the partition is a non-empty subset of G and any two distinct members of the partition are disjoint.

If $a \in G$, we denote the equivalent class generated by 'a' by $[a]$.

i.e. $[a] = \{x \in G \mid x \equiv a \text{ (mod } H)\} = \{x \in G \mid xa^{-1} \in H\}$

Definition : Let H be a subgroup of a group $(G, *)$ and a be any element of G.

The set Ha defined by $H\,a = \{ha \mid h \in H\}$ is called the right coset of H in G w.r.t. 'a'

The set defined by $aH = \{a\,h\}\ h \in H\}$ is called the left coset of H in G w.r.t. 'a'.

Note :

1. $Ha \neq \phi$ - for $e \in H$ and $e\,.\,a = a \in Ha \Rightarrow a \in Ha$
   Similarly $aH \neq \phi$ - as $a \in a\,H$
2. $He = \{he \mid h \in H\} = \{h \mid h \in H\} = H$
   Similarly $eH = H$. Thus every subgroup H itself is a right coset and as well left coset of H in G.
3. If G is abelian, then every right coset Ha is equal to corresponding left coset aH - i.e., if G is abelian $Ha = aH$. But in general $Ha \neq aH$.

Example 1. Find all the right cosets of the subgroup $H = \{4n \mid n \in Z\} = 4Z$, in the additive group of integers Z.

Solution : The right coset of 4Z are of the form $4Z + a, a \in Z$.

By division algorithm, there exists two unique integers q and r such that $a = 4q + r, 0 < r < 4$

$$\therefore \quad 4Z + a = \{4n + a \mid n \in Z\}$$
$$= \{4n + 4q + r \mid n \in Z, 0 < r < 4\}$$
$$= \{4m + r \mid m \in Z, 0 < r < 4\} \quad (n + q = m).$$

$\therefore$ the distinct right cosets of 4Z are

$$4Z+0, \quad 4Z+1, \quad 4Z+2, \quad 4Z+3. \text{ i.e. } 4Z, \quad 4Z+1, \quad 4Z+2, \quad 4Z+3$$

For example,

$$4Z + 2 = \{4n + 2 \mid n \in Z\}$$
$$= \{\ldots\ldots\ldots -6, -2, 2, 6, 10, \ldots\ldots\ldots\}$$

Theorem 2. If H is a subgroup of a group G, then for all $a \in G$, $[a] = Ha$.

i.e. equivalence class generated by the element "a" under the equivalence relation $a \equiv b$ (mod H) is precisely the right coset of H in G w.r.t. 'a'.

Proof : We have, $Ha = \{ha \mid \forall\, h \in H\}$ and $[a] = \{x \in G \mid x \equiv a \text{ (mod } H)\}$

Let, $x \in [a] \Rightarrow x \equiv a$ (mod H)

$\Rightarrow xa^{-1} \in H$

$$\Rightarrow \quad (xa^{-1})a \in Ha \quad \Rightarrow xe \in Ha \quad \Rightarrow x \in Ha$$

$\therefore \quad [a] \subset Ha$

Again, let $y \in Ha \quad \Rightarrow \quad y = ha$ for some $h \in H$

$$\Rightarrow \quad ya^{-1} = (ha)a^{-1}$$
$$\Rightarrow \quad ya^{-1} = h(aa^{-1})$$
$$\Rightarrow \quad ya^{-1} = he \Rightarrow ya^{-1} = h \Rightarrow y \equiv a \pmod{h} \Rightarrow y \in [a]$$

$\therefore \quad Ha \subset [a]$. **Hence, $[a] = Ha$.**

From the above theorem, we have $\forall\, a \in G, Ha = [a]$. Thus any two distinct cosets of H in G are disjoint and the union of all the right cosets of H is whole of G. In other words the collection of all distinct right cosets of H in G is called the **"right coset composition"** of the group G, w.r.t. the subgroup H.

In the above manner, if one defines a relation called **"congruence modulo H"** on G by $a \equiv b \pmod{H}$ if and only if $b^{-1}a \in H$. It can be shown that $[a] = aH$ and hence the collection of all distinct left cosets of H in G form a **partition** of the group G. This decomposition of the group G is called **"left coset decomposition"** of the group w.r.t. the subgroup H.

The following two theorems assert that the number of elements in any two distinct right (left) cosets of H in G are same and also the total number of right cosets of H in G is same as the total number of left cosets of H in G.

Theorem 3. If H is a subgroup of G, then there exists a one-to-one correspondence between any two right (left) cosets of H in G.

Proof : Let $a, b \in G$ and Ha and Hb be two right cosets of H w.r.t. a and b.

If $Ha = Hb$, then there is nothing to prove. Let, $Ha \cap Hb = \phi$.

Define a mapping, $f : Ha \to Hb$

by $f(ha) = h\,b$, for all $h \in H$

(i) f is well defined and one-one :

Let, $h_1 a = h_2 a \iff h_1 = h_2$ [By cancellation law]

$$\iff h_1 b = h_2 b \iff f(h_1 a) = f(h_2 a)$$

$\therefore f$ is well defined and one-one.

(ii) f is onto :

Let $h\,b \in Hb$ be arbitrary. As $h \in H$, there exists $ha \in Ha$ such that $f(ha) = hb$. Thus f is onto.

Hence **f is a bijection and hence the theorem.**

Theorem 4. There is one-to-one correspondence between the set of all right cosets and the set of all left cosets of a subgroup of a group.

Proof : Let R and L be the sets of all distinct right cosets and left cosets of a subgroup H of a group G, respectively.

Define a map, $f : R \to L$ by $f(Ha) = a^{-1}H$, for all $a \in G$

(i) f is well defined and one-one :

Let, $Ha = Hb \iff [a] = [b]$

$$\iff a \in [b]$$
$$\iff ab^{-1} \in H \quad \text{[Right coset decomposition]}$$
$$\iff a = hb$$
$$\iff a^{-1} = b^{-1}h^{-1}$$
$$\iff ba^{-1} = h^{-1}$$

$\Leftrightarrow \quad (b^{-1})^{-1} a^{-1} = h^{-1} \in H$ [Since H is a subgroup]

$\Leftrightarrow \quad a^{-1} \equiv b^{-1} \pmod H$

$\Leftrightarrow \quad [a^{-1}] \equiv [b^{-1}]$ [Left coset decomposition]

$\Leftrightarrow \quad a^{-1}H \equiv b^{-1}H$

$\Leftrightarrow \quad f(Ha) \equiv f(Hb)$

$\therefore$ f is well defined and one-one.

(ii) *f* is onto :

Consider, $aH \in L$. Thus, $Ha^{-1} \in R$

Further, $\because f(Ha^{-1}) = (a^{-1})^{-1} H = aH$

$\therefore$ for every $aH \in L$, there exists $Ha^{-1} \in R$ such that $f(Ha^{-1}) = aH$. Thus, f is onto.

Thus *f* is a bijection and hence the theorem.

Definition [Index of a subgroup] : The number of distinct right (left) coset of a subgroup *H* of a group *G* is called the index of the subgroup *H* in *G* and is denoted by [*G* : *H*] and read as "the index of *H* in *G*".

1.17 Lagrange's Theorem

In this section we shall establish the well known theorem due to Lagrange and prove some of the famous results about finite group.

Theorem (Lagrange) : If *G* is any finite group and *H* is any subgroup of *G*, then *O* (*H*) divides *O* (*G*).

Proof : Let $O(G) = n$, and $O(H) = m$.

We shall show that $m \mid n$. i.e. $n = km$, for some integer k.

Since the group G is finite, there are finite number of right (left) cosets of H in G.

Let G have k distinct right (left) cosets of H and these be are.

$$Ha_1, Ha_2, Ha_3, \ldots\ldots\ldots, Ha_k \Rightarrow [G:H] = k$$

We know that any two right (left) cosets of H have the same number of elements.

In particular H itself is a right (left) coset of H in G. Since $O(H) = m$, each right (left) coset of H will have m elements.

Also we have, $G = Ha_1 \cup \ldots\ldots\ldots \cup Ha_k$

$$\therefore \quad n = O(G) = O(Ha_1) + O(Ha_2) + \ldots\ldots\ldots O(Ha_k)$$
$$= O(H) + O(H) + \ldots\ldots\ldots O(H)$$
$$= k \cdot O(H) = km$$

$\therefore \quad n = km \Rightarrow m \mid n \Rightarrow$ **$O(H)$ divides $O(G)$**

Corollary : The index of any subgroup of a finite group is a divisor of the order of the group.

Proof : Index of the subgroup H in the group G is the number of distinct right (left) coset of H in G. Let the index of H in G be k.

i.e. $\quad [G:H] = k \Rightarrow [G:H] = \dfrac{n}{m}$ [Since $n = km$, from Lagrange's theorem]

$\Rightarrow \quad [G:H] = \dfrac{O(G)}{O(H)} \Rightarrow O(G) = O(H)\,[G:H]$

Hence the index of the subgroup divides the order of the group.

Note :

1. The converse of Lagrange's theorem is not true in general - i.e., if G is a finite group of order n and m is a positive integral divisor of n, then G need not have a subgroup of order m.

2. It follows from the corollary that the number of right (left) coset of H in G is equal to $\frac{O(G)}{O(H)}$.

1.18 Cosequences of Lagrange's Theorem

1. If G is a finite group and $a \in G$, then $O(a)$ divides $O(G)$ i.e. the order of an element of a finite group G divides the order of G.

Proof : Let $O(G) = n,\ a \in G$ and $O(a) = m.$ Clearly m is finite as G is a finite group.

Let H be a subgroup generated by the element 'a'. i.e. $G = < a >$ and $O(H) = O(a) = m$

By Legrange's theorem $O(H) \mid O(G)$. i.e. $m \mid n,$ hence $\mathbf{O(a) \mid O(G)}$.

Corollary.

1. If G is a finite group of orer n then $\forall\ a \in G,\ a^n = e,$ where e is the identity element of G.

Proof : Let $O(a) = m \quad \Rightarrow \quad O(a) \mid O(G) = m \mid n$

$\therefore \quad \exists$ a positive integer k, such that $n = mk$

Now, $\quad a^n = a^{mk} \quad \Rightarrow \quad a^n = e \quad$ [Since $a^m = e$]

Hence, $\quad a^n = e.$

The above result can also be written as $\mathbf{a^{O(G)} = e}$.

2. Euler's theorem. If n is any positive integer and 'a' is relatively prime to n, then $a^{\varphi(n)} \equiv 1 \pmod n$.

Proof : We know that the set G_n of all positive integers less than n and prime to n is an abelian group under multiplication modulo n and its order is $\varphi(n)$.

Let, $a = qn + r,\ q,\ r \in Z, 0 \le r < n$

By data $(a, n) = 1 \quad \Rightarrow \quad (r, n) = 1$

$\Rightarrow \quad r \in G_n \quad$ [as $r < n$]

By the above corollary, we have $r^{\varphi(n)}$ = identity of $G_n = 1$

Now, $\quad a = qn + r \quad \Rightarrow \quad a \equiv r \pmod n$

$\Rightarrow \quad a^{\varphi(n)} \equiv r^{\varphi(n)} \pmod n$

$\Rightarrow \quad a^{\varphi(n)} \equiv 1 \pmod n \quad$ [Since $r^{\varphi(n)} = 1$]

Hence the theorem.

3. Fermat's theorem. If a is any integer and p is any positive prime, then $a^p \equiv a \pmod p$.

Proof : Here two cases arise.

Case (i) : Let $(a, p) = 1$. i.e. a and p are relatively prime. Then by Euler's theorem, we have

$$a^{\varphi(p)} \equiv 1 \pmod p$$

But, $\varphi(p) = p - 1$, as p is prime.

$\therefore\ a^{p-1} \equiv 1 \pmod p \quad \Rightarrow \quad a \,.\, a^{p-1} \equiv a \,.\, 1 \pmod p$

$\Rightarrow \quad a^p \equiv a \pmod p$

Hence the theorem.

Case (ii) : Let $(a, p) \neq 1$.

Now $(a, p) \neq 1$ and p is prime $\Rightarrow p \mid a$

$\Rightarrow \quad a \equiv 0 \pmod p \Rightarrow a^p \equiv 0^p \pmod p \Rightarrow a^p \equiv 0 \pmod p$

Also we have $\quad 0 \equiv a \pmod p \Rightarrow a^p \equiv a \pmod p \quad$ [Since $p \mid a \Rightarrow p \mid -a$]

Hence the theorem.

4. A finite group of prime order is cyclic and hence abelian.

Proof : Let G be a finite group and $O(G) = p$, where p is prime.

Let, $a (\neq e) \in G$ and $H = \langle a \rangle$.

By Legrange's theorem $O(H) | O(G)$. i.e. $O(H) | p$

Since p is prime, p has no divisors other than p and 1.

$\therefore$ $O(H) = p$ [Since $O(H) \neq 1$ as $a \neq e$]

$\therefore$ $G = H = \langle a \rangle$ [Since $O(G) = p$ and $H \not\subset G$]

$\therefore$ **G is cyclic and hence abelian.**

Note : If G is a finite group of prime order, then every element other than the identity element is a generator of G.

5. If G is a cyclic group of prime order then G has no proper subgroup.

Proof : Let $O(G) = p$, where p is prime. Let H be any subgroup of G.

Then by Legrange's theorem $O(H) | O(G) \Rightarrow O(H) | p$

Since p is prime, we must have $O(H) = 1$ or p

$\Rightarrow H = \{e\}$ or $H = G$ $\Rightarrow$ **G has no proper subgroup.**

6. Every finite group of composite order has proper subgroups.

Proof : Let G be a finite group and $O(G) = n$, where n is a compositie number.

Case (i) : Let G be a cyclic group and $g \in G$ be a generator of G.

$\therefore$ $O(g) = n,$ i.e. $g^n = e$

Since n is a composite integer, n can be written as $n = l \cdot m$, where l and m are integers such that $l \neq 1$ and $m \neq 1$

Now, $g^n = e \Rightarrow g^{lm} = e \Rightarrow (g^l)^m = e$

Since n is the smallest positive integer such that $g^n = e$, the integer m is the smallest such that $(g^l)^m = e$.

Thus the subset generated by the element g^l is a subgroup of G and it is a proper subgroup.

Case (ii) : Let G be any other group and $g \in G$ be arbitrary.

Let, $O(g) = p$. By Legrange's theorem $p | n$.

As n is not prime, we have $p < n$. Thus the subgroup $\langle g \rangle$ generated by g is a proper subgroup of G [Since $O(\langle g \rangle) = p < O(G) = n$]

Hence the theorem.

7. Any group of order five or less than five is abelian.

Proof : Clearly the group of order 1 is abelian.

We know that the group of prime order is cyclic and hence abelian.

Therefore the groups of orders 2, 3 and 5 are abelian.

Now we shall consider a group G of order 4.

If G has an element of order 4, then G is cyclic and abelian.

Supposing G has no element of order 4, then the order of every element other than the identity element, should be 2. This follows from the fact that the order of every element of the group is a divisor of the order of G.

$$\therefore \quad \forall \ a \in G, \ a^2 = e \Rightarrow a^{-1} = a \ \forall \ a \in G$$

Now let $a, b \in G$. Then, $ab = (ab)^{-1} = b^{-1} . a^{-1} = ba$

That is $\forall \ a, b \in G$, $ab = ba$. Hence, **G is abelian.**

Now we shall show that the converse of Lagrange's theorem holds for a finite cyclic group.

Theorem, If G is a finite cyclic group of order k, then for every divisor n of k, there exists a unique subgroup of G of order n.

Proof : Let $G = \langle a \rangle$, then $O(a) = O(G) = k$. By data n is a divisor of k.

$\therefore$ $\exists$ a positive integer m such that $k = m \cdot n$.

We shall show that the required subgroup is $H = \langle a^m \rangle$.

To this end we shall prove $O(a^m) = n$.

Let $O(a^m) = s$, i.e. s is the least positive integer such that $(a^m)^s = e$.

Again, $(a^m)^n = a^{mn} = a^k = e$ [Since $O(a) = k$]

Hence, $s \le n$

Supposing $s < n$, then $ms < mn = k \Rightarrow a^{ms} = e$ with $ms < k$.

This contradicts the fact that $O(a) = k$.

Hence, $s \le n \Rightarrow s = n$. i.e. $O(a^m) = n$.

Thus there exists a subgroup $H = \langle a^m \rangle$ of the group G of order n and a^l be its generator.

$\therefore$ $O(a^m) = O(a^l) = n$.

By division algorithm there exists two unique integers q and s such that $l = mq + r,\ 0 \le r < m$.

Now, $e = a^{l\,n} = a^{nmq+nr} = (a^{mn})^q \cdot a^{nr} = e \cdot a^{nr} = a^{nr}$

Now, if $0 \le r < m$, then $0 \le r\,n < m\,n = k \therefore a^{rn} = e \Rightarrow r = 0.$ i.e. $l = mq$

$\therefore$ $a^l = (a^m)^q \Rightarrow$ subgroup H generated by a^l is contained in subgroup H' generated by a^m.

Also, $O(H') = O(H) \Rightarrow H' = H$. Thus the uniqueness of the subgroup.

Example 1. Consider the multiplicative group $G = \{1, -1, i, -i\}$ of fourth roots of unity and $H = [-1, 1]$ be a subgroup of G. Write all the right cosets of H in G.

Solution : Hence $O(G) = 4$ and $O(H) = 2$.

We have $\dfrac{O(G)}{O(H)}$ = number of distinct right (left) cosets of H in G.

$\therefore$ number of right cosets $= \dfrac{4}{2} = 2$

These are $H.1 = \{-1, 1\}$, and $H.i = \{-i, i\}$

Example 2. Find all the left cosets of $H = \{0, 4, 8\}$ in $(Z_{12}, +_{12})$

Solution : Here $O(Z_{12}) = 12$ and $O(H) = 3$. Number of distinct left cosets $= H = \dfrac{12}{3} = 4$

These are,

$0 +_{12} H = \{0, 4, 8\} = H,\quad 1 +_{12} H = \{1, 5, 9\},\quad 2 +_{12} H = \{2, 6, 10\},\quad 3 +_{12} H = \{3, 7, 11\}$

Example 3. List all the subgroups of the cyclic group of fourth roots of unity.

Solution : Let $G = \{1, -1, i, -i\}$. Here $O(G) = 4$ and i is a generator.

Now the divisors of 4 are only 1, 2, and 4.

$\therefore$ the subgroups of G are the only subgroups generated by $i, i^2 = -1$ and $i^4 = 1$

The subgroups are,

$$<i> = G, \quad <-1> = \{1, -1\} \quad \text{and} \quad <1> = \{1\}$$

Example 4. List all the subgroups of a cyclic group G of order 24 and whose generator is g.

Solution : By data $O(G) = 24$, $\therefore$ $O(g) = 24$.

The divisors of 24 are 1, 2, 3, 4, 6, 8, 12 and 24.

$\therefore$ the subgroups of G are the only subgroups generatred by,

$$g, g^2, g^3, g^4, g^6, g^8, g^{12} \text{ and } g^{24} = e$$

Example 5. Find the order of the cyclic subgroup of $(Z_{30}, +_{30})$ generated by 25.

Solution : If g is a generator of a cyclic group G of order n, then g^m generates a subgroup H of G or order $\frac{n}{d}$ when $d = (n, m)$.

Here 1 is a generator of Z_{30}. Also 25 = 1.25 and (25, 30) = 5

$\therefore$ $O(H) = \frac{30}{5} = 6$, where H is the subgroup generated by 25.

Example 6. Find all the subgroups of $(Z_{18}, +_{18})$.

Solution : 1 is a generator of Z_{18}. The divisors of 18 are, 1, 2, 3, 6, 9, 18.

$\therefore$ The generators of the subgroups are

1.1 = 1, 1.2 = 2, 1.3 = 3, 1.6 = 6, 1.9 = 9, 1.18 = 0

Subgroups are

$<1> = Z_{18}$

$<2> = \{0, 2, 4, 6, 8, 10, 12, 14, 16\}$

$<3> = \{0, 3, 6, 9, 12, 15\}$

$<6> = \{0, 6, 12\}$

$<9> = \{0, 9\}$

$<0> = \{0\}$

EXERCISES

1. Find all the right cosets of the subgroup $H = \{0, 3\}$ in the group $(Z_6, +_6)$.
2. Find all the right cosets of the subgroup $H = \{1, 3, 9\}$ in the group $(Z_{13} - \{0\}, \times_{13})$.
3. List all the subgroups of $(Z_9, +_9)$.
4. Find the number of elements in the cyclic subgroup $(Z_{12}, +_{12})$ generated by 3 and 8.
5. Find all the cosets of $H = \{0, 3, 6, 2\}$ in $(Z_{12}, +_{12})$ and verify Lagrange's theorem in each case.
6. Find all the subgroups of $(Z_6, +_6)$.

1.19 Normal Subgroups

A special type of subgroups are called **normal subgroups**. For such subgroup H of a group G the right (left) cosets will form a group under the induced operation called **"coset multiplication"**.

Definition (Normal subgroup) : A subgroup H of a group G is said to be normal subgroup of G if $ghg^{-1} \in H$ for every $g \in G$ and $h \in H$.

If H is a normal subgroup of G, then we denote this by $H \Delta G$ and read as **"H is normal in G"**.

Note :

1. Every group G is normal in itself and the subgroup containing identity element alone is normal in G. These two normal subgroups are called *trivial* or *improper* normal subgroups of the group G.
2. Every group of prime order has no proper normal subgroups. This follows from the fact that a group of prime order has no proper subgroups (by Lagrange's theorem).

The following theorems gives us the equivalent definitions of normal subgroups.

Theorem 1. A subgroup H of a group G is normal if and only if $gHg^{-1} = H, \forall g \in G$.

Proof : Let H be a normal subgroup of G. That is, $H \Delta G$.

Now, $\quad gHg^{-1} = \{ghg^{-1} \mid h \in H\}$

Let, $\quad x \in gHg^{-1}$

$\Rightarrow \quad x = ghg^{-1}$ for some $h \in H$

$\Rightarrow \quad x \in H \quad [\because H \Delta G]$

$\Rightarrow \quad gHg^{-1} \subseteq H, \ \forall g \in G \quad$ .... (1)

Now replacing g by g^{-1} in (1), we have,

$$g^{-1}H(g^{-1})^{-1} \subseteq H \Rightarrow g^{-1}Hg \subseteq H$$

Now, $\quad H = (gg^{-1})\, H\, (gg^{-1})$

$\Rightarrow \quad H = g(g^{-1}Hg)g^{-1} \subseteq gHg^{-1} \quad [\because g^{-1}Hg \subseteq H]$

$\Rightarrow \quad H \subseteq gHg^{-1} \quad$ .... (2)

Thus from (1) and (2) we have, $\quad gHg^{-1} = H$

Conversely, let $\quad gHg^{-1} = H, \ \forall g \in G$

Now, $\quad ghg^{-1} \in gHg^{-1}, \ \forall h \in H$

But by data, $\quad gHg^{-1} = H, \ \forall g \in G \Rightarrow ghg^{-1} \in H, \ \forall h \in H$ and $g \in G$

Hence $H \Delta G$.

Theorem 2. A subgroup H of a group G is normal if and only if every right coset of H in G is a left coset of H in G.

Proof : Let H be a normal subgroup of G. By theorem 1,

$$gHg^{-1} = H, \ \forall g \in G$$

$$\Rightarrow \quad (gHg^{-1})g = Hg, \ \forall g \in G$$

$$\Rightarrow \quad (gH)(g^{-1}g) = Hg, \ \forall g \in G$$

$$\Rightarrow \quad gH = Hg, \ \forall g \in G$$

Thus, every right coset of H in G is a left coset of H in G.

Conversely, let every right coset of H in G be equal to the left coset of H in G.

Let the right coset Hg, for $g \in G$, be equal to the left coset, say xH for $x \in G$.

Now, $g \in Hg = xH$. Also, $g \in gH$

$\Rightarrow$ $g \in xH$ and $g \in gH$

$\Rightarrow$ $gH = xH$ ($\because$ any two left cosets are either identical or disjoint)

$\Rightarrow$ $gH = Hg, \forall g \in G$ ($\because xH = Hg$)

$\Rightarrow$ $gHg^{-1} = H \Rightarrow H \Delta G$

Corollary : Every subgroup of an abelian group is a normal subgroup.

Proof : If a group G is abelian, then every right coset Ha of subgroup H is equal to the corresponding left coset aH of H. Hence, by the theorem 2, H is normal in G.

Theorem 3. A subgroup H of the group G is normal in G, if and only if the product of any two right (or left) cosets of H in G is again a right (or left) coset of H in G.

Proof : Let H be normal in G and Ha and Hb be any two right cosets of H.

Consider, $Ha \cdot Hb = H(aH)b$

$= H(Ha)b$ ($\because$ H is normal $\Rightarrow Ha = aH$)

$= (HH)ab$

$= Hab$ ($\because$ H is subgroup $\Rightarrow HH = H$)

Thus, the product of two right cosets of H in G is again a right coset of H in G.

Conversely, let the product of any right cosets of H in G be again a right coset. Then, Hx and Hx^{-1} will be two right cosets, for $x \in G$.

By data $Hx \cdot Hx^{-1}$ is again a right coset.

We have, $x \in Hx$ and $x^{-1} \in Hx^{-1} \Rightarrow e = xx^{-1} \in Hx \cdot Hx^{-1}$

Further $e \in He$, which is a right coset of H.

Thus, two right cosets, He and $Hx \cdot Hx^{-1}$ has an element in common.

$\Rightarrow$ $Hx \cdot Hx^{-1} = H$ ($\because$ two right cosets of H are either equal to disjoint)

Now, $Hx \cdot Hx^{-1} = H$

$\Rightarrow$ $h_1x \cdot h_2x^{-1} \in H, \forall x \in G$ and $h_1, h_2 \in H$

$\Rightarrow$ $x h_2x^{-1} \in (h_1)^{-1}H, \forall x \in G$ and $h_1, h_2 \in H$

$\Rightarrow$ $x h_2x^{-1} \in H \Rightarrow H \Delta G.$

1.20 Results Related to Normal Subgroups

1. **The intersection of any two normal subgroups of a group is also a normal subgroup.**

   Proof : Let H and K be any two normal subgroups of the group G.

   Since H and K are subgroups. $H \cap K$ is also a subgroup of G.

   Let, $x \in G$ and $h \in H \cap K$ be arbitrary

   $\Rightarrow$ $x \in G$ and $h \in H$, $h \in K$

   $\Rightarrow$ $xhx^{-1} \in H$ and $xhx^{-1} \in K$ [$\because H \Delta G$ and $K \Delta G$]

   $\Rightarrow$ $xhx^{-1} \in H \cap K$, $\forall x \in G$ and $h \in H \cap K$

   Hence $H \cap K$ is normal in G.

2. **The product of any two normal subgroups of a group is a subgroup of the group.**

   Proof : Let H and K be any two normal subgroups of the group G.

   $$HK = \{h_1k_1 \mid \forall h_1 \in H, k_1 \in K\}$$

Since $e \in H$ and $e \in K$, we have, $e \cdot e = e \in HK$. Thus, HK is non empty.

Let, $x, y \in HK$ be arbitrary.

Consider, $xy^{-1} = (h_1k_1)(h_2k_2)^{-1}$ for some $h_1, h_2 \in H$ and $k_1, k_2 \in K$

$\Rightarrow$ $xy^{-1} = h_1\left(h_2^{-1}h_2\right)\left(k_1k_2^{-1}h_2^{-1}\right)$ $(\because h_2^{-1}h_2 = e)$

$\Rightarrow$ $xy^{-1} = \left(h_1h_2^{-1}\right)\left[h_2\left(k_1k_2^{-1}\right)h_2^{-1}\right]$

$\Rightarrow$ $xy^{-1} = h \cdot k$ [where $h = h_1h_2^{-1}$ and $k = h_2(k_1k_2^{-1})h_2^{-1}$]

Now, $h_1h_2 \in H \Rightarrow h = h_1h_2^{-1} \in H$ [$\because$ H is a subgroup]

Again, $h_2 \in H \Rightarrow h_2 \in G$ and $k_1k_2^{-1} \in K$

$\Rightarrow$ $h_2(k_1k_2^{-1})h_2^{-1} \in K$ ($\because$ K is a normal)

$\Rightarrow$ $k \in K \Rightarrow xy^{-1} = h \cdot k \in HK$

Hence HK is a subgroup of G.

3. Let H be a subgroup and K be a normal subgroup of the group G. Then $H \cap K$ is normal in H.

Proof : Since H and K are subgroups of G, $H \cap K$ is also a subgroup of G. Further $H \cap K \subset H$. Thus, $H \cap K$ is a subgroup of H.

Let $g \in H$ and $h \in H \cap K$ be arbitrary.

Now $h \in H \cap K \Rightarrow h \in H$ and $h \in K$. Also, as $K \Delta G$, we have, $ghg^{-1} \in K$ .... (1)

Further $hg^{-1} \in H$ as H is a subgroup $\Rightarrow ghg^{-1} \in H$ [$\because g \in H, hg^{-1} \in H$] .... (2)

From (1) and (2) we have,

$$ghg^{-1} \in H \cap K, \ \forall g \in H \text{ and } h \in H \cap K$$

Hence $H \cap K$ is normal in H.

4. If N is a normal subgroup of G and H is any subgroup of G, then NH is a subgroup of G.

Proof : $NH = \{nh \mid n \in N, h \in H\}$. $e \in N$ and $e \in H \Rightarrow e \in NH \Rightarrow NH \neq \phi$.

Let $x, y \in NH$ be two arbitrary elements.

$\Rightarrow$ $x = n_1h_1$ and $y = n_2h_2$ for $n_1, n_2 \in N$ and $h_1, h_2 \in H$.

We shall show that, $xy^{-1} \in NH$

Consider, $xy^{-1} = n_1h_1(n_2h_2)^{-1} = n_1h_1h_2^{-1}n_2^{-1}$

$= n_1(h_1h_2^{-1})n_2^{-1}$

$= n_1h_3n_2^{-1}$(where $h_1h_2^{-1} = h_3$)

$= (n_1h_3n_2^{-1})h_3^{-1}\, h_3 = n_1(h_3n_2^{-1}h_3^{-1})\, h_3$

Since N is normal in G, the element $(h_3n_2^{-1}h_3^{-1}) \in N$. Let, $h_3n_2^{-1}h_3^{-1} = n_3$

Then, $xy^{-1} = (n_1n_3)h_3$

$= n_4h_3$ (where $(n_1n_2) = n_4 \in N$)

Therefore $xy^{-1} \in NH$ and hence NH is a subgroup of G.

5. The centre Z of a group G is a normal subgroup of G.

Proof : The centre Z of a group G is the set of those elements of G which commute with every element of G. That is,

$$Z = |z \in G| \ zx = xz \ \forall x \in G|$$

Clearly Z is a subgroup of G (exercise)

To show that Z is normal in G, let $x \in G$ and $z \in Z$ be arbitrary.

Now,
$$\begin{aligned} xzx^{-1} &= (xz)x^{-1} \\ &= (zx)x^{-1} && [\because z \in Z] \\ &= z(xx^{-1}) = z && [\because xx^{-1} = e] \end{aligned}$$

$$\Rightarrow \quad xzx^{-1} = z \in Z, \ \forall \ x \in G \text{ and } z \in G$$

Hence Z is a normal subgroup of G.

6. **The normalizer of any element of a group is a subgroup of the group. The centre Z of a group is a normal subgroup of a normalizer of an element.**

Proof : If $a \in G$, then the normalizer of a in G is the set of all those elements of G which commutes with a. The normalizer of a is denoted by N_a. That is,

$$N_a = \{x \in G \mid ax = xa\}$$

N_a is a subgroup of G(exercise).

We shall show that the centre Z of the group G is normal in N_a.

We have, $\forall \ z \in Z, \ zx = xz \ \forall \ x \in G$. In particular, $za = ax$ (as $a \in G$)

$$\Rightarrow \quad z \in N_a \ \Rightarrow \ Z \subset N_a$$

As Z is a group by itself, Z is a subgroup of Na. Also we know that Z is normal in G. Hence Z is normal in N_a.

1.21 Factor Group (Quotient Group)

Let G be a group and H be a normal subgroup of G. As H is a normal subgroup, there is no distinction between the right and left cosets. Therefore we shall simply say H_a is a coset of H in G. Let the set of all cosets of H in G be denoted by $G|H$. That is,

$$G|H = \{H_a | a \in G\}$$

We shall define a binary composition in the set $G|H$ by,

$$H_a \cdot H_b = H_{ab} \quad \forall \, H_a, \, H_b \in G|H$$

Clearly, H_{ab} is also a coset of H in G - for H is a normal subgroup of G, if and only if the product of two cosets is again a coset of H in G. Thus $G|H$ is closed under the binary composition of cosets of H.

Now we shall show that the above binary composition induces a group structure on $G|H$.

Theorem. The set $G|H$ of all cosets of a normal subgroup H, of the group G, is a group under the binary operation defined by

$$\boldsymbol{H_a \cdot H_b = H_{ab} \quad \forall \, H_a, H_b \in G|H}$$

Proof :

(i) Associative property

Let, $H_a, H_b, H_c \in G|H$ be arbitrary elements.

Consider,
$$\begin{aligned} H_a \cdot [H_b \cdot H_c] = H_a \cdot H_{bc} &= H_{a(bc)} \\ &= H_{(ab)c} = H_{ab} \cdot H_c = [H_a \cdot H_b] \cdot H_c \end{aligned}$$

Thus, the binary composition is associative.

(ii) Existence of identity

Corresponding to the identity element e of G, the coset, $H_e = H \in G|H$

Consider, $H_a \cdot H = H_a \cdot H_e \Rightarrow H_a \cdot H = H_{ae} \Rightarrow H_a \cdot H = H_a$

Similarly, $H \cdot H_a = H_a$

Thus, H acts as the identity element of $G|H$, w.r.t. the binary operation defined.

(iii) Existence of Inverse.

Let $a \in G$, then H_a and $H_{a^{-1}}$ are the elements of $G|H$.

Consider, $H_a \cdot H_{a^{-1}} = H_{aa^{-1}} = H_e = H = H_a$. Similarly, $H_{a^{-1}} \cdot H_a = H$.

Thus, the inverse of H_a is $H_{a^{-1}}$.

Hence $G|H$ is a group under the binary operation defined.

Definition : (Factor group or Quotient group) : The group of $G|H$ of all cosets of a normal subgroup H in G, together with the binary composition defined by $H_a \cdot H_b = H_{ab}$ $\forall\, H_a, H_b \in G|H$ is called the factor group of G w.r.t. H or the Quotient group of G w.r.t H.

Note :

1. The factor group $G|H$ of an abelian group G, w.r.t. a subgroup H, is itself an abelian group.
2. If the group operation in G is in additive notation, then the cosets of H are of the form

$$H_{+a} = \{h + a \mid h \in H\}$$

Also if H_{+a} and H_{+b} are any two elements of $G|H$, then their composition in $G|H$ is denoted by + and is defined by

$$(H_a) + (H_b) = H_{+(a+b)}$$

Further the identity element of the factor group $G|H$ is H and the inverse of H_{+a} is $H_{+(-a)}$.

Theorem. If H is a normal subgroup of finite order, then $O(G|H) = \dfrac{O(G)}{O(H)}$.

Proof : By definition, $O(G|H)$ = number of distinct cosets of H

= Index of H in G

$= \dfrac{O(G)}{O(H)}$ (By Lagrange's theorem)

Example 1. If G is a group and H is a subgroup of index 2 in G, show that H is normal subgroup of G.

Solution : Since the index of H in G is 2, there are only two left (or right) cosets. Let $x \in G$ be arbitrary.

If $x \in H$, then $Hx = H$ and $xH = H \Rightarrow Hx = xH \Rightarrow H$ is normal in G

Supposing $x \notin H$, then, $Hx \neq H$ and $xH \neq H$.

Since the index of H is 2, the cosets H, Hx and xH are such that,

$$G = H \cup Hx = H \cup xH \text{ and } \phi = H \cap Hx = H \cap xH$$

$$\Rightarrow \quad Hx = xH \quad \forall\, x \in G$$

Hence H is a normal subgroup of G.

Example 2. Let H and K be two normal subgroups of a group G, such that $H \cup K = \{e\}$, where e is the identity element of G. Show that for every $h \in H$, $k \in H$, $hk = kh$.

Solution : We shall show that,

$$hkh^{-1}k^{-1} = e \text{ which implies } hk = kh$$

Since H is normal, $kh^{-1}k^{-1} \in H, \ \forall \ h \in H$ and $k \in K \subset G$

Now, $h \in G$ and $kh^{-1}k^{-1} \in H \Rightarrow hkh^{-1}k^{-1} \in H$ ($\because$ H is a subgroup)

Again as K is normal,

$$hkh^{-1} \in K, \forall \ k \in K \text{ and } h \in H \subset G$$

Now, $k^{-1} \in K$ and $hkh^{-1} \cdot K \Rightarrow hkh^{-1}k^{-1} \in K$ ($\because$ K is a subgroup)

Thus, $hkh^{-1}k^{-1} \in H \cap K = \{e\}$ (by data)

$\Rightarrow$ $hkh^{-1}k^{-1} = e \Rightarrow hk = kh, \ \forall \ h \in H$ and $h \in K$

Example 3. If H is the only subgroup of finite order m in the group G, then show that H is normal in G.

Solution : H is a subgroup of G and $O(H) = m$. To show that H is normal in G, we shall show that,

$$\forall \ x \in G, \ xHx^{-1} = H,$$

Clearly, xHx^{-1} is a subgroup of G.

Let, $H = \{h_1, h_2, h_3, \ldots\ldots, h_m\}$, where h_i, $i = 1, 2, \ldots\ldots, m$ are all distinct.

Then, $xHx^{-1} = \{xh_1x^{-1}, \ xh_2x^{-1}, \ldots\ldots, xh_mx^{-1}\}$

Clearly, the elements of xHx^{-1} are all distinct for if,

$$xh_ix^{-1} = xh_jx^{-1} \Rightarrow h_i = h_j \text{ which is a contradiction.}$$

Thus, $O(xHx^{-1}) = m = O(H)$

But H is the only subgroup of G of order $m \Rightarrow xHx^{-1} = H, \ \forall \ x \in G$.

Hence H is normal in G.

Example 4. Give an example of a non-abelian group all of whose subgroups are normal.

Solution : Consider the group, $G = \{\pm 1, \pm i, \pm j, \pm k\}$, under multiplication defined by,

$$i^2 = j^2 = k^2 = -1, \ \ i \cdot j = k, \ \ j \cdot k = i, \ \ k \cdot i = j, \ \ j \cdot i = -k, \ k \cdot j = -i, \ \ i \cdot k = -j$$

Clearly G is abelian and is called **Quaternion group**.

Now the subgroup $(1, -1)$ is the only subgroup of order 2. Thus by example 3, it is normal in G.

If H is any subgroup of order 4, then the index of H in G is $\dfrac{O(G)}{O(H)} = \dfrac{8}{4} = 2$. Thus by example 1, H is normal in G.

Hence all the subgroup of G is normal.

Example 5. Show that every quotient group of an abelian group is abelian.

Solution : Let, G be an abelian group and N be a normal subgroup of G.

Let, $Na, Nb \in G|N$ be arbitrary

Now, $(Na)(Nb) = Nab \Rightarrow (Na)(Nb) = Nba$ ($\because ab = ba$)

$\Rightarrow (Na)(Nb) = (Nb)(Na)$

$\Rightarrow G|N$ is abelian.

Example 6. Show that every factor group of cyclic group is cyclic.

Solution : Let G be a cyclic group with generator 'a' and N be a normal subgroup of G.

We have to show that the factor group $G|N$ is cyclic.

To this end we shall show that the elements of $G|N$ are generated by the coset Na.

First we shall prove that, $Na^m = (Na)^m$ .... (1)

Case (i) : Let $m = 0$.

Then, $Na^m = Na^0 = Ne = N$, $(Na)^m = (Na)^0 =$ identity of $G|N = N \Rightarrow Na^m = (Na)^m$.

Case (ii) : Let m be a positive integer.

Then,
$$Na^m = Na \cdot a \cdot ...... a = Na \cdot Na \cdot Na ..... Na \quad (\because Nab = Na \cdot Nb)$$
$$= (Na)^m$$

Case (iii) : Let m be a negative integer.

Let, $m = -m$ where n is a positive integer.

$$Na^m = Na^{-n} = (Na^{-1})^n = Na^{-1} \cdot Na^{-1} \cdot Na^{-1} .... Na^{-1}$$
$$= (Na^{-1})^n = (a^{-1}N)^n \quad (\because N \Delta G)$$
$$= [(Na)^{-1}]^n = (Na)^{-n} = (Na)^m$$

Hence (1) is true in all the cases.

Let $Ng \in G|N$ be arbitrary.

Now, $g \in G \Rightarrow g = a^m$ for some $m \in Z$

$\Rightarrow$ $Ng = Na^m = (Na)^m \Rightarrow$ every element of $G|N$ is some power of Na.

Hence, $G|N$ is cyclic with Na as generator.

Example 7. If N is a normal subgroup of G and a is an element of finite order in G, then show that the element Na of the quotient group $G|N$ is also of finite order and the order of Na divides the order of a.

Solution : Let $O(a) = n$. That is, n is the least positive integer such that $a^n = e$, e is the identity element of G.

Consider, $(Na)^n = Na^n = Ne = N =$ identity element of $G|N$

$\Rightarrow$ Na is of finite order and $O(Na) \leq n$

Let $O(Na) = m$. That is m is the least positive integer such that $(Na)^m = N =$ identity element of $G|N$. Further, $m \leq n$.

By division algorithm, there exists q and r such that, $n = qm + r$, $0 \leq r < m$.

Now,
$$N = (Na)^n = (Na)^{qm+r}$$
$$= [(Na)^m]^q \cdot (Na)^r$$
$$= N^q \cdot (Na)^r \quad (\because (Na)^m = N)$$
$$= N \cdot (Na)^r \quad (\because N \text{ is a subgroup})$$
$$= (Na)^r \quad (\because N \text{ is the identity element})$$
$$\Rightarrow \quad N = (Na)^r \text{ for } r < m$$

This contradicts the fact that m is the least positive integer such that $(Na)^m = N$.

Thus, $r = 0 \Rightarrow n = qm \Rightarrow m|n \Rightarrow O(Na)|O(a)$

EXERCISES

1. If a cyclic subgroups S of a group G is normal in G, show that every subgroup of S is also normal in G.
2. Let M and N be two normal subgroups of a group G such that $M \cap N = \{e\}$. Show that $mn = nm \ \forall \ m \in M$ and $n \in N$.
3. Show that $\{1, -1\}$ is a normal subgroup of the multiplicative group $G = \{1, i, -i, -1\}$.

4. Let G be the multiplicative group of all $n \times n$ non-singular matrices over the set of real numbers. H be the subset of G consisting of all $n \times n$ non-singular matrices whose determinants are equal to 1. Show that H is normal in G.
5. Let Z be the additive group of integers. For each $n > 1$, let nZ be the subset consisting of all multiplies of n. Show that nZ is normal in Z.

1.22 Homomorphism of Groups

In this section we shall consider mapping defined from one group into another which "preserves" the binary composition. Such mapings are called **Homomorphisms**. If two groups are homomorphic, then they have the similar structure.

Definition : Let (G, ·) and (G′, *) be two groups. A mapping, $f : G \to G'$ is said to be a homomorphism if

$$f(g_1 \cdot g_2) = f(g_1) * f(g_2), \quad \forall\, g_1, g_2 \in G$$

That is the mapping f is said to be a homomorphism, if f maps the composition of any two elements of G into the composition of the images of the elements.

This we say by f preserves composition. If f is a homomorphism, from the group G into G′, we say **"G is homomorphic to the group of G′"**.

Definition : A homomorphism f from a group G into G′ is said to be isomorphism if f is a bijection. i.e., f is both one-one and onto.

Two groups G and G′ are said to be "isomorphic", if there exists an isomorphism $f : G \to G'$. This we denote it by $G \approx G'$.

Definition : A homomorphism f from a group G into itself is called Endomorphism.

Definition : An isomorphism f from a group G onto itself is called automorphism.

Examples of homomorphisms.

1. The identify map $I_g : G \to G$ from the group G onto itself, defined by $I_g(g) = g$ is a homomorphism. This shows that there exists atleast one homomorphism from G into itself. Further $I_g(g) = g$ is an **automorphism**.
2. Let (G, *) and (G′, *) be two groups and e' be the identity element of G'. Define, $f : G \to G'$
   by, $f(g) = e'$, $\forall\, g \in G$.
   Now, $f(g_1 \cdot g_2) = e'$, $g_1, g_2 \in G$. Also, $f(g_1) * f(g_2) = e' * e' = e'$
   Thus, $f(g_1 \cdot g_2) = f(g_1) * f(g_2)$.
   Hence f is homomorphism. This homomorphism is called a **trivial homomorphism.**
3. Consider the additive group (Z, +) of all integers and the group $H = \{1, -1\}$ under multiplication. Define,

$$\varphi = Z \to H$$

by
$$\varphi(n) = \begin{cases} 1 & \text{if } n \text{ is even or zero} \\ -1 & \text{if } n \text{ is odd} \end{cases}$$

We shall show that φ is a homomorphism. To this end we shall verify

$$\varphi(m + n) = \varphi(m) + \varphi(n), \text{ for all } m, n \in Z$$

Case (i) : Let m and n both even.

Then, $m + n$ is again even $\Rightarrow \varphi(m + n) = 1$

Also, $\varphi(m) = 1$ and $\varphi(n) = 1$. Thus, $\varphi(m + n) = \varphi(m) \cdot \varphi(n)$.

Case (ii) : Let one of m and n be even, say m be even and n be odd.

Then $m + n$ is odd. Then,

$$\varphi(m+n) = -1, \ \varphi(m) = 1 \text{ and } \varphi(n) = -1$$

$$\Rightarrow \quad \varphi(m+n) = \varphi(m) \cdot \varphi(n)$$

Case (iii) : Let m and n both odd. Then $m + n$ is even.

$$\Rightarrow \quad \varphi(m+n) = 1, \ \varphi(m) = -1 \text{ and } \varphi(n) = -1$$

$$\Rightarrow \quad \varphi(m+n) = \varphi(m) \cdot \varphi(n)$$

Hence φ is homomorphism.

4. Let $M_n(R)$ be the group of all non-singular $n \times n$ matrices, with real entries under matrix multiplication. Let R^* be the multiplicative group of non-zero real numbers. Define,

$$\varphi \cdot M_n(R) \to R^*$$

by, $\varphi(A) = |A|, \forall A \in M_n(R)$

Consider, $\varphi(AB) = |AB|, \ A, B \in M_n(R) \ \Rightarrow \ \varphi(AB) = |A| \cdot |B| = \varphi(A) \cdot \varphi(B)$

Hence φ is homomorphism.

5. Consider the multiplication group $(R^*, \cdot)$ of all non zero real numbers. Define, $f: R^* \to R^*$

by $f(x) = 2x$ for every $x \in R^*$.

Consider, $f(x \cdot y) = 2xy$ for every $x, y \in R^*$. But, $f(x) \cdot f(y) = 2x \cdot 2y \neq 2xy$.

Hence f is not a homomorphism.

1.23 Elementary Properties of Homomorphism

We shall see few properties of homomorphism in the form of theorems.

Theorem 1. Let $f: G \to G'$ be a homomorphism from the group $(G, \cdot)$ into the group $(G', *)$. Then

(i) $f(e) = e'$ where e and e' are the identity elements of the groups G and G' respectively.

(ii) $f(a^{-1}) = [f(a)]^{-1}, \forall a \in G$.

Proof : The map $f: G \to G'$ is a homomorphism. Thus, $\forall a, b \in G, \ f(a, b) = f(a) * f(b)$.

(i) Now,

$$\begin{aligned} f(a) * e' = f(a) &= f(a \cdot e) \\ &= f(a) * f(e) \qquad (\because f \text{ is a homomorphism}) \\ \Rightarrow \quad f(a) * e' &= f(a) * f(e) \\ \Rightarrow \quad f(e) &= e' \qquad \text{(by cancellation law)} \end{aligned}$$

(ii) Consider, $\quad f(a) * f(a^{-1}) = f(a \cdot a^{-1}) = f(e) = e'$

Again consider, $\quad f(a^{-1}) * f(a) = f(a^{-1} \cdot a) = f(e) = e'$

Thus, $\quad f(a) * f(a^{-1}) = f(a^{-1} * f(a) = e' \Rightarrow f(a^{-1}) = \text{inverse of } f(a)$

That is, $\quad f(a^{-1}) = [f(a)]^{-1}$

Theorem 2. If f is a homomorphism of a group G into G', then the range $f(G) = \{f(g) \mid \forall g \in G\}$ is a subgroup of G'.

Proof : Clearly, $\quad f(G) = \{f(g) \mid \forall g \in G\} \subset G'$

To show that $f(G)$ is a subgroup of G', we shall show that

$$x' \cdot (y')^{-1} \in f(G), \ \forall x', y' \in f(G)$$

Since, $x', y' \in f(G)$, $\exists\, x, y \in G$ such that, $f(x) = x'$, $f(y) = y'$

Consider,
$$x' \cdot [y']^{-1} = f(x) \cdot [f(y)]^{-1}$$
$$= f(x) \cdot f(y^{-1}) \quad \text{(by theorem 1)}$$
$$= f(x \cdot y^{-1}) \quad (\because f \text{ is homomorphism})$$

Thus, $x' \cdot [y']^{-1}$ is the image of $x \cdot y^{-1} \in G \Rightarrow x' \cdot [y']^{-1} \in f(G)$.

Hence $f(G)$ is a subgroup of G'.

Definition : If f is a homomorphism of a group G into G', then the set

$$f(G) = \{f(g) \mid g \in G\}$$

is called the homomorphic image of G in G'.

Theorem 4. If $f : G \to G'$ is a homomorphism of a group G into G' and H is a subgroup of G, then $f(H)$ is again a subgroup of G'.

Proof : By data, $f : G \to G'$ is a homomorphism. Now, $f(H) = \{f(h) \mid h \in H\}$

Let h_1' and $h_2' \in f(H)$ be arbitrary.

Then, $\exists\, h_1, h_2 \in H$ such that $f(h_1) = h_1'$ and $f(h_2) = h_2'$.

Consider,
$$h_1' \cdot (h_2')^{-1} = f(h_1) \cdot [f(h_2)]^{-1}$$
$$= f(h_1) \cdot f(h_2^{-1})$$
$$= f(h_1 \cdot h_2^{-1}) \quad (\because f \text{ is a homomorphism})$$
$$= f(h) \quad (\text{where } h = h_1 \cdot h_2^{-1})$$

Now, $h_1 \cdot h_2 \in H \Rightarrow h_1 \cdot h_2^{-1} \in H$ as H is a subgroup.

Thus, $\exists$ an element, $h = h_1 \cdot h_2^{-1}$ such that, $f(h) = h_1' \cdot (h_2')^{-1}$

$$\Rightarrow \quad h_1' \cdot (h_2')^{-1} \in f(H)$$

Hence $f(H)$ is a subgroup of G'.

Theorem 5. If $f : G \to G'$ be a homomorphism from the group G into itself and H is a cyclic subgroup of G, then $f(H)$ is again a cyclic subgroup of G.

Proof : Let H be a cyclic subgroup of G, generated by the element $a \in G$.

Clearly, $f(H)$ is a subgroup of G' (by theorem 4). We shall show that $f(H)$ is cyclic.

Let, $x \in f(H)$. Then there exists $y \in H$, such that $f(y) = x$. Now,

$$y \in H \Rightarrow y = a^m \quad \text{for some integer } m.$$

Thus,
$$x = f(y) \Rightarrow x = f(a^m)$$

Here two cases will arise.

Case (i) : Let m be a positive integer.

Now,
$$x = f(a^m)$$
$$\Rightarrow \quad x = f(a \cdot a \cdot a \cdot \ldots\ldots \cdot m \text{ times})$$
$$\Rightarrow \quad x = f(a) \cdot f(a) \ldots\ldots \cdot f(a)\ m \text{ times} \quad (\because f \text{ is a homomorphism})$$
$$\Rightarrow \quad x = [f(a)]^m$$

Case (ii) : Let m be a negative integer.

Let $m = -n$, where n is a positive integer.

Now,
$$x = f(a^m)$$
$$\Rightarrow \quad x = f(a^{-n})$$
$$\Rightarrow \quad x = [f(a^{-1})^n]$$

$\Rightarrow \quad x = [f(a^{-1})]^n$ (∵ f is a homomorphism and n is +ve)

$\Rightarrow \quad x = [\{f(a)^{-1}\}]^n$

$\Rightarrow \quad x = [f(a)]^{-n}$

$\Rightarrow \quad x = [f(a)]^m$ (∵ $m = -n$)

Thus in both the cases, for $x \in H$, there exists an integer m such that $x = [f(a)]^m$. Thus $f(H)$ is a cyclic sub group of G, generated by $f(a)$.

1.24 Kernel of a Homomorphism

Definition (Kenel of a Homomorphism) : Let $f : G \to G'$ be a homomorphism from the group G into G' and e' be the identity element of G'. The subset K of G defined by

$$K = \{a \in G \mid f(a) = e'\}$$

is called the Kernel of a Homomorphism f and is denoted by Ker f.

1. Consider the multiplicative group C^* of all non zero complex numbers and the mapping

$$f : C^* \to C^*$$

defined by, $f(a + ib) = a - ib \ \forall \ a + ib \in C$

Clearly f is a homomorphism (exercise).

Now, $\text{Ker} f = \{x \in C^* : f(x) = e' = 1\}$

$= \{a + ib : f(a + ib) = 1\}$

$= \{a + ib : a - ib = 1\} = \{a + ib : a = 1, b = 0\} = \{1\}$

2. Consider the homomorphism

$$f : R^* \to R$$

defined by, $f(x) = \log_{10} x$

Here $(R^+, .)$ is the multiplicative group of all positive real numbers and $(R, +)$ is the additive group of all real numbers. Clearly f is a homomorphism.

Now, $\text{Ker} f = \{x \in R^+ : f(x) = 0\}$

$= \{x \in R^+ : \log_{10} x = 0\} = \{x \in R^+ : x = 1\} = \{1\}$

Theorem 1. Let $f : G \to G'$ be a homomorphism from the group G into G' with Kernel K. Then K is a normal subgroup of G.

Proof : We shall prove this theorem in two steps.

(i) K is a subgroup of G and **(ii)** K is normal in G

(i) By the definition of Ker f, K is a subset of G.

Further $f(e) = e'$ where e and e' are identities of G and G' respectively. Hence $e \in K$. Thus $K \neq \phi$.

Now, let $a, b \in K$. Then, $f(a) = e'$ and $f(b) = e'$

$f(a \cdot b^{-1}) = f(a) \cdot f(b^{-1})$ (∵ f is a homomorphism)

$= e' \cdot (e')^{-1}$

$= e' \cdot e' = e'$

Hence $a \cdot b^{-1} \in K$ and hence K is a subgroup of G.

(ii) To show that K is normal in G, we shall show that $gkg^{-1} \in K$ for every $g \in G$ and $k \in K$.

Consider, $f(gkg^{-1}) = f(g) \cdot f(k) \cdot f(g^{-1})$ (∵ f is in homomorphism)

$= f(g) \cdot e' \cdot f(g^{-1})$ (∵ $k \in K$)

$$= f(g) \cdot f(g^{-1})$$
$$= f(g \cdot g^{-1})$$
$$= f(e) = e'$$
$$\Rightarrow \quad gkg^{-1} \in K, \ \forall g \in G \text{ and } k \in K$$

Theorem 2. Let $f: G \to G'$ be a homomorphism from the group G into G' with Kernel K. Then f is one-one if and only if $K = \{e\}$ where e is the identity element of G.

Proof : Let $f: G \to G'$ be one-one. That is, $f(x_1) = f(x_2) \Rightarrow x_1 = x_2$

Let, $a \in K \Rightarrow f(a) = e'$ where e' is the identity element of G'

$$\Rightarrow f(a) = f(e) \qquad (\because f(e) = e')$$
$$\Rightarrow a = e \qquad (\because f \text{ is one-one})$$
$$\Rightarrow K = \{e\}$$

Conversely, let, $K = \{e\}$. We shall show f is one-one.

Let, $f(a) = f(b)$

$$f(a) \cdot [f(b)]^{-1} = f(b) \cdot [f(b)]^{-1}$$
$$\Rightarrow f(a \cdot b^{-1}) = e' \qquad (\because K = \{e\})$$
$$\Rightarrow a \cdot b^{-1} = e$$
$$\Rightarrow (a \cdot b^{-1})b = e \cdot b$$
$$\Rightarrow a \cdot (b^{-1} \cdot b) = b \Rightarrow a = b$$

Hence f is one-one.

Corollary : A homomorphism f from group G onto the group G' is an isomorophism if and only if Ker $f = \{e\}$.

Proof : By data f is onto. Further, Ker $f = \{e\} \Rightarrow f$ is one-one (by theorem 2)

Thus, f is a homomorphism, which is both one-one and onto. Hence f is an isomorphism.

Theorem 3. Let G be a group and H be a normal subgroup of G. Then $G|H$ is a homomorphic image of G with H as its Kerner.

Proof : We shall define a mapping from G onto $G|H$ and show that this map is a homomorphism and its Kernel is H.

Since $G|H = \{Ha \mid a \in G\}$, it suggests us to define the mapping

$$f: G \to G|H$$

by $f(x) = Hx$ for every $x \in G$

Clearly f is well defined and further f is onto.

We shall show that f is homomorphism.

Consider, $f(x \cdot y) = Hxy$ where $y \in G \Rightarrow f(x \cdot y) = Hx \cdot Hy = f(x) \cdot f(y)$

Thus, f is homomorphism.

Now,
$$\text{Ker} f = \{x \in G \mid f(x) = \text{identity of } G|H\}$$
$$= \{x \in G \mid Hx = H\}$$
$$= \{x \in G \mid x \in H\} \qquad (\because Hx = H \Leftrightarrow x \in H)$$
$$\Rightarrow \text{Ker} f = H$$

The map f defined above is called the **natural homomorphism** or **canonical homomorphism** from G into $G|H$.

The above theorem shows that every normal subgroup H of a group G, gives rise to a homomorphism f from G, such that Ker $f = H$. The following theorem tells us about the converse of this statement. This theorem is called **Fundamental Homomorphism theorem**.

Theorem 4 (Fundamental Homomorphism theorem) : Let f be a homomorphism of a group G into a group G', with Kernel K. Then $f(G)$ is isomorphic to factor group $G|K$.

Proof : Since, $f: G \to G'$ is a homomorphism, the set $f(G)$ is a subgroup of G' hence a group by itself.

We shall define the map, $\varphi : G|K \to f(G)$

by $\varphi(Ka) = f(a), \ \forall\, a \in G, \ Ka \in G|K$

1. **φ is well defined**

We have to show that φ is independent of choice of representative.

Let, $b \in Ka$ we must show that $f(a) = f(b)$

Now, $b \in Ka \Rightarrow b = k_1 a$, for $k_1 \in K$

$\Rightarrow ba^{-1} = k_1$

Now, $f(k_1) = e'$ $\quad (\because k_1 \in \text{Ker} f = K)$

$\Rightarrow f(ba^{-1}) = e'$

$\Rightarrow f(b)\cdot f(a^{-1}) = e'$

$\Rightarrow f(b)\cdot [f(a)]^{-1} = e' \Rightarrow f(b) = [f(a)]\cdot e' \Rightarrow f(b) = f(a)$

Hence φ is well defined.

2. **φ is one to one.**

Let, $\varphi(Ka) = f(Kb) \Rightarrow f(b) = f(a)$

$\Rightarrow f(a)\cdot [f(b)]^{-1} = e'$

$\Rightarrow f(ab^{-1}) = e'$

$\Rightarrow ab^{-1} \in K \Rightarrow a \in Kb \Rightarrow Ka = Kb$

Hence φ is one to one.

3. **φ is onto.**

Let, $b \in f(G)$, be arbitrary. Then, $\exists\, a \in G$ such that $f(a) = b$.

Thus, $\exists\, Ka \in G|K$ such that $\varphi(Ka) = f(a) = b$.

Hence φ is onto.

4. **φ is a homomorphism**

Consider, $\varphi(Ka \cdot Kb) = \varphi(Kab) = f(ab)$

$= f(a)\cdot f(ab)$ $\quad [\because f$ is homomorphism$]$

$= \varphi(Ka)\cdot \varphi(Kb)$

Hence φ is a homomorphism.

Hence φ is isomorphism and hence $f(G) \approx G|K$.

1.25 Some Results on Isomorphism

Theorem 1. If $f: G \to G'$ be an isomorphism of a group G onto a group G' and a is any element of G, then the order of $f(a)$ equals the order of a.

Proof : Let e and e' be the identity elements of G and G'.

Case (i) : Let a be of finite order, say n. Thus, n is the least positive integer such that $a^n = e$.

Now, $a^n = e \Rightarrow f(a^n) = f(e)$

$\Rightarrow [f(a)]^n = e'$ $\quad (\because f$ is homomorphism$)$

$\Rightarrow f(a)$ is also of finite order.

Let $O[f(a)] = m$. That is, m is the least positive integer such that $[f(a)]^m = e'$.

Also we have, $[f(a)]^n = e'$. Thus, m divides n.

Now, $[f(a)]^m = e' \Rightarrow f(a^m) = f(e) \Rightarrow a^m = e$ $(\because f$ is one-one$)$

But, $O(a) = n$. Thus n dives m.

Hence $m = n$. That is, $O[f(a)] = O(a)$

Case (ii) : Let a be of infinite order. Let $f(a)$ be of finite order say m.

That m is the least positive integer such that $[f(a)]^m = e'$. Thus, $f(a^m) = e'$.

Now, $f(a^m) = e' \Rightarrow f(a^m) = f(e) \Rightarrow (a^m) = e$

Thus, a is of finite order. This is a contradiction. Hence, $O[f(a)]$ is infinite.

Theorem 2. Let $f : G \to G'$ be an isomorphism. If G is abelian, then G' is also abelian.

Proof : Let, $a'b' \in G'$ be arbitrary. Since f is onto, $\exists\, a, b \in G$ such that, $f(a) = a'$, $f(b) = b'$.

Consider, $a' \cdot b' = f(a) \cdot f(b) = f(ab) = f(ba)$ $(\because G$ is abelian$)$

$= f(b) \cdot f(a) = b' \cdot a'$

$\Rightarrow a'b' = b'a' \ \forall\, a', b' \in G$

Hence G' is abelian.

Theorem 3. Any infinite cyclic group is isomorphic to the group Z of integers, under addition.

Proof : Let G be an infinite cyclic group having generator 'a'. That is, $G = \{a^n : n \in \mathbf{Z}\}$.

Define, $f : G \to \mathbf{Z}$ by $f(a^n) = n, \ \forall\, a^n \in G$

(i) f is homomorphism.

Consider, $f(a^m \cdot a^n) = f(a^{m+n}) \Rightarrow f(a^m \cdot a^m) = m + n = f(a^m) + f(a^n)$

Thus, $f(a^m \cdot a^n) = f(a^m) + f(a^n)$. Hence, f is homomorphism.

(ii) f is onto.

Now, $\forall\, n \in \mathbf{Z}$, $a^n \in G$, G being infinite group

$\Rightarrow \forall\, n \in \mathbf{Z}$, $\exists\, a^n \in G$ such that $f(a^n) = n$. Hence, f is onto.

(iii) f is one -one.

Let, $f(a^n) = f(a^m) \Rightarrow n = m \Rightarrow a^n = a^m$. Hence, f is one-one.

Theorem 4. Any finite cyclic group of order n is isomorphic to additive group of integers modulo n.

Proof : Exercise.

Example 1. Define $f : R \to R$, by $f(x)$ = greatest integer $\leq x$. Is f is a homomorphism.

Solution : By definition of f, $f(1) = 1$, $f\left(\frac{1}{2}\right) = 0$, $f\left(\frac{3}{2}\right) = 1$ etc.

Consider, $f\left(\frac{3}{2}\right) = 1$ and $f\left(\frac{5}{2}\right) = 2$.

Now, $f\left(\frac{3}{2} + \frac{5}{2}\right) = f(4) = 4$. But, $f\left(\frac{3}{2}\right) + f\left(\frac{5}{2}\right) = 1 + 2 = 3$

Thus, $f\left(\frac{3}{2} + \frac{5}{2}\right) \neq f\left(\frac{3}{2}\right) + f\left(\frac{5}{2}\right)$. Hence, f is not a homomorphism.

Example 2. Show that the additive group $(R, +)$ of all real numbers and the multiplicative group $(R^+, \cdot)$ of all positive real numbers are isomorphic.

Solution : Define, $f : (R, +) \to (R^+, \cdot)$, by $f(x) = e^x, \forall x \in R$

Now, $f(x + y) = e^{x+y} \Rightarrow f(x + y) = e^x \cdot e^y \Rightarrow f(x + y) = f(x) \cdot f(y)$

Thus, f is homomorphism.

Now, $f(x) = f(y) \Rightarrow e^z = e^y$

$\Rightarrow e^{x-y} = 1 = e^0 \Rightarrow x - y = 0 \Rightarrow x = y$

Thus, f is one-one.

Let, $y \in R^+$

Consider, $f(\log_e y) = e^{\log ey} = y \Rightarrow \forall y \in R^+, \exists \log_e y \in R$ such that $f(\log_e y) = y$

Thus, f is onto. Hence f is isomorphism.

EXERCISES

1. Which of the following maps are homomorphism? If the map is a homomorphism, find its kernel.
   (i) $f : (Z, +) \to (2Z, +)$ defined by $(x) = 2x, \forall x \in \mathbf{Z}$.
   (ii) $f : (R, +) \to (R^+, \cdot)$ defined by $f(x) = e^x, \forall x \in \mathbf{Z}$.
   (iii) $f : (\mathbf{Z}, +) \to (\mathbf{Z}_n, +_n)$ defined by $f(a) = r, \forall a \in Z$, where r is the unique least non negative remainder when a is divided by n.
   (iv) $f : (\mathbf{R}, +) \to (\mathbf{C}^*, \cdot)$ defined by $f(x) = \cos x + i \sin x, \forall x \in R$, where C^* is the set of all non-zero complex numbers.
   (v) $g : (\mathbf{Q}^*, \cdot) \to (\mathbf{Q}^*, \cdot)$ defined by $g(x) = |x| = x, \forall x \in \mathbf{Q}^*$, where $\mathbf{Q}^*$ is the set of all non-zero rationals.
   (vi) $f : (\mathbf{R}, +) \to (\mathbf{R}, +)$ defined by $f(x) = 3x + 1, \forall x \in \mathbf{R}$.
   (vii) $h : (\mathbf{Z}_6, +_6) \to (\mathbf{Z}_2, +_2)$ defined by $h(x) = r, \forall a \in \mathbf{Z}$, where r is remainder when x is divided by 2.
2. Let $G = |a + b\sqrt{2} \mid a, b \in Q|$, show that $T : (G, +) \to (G, +)$ defined

   $T(a + b\sqrt{2}) = a - b\sqrt{2}$ is an isomorophism.
3. Let $G = \left\{\begin{pmatrix} a & 0 \\ 0 & 0 \end{pmatrix} \middle| a \in \mathbf{R}^*\right\}$ show that, $f : (G, \cdot) \to \mathbf{R}^*$ defined by $f\begin{pmatrix} a & 0 \\ 0 & 0 \end{pmatrix} = a$ is an

   isomorphism.
4. G is any group and g a fixed element of G. Define $T : G \to G$ by $T(x) = gxg^{-1}$. Show that T is an isomorphism.
5. Let G be an abelian group and define $f : G \to G$ by $f(x) = x^{-1}, \forall x \in G$. Show that f is isomorphism. Conversely, if G is a group and $f : G \to G$, defined by $f(x) = x^{-1}$ is an isomorphism, show that G is abelian.

Answer

1. **(i)** Ker $f = (0)$ **(ii)** Ker $f = (f)$ **(iii)** Ker $f = \{kn \mid k \in Z\}$
   (iv) Ker $f = \{2n\pi \mid n$ is an integer$\}$ **(v)** Ker $g = \{1, -1\}$
   (vii) Ker $h = \{0, 2, 4\}$

1.26 Permutation Group of Degree n-Cayley's Theorem

Consider a finite set $S = \{a_1, a_2, a_3, ... a_n\}$ of n distinct elements. Let $f : S \to S$ be a bijection, That is, f is one-one and onto function. Now $f(a_1), f(a_2), f(a_3), ..., f(a_n)$ are also elements of S and also these $f(a_1), f(a_2, f(a_3), ..., f(a_n)$ are $a_1, a_2, a_3, ..., a_n$ in some order. We have,

$$S = \{f(a_1), f(a_2), f(a_3), ..., f(a_n)\}$$

Let, $f(a_1) = b_1,\ f(a_2) = b_2,\ f(a_3) = b_3, ..., f(a_n) = b_n$.

Then the mapping f can be denoted by a two row notation as

$$f = \begin{pmatrix} a_1 & a_2 & a_3 .... & a_n \\ b_1 & b_2 & b_3 .... & b_n \end{pmatrix}$$

In this notation, we put the elements of S in the first row (in any order) and under each element of this row we put down its image under the mapping f. Such a function f is called a **permutation** of degree n.

Definition : A one-one mapping of a finite set S onto itself is called a permutation. The number of elements in the set is called degree of permutation.

As we are dealing with only finite sets and considering only bijections, we can denote the elements of S by 1, 2, 3, ..., n instead of $a_1, a_2, a_3, ..., a_n$.

Equality of Two Permutations

Two permutations f and g of degree n on S are said to be equal if

$$f(x) = g(y),\ \forall\ x \in S$$

For example, consider the permutations

$$f = \begin{pmatrix} 1 & 2 & 3 \\ 3 & 1 & 2 \end{pmatrix} \text{ and } g = \begin{pmatrix} 2 & 1 & 3 \\ 1 & 3 & 2 \end{pmatrix}$$

In both these permutations, 1 is mapped onto 3, 2 is mapped onto 1 and 3 is mapped onto 2. Thus $f = g$.

Identity Permutations

The identity mapping I on a set S consisting of a distinct elements to itself is called an **identity permutation of degree *n*.**

That is the permutation $$I = \begin{pmatrix} 1 & 2 & 3 & .... & n \\ 1 & 2 & 3 & ...... & n \end{pmatrix}$$

is called the **identity permutation of degree *n*.**

Permutation Multiplication (Composition of Permutation)

Let f and g be two permutation of degree n defined on the set S. Since both f and g are bijections from S onto itself, the composition gof is also a bijection from S onto S. Similarly the composition fog is also a bijection from S onto S. Thus gof and fog are both permutations of degree n on S. The permutations gof and fog are called the **product** or **composite** of the permutations g and f and that of f and g respectively. These are defined by

$$(gof)\ (x) = f[g(x)],\ \forall\ x \in S$$
$$(fog)\ (x) = g[f(x)],\ \forall\ x \in S$$

Thus gof means the permutation obtained by applying first g to each element of S and then applying f to the element so obtained.

Let, $S = \{1\ \ 2\ \ 3\ \ 4\}$ and $f = \begin{pmatrix} 1 & 2 & 3 & 4 \\ 3 & 1 & 2 & 4 \end{pmatrix},\ g = \begin{pmatrix} 1 & 2 & 3 & 4 \\ 4 & 1 & 2 & 3 \end{pmatrix}$

be two permutations of degree 4 on S.

$$gof(1) = f[g(1)] = f(4) = 4$$
$$gof(2) = f[g(2)] = f(1) = 3$$
$$gof(3) = f[g(3)] = f(2) = 1$$
$$gof(4) = f[g(4)] = f(3) = 2$$

Thus, $gof = \begin{pmatrix} 1 & 2 & 3 & 4 \\ 4 & 3 & 1 & 2 \end{pmatrix}$. Similarly, $fog = \begin{pmatrix} 1 & 2 & 3 & 4 \\ 2 & 4 & 1 & 3 \end{pmatrix}$.

Note that, $gof \neq fog$

The product of two permutation can be computed as shown below.

Computation of gof

$$\begin{array}{cccccc} g & : & 1 & 2 & 3 & 4 \\ & & \downarrow & \downarrow & \downarrow & \downarrow \\ & & 4 & 1 & 2 & 3 \\ & & \downarrow & \downarrow & \downarrow & \downarrow \\ f & : & 4 & 3 & 1 & 2 \end{array}$$

Thus, $gof = \begin{pmatrix} 1 & 2 & 3 & 4 \\ 4 & 3 & 1 & 2 \end{pmatrix}$.

Inverse of a Permutation

Consider a permutation f of degree n on S. Thus f is a bijection on S onto S. Therefore the inverse of the mapping f exists and is also a permutation of degree n on S. This inverse of f is called **inverse permutation** of f and is denoted by f^{-1}.

To obtain the inverse permutation of f^{-1} of f interchange the rows of the permutation. For example, if,

$$f = \begin{pmatrix} 1 & 2 & 3 & 4 \\ 3 & 1 & 2 & 4 \end{pmatrix}, \text{ then } f^{-1} = \begin{pmatrix} 3 & 1 & 2 & 4 \\ 1 & 2 & 3 & 4 \end{pmatrix}. \text{ That is, } f^{-1} = \begin{pmatrix} 1 & 2 & 3 & 4 \\ 2 & 3 & 1 & 4 \end{pmatrix}$$

Note : $fof^{-1} = f^{-1}of = I$ (Identity permutations)

Let S be a finite set of n distinct elements. Then clearly there are $n!$ different ways of arranging the elements of S. Thus the total number of distinct permutations of degree n defined on S is $n!$.

For example, if $S = (1, 2, 3)$, then there are 3! permutations of order 3, which are

$$f_1 = \begin{pmatrix} 1 & 2 & 3 \\ 1 & 2 & 3 \end{pmatrix}, \; f_2 = \begin{pmatrix} 1 & 2 & 3 \\ 2 & 1 & 3 \end{pmatrix}, \; f_3 = \begin{pmatrix} 1 & 2 & 3 \\ 3 & 1 & 2 \end{pmatrix},$$

$$f_4 = \begin{pmatrix} 1 & 2 & 3 \\ 1 & 3 & 2 \end{pmatrix}, \; f_5 = \begin{pmatrix} 1 & 2 & 3 \\ 2 & 3 & 1 \end{pmatrix}, \; f_6 = \begin{pmatrix} 1 & 2 & 3 \\ 3 & 2 & 1 \end{pmatrix}$$

Observe that, $f_4of_2 \neq f_2of_4$.

Theorem. Let S be a non empty finite set of n elements. The set S_n of all permutations of degree n defined on S, is a finite group of order n!, under the permutation multiplication.

Proof : Since S contain a distinct elements, the set S_n contain $n!$ elements.

We shall show that the permutation multiplication defined on S_n satisfies the axioms of a group.

1. Closure Property

If f and g are any two elements of S_n, then clearly fog is also a permutation of degree n on S. Thus, $fog \in S_n$. That is, S_{n31} is closed under permutation multiplication.

2. Associative Law

The permutation multiplication of two permutation is defined as composition of two bijective mapping from S onto S. Since the composition of mapping is associative, it follows that permutation multiplication is associative.

3. Existence of Identity

The identity permutation I on S acts as the identity element of S_n. For, let

$$f = \begin{pmatrix} a_1 & a_2 & a_3 & \cdots & a_n \\ b_1 & b_2 & b_3 & \cdots & b_n \end{pmatrix}$$

be arbitrary element of S_n.

Consider, $Iof = \begin{pmatrix} a_1 & a_2 & \cdots\cdots & a_n \\ a_1 & a_2 & \cdots\cdots & a_n \end{pmatrix} o \begin{pmatrix} a_1 & a_2 & \cdots\cdots & a_n \\ b_1 & b_2 & \cdots\cdots & b_n \end{pmatrix} = \begin{pmatrix} a_1 & a_2 & \cdots\cdots & a_n \\ b_1 & b_2 & \cdots\cdots & b_n \end{pmatrix} = f$

Similarly, $foI = f$.

4. Existence of Inverse

Clearly for $f \in S_n$, the function $f^{-1} \in S_n$ acts as inverse of f.

Hence S_n is a finite group of order $n!$

Definition : The group S_n of all permutation of a finite set S of n distinct elements is called the symmetric group of degree n.

Note :

1. The order set S_n is $n!$
2. The group S_n is a non-abelian group.

Definition : Any subgroup of S_n is called a permutation group of degree n.

Theorem (Cayley's theorem) : Every finite group is isomorphic to a group of permutation.

Proof : Let G be a finite group of order n. Let $g \in G$. We shall define a mapping

$$T_g : G \to G$$

by
$$T_g(x) = xg, \forall x \in G$$

The map T_g is well defined.

Now, $T_g(x) = T_g(y) \Rightarrow xg = yg \Rightarrow x = y$. Thus, T_g is one-one.

If $y \in G$, then, $T_g(yg^{-1}) = (yg^{-1})g = y(g^{-1}g) = y$. Thus, $\forall y \in G, \exists yg^{-1} \in G$ such that

$$T_g(yg^{-1}) = y$$

Hence, T_g is onto.

Thus T_g is a permutation of n elements of G.

Hence, $\forall y \in G$ we have a permutation T_g of degree n.

Let, $G' = \{T_g \mid g \in G\}$ and $T_g, T_h \in G'$ be arbitrary.

Consider, $(T_g \cdot T_h)x = T_h[T_g(x)] = T_h(xg)$

$= (xg)h = x(gh) = T_{gh}(x)$

Thus, $\forall x \in G \ (T_g \cdot T_h)(x) = T_{gh}(x)$

Since $g, h \in G$, the composition $gh \in G$ and therefore $T_{gh} \in G'$.

Thus, G' is closed under permutation multiplication.

Further G' is a group under permutation multiplication.

Now we shall prove $G \approx G'$. Define a mapping,

$$\varphi : G \to G'$$

by $$\varphi(g) = T_g \quad \forall\, g \in G$$

Let, $\varphi(g) = \varphi(h)$

$$\Rightarrow \quad T_g = T_h \quad \Rightarrow \quad T_g(x) = T_h(x) \quad \forall\, x \in G$$

$$\Rightarrow \quad xg = xh \quad \forall\, x \in G$$

$$\Rightarrow \quad g = h$$

Thus, φ is one-one.

Also, $\forall$ Tg $\in$ G′, $\exists\ g \in G$ such that $\varphi(g) = T_g$. Thus, φ is onto.

Now consider, $\varphi(gh) = T_{gh}$

$$= T_g \cdot T_h = \varphi(g) \cdot \varphi(h) \quad \forall\, g, h \in G$$

Thus, φ is homomorphism.

Hence φ is a homomorphism and therefore $G \approx G'$.

Definition : The group of permutations G' in the proof of the above theorem is called right regular representation of G.

Example 1. Find the right regular representation of the multiplicative group {1, –1, *i*, –*i*}.

Solution : Let, $G = \{1, -1, i, -i\}$, $G' = \{T_g \mid g \in G\}$, $G' = \{T_1, T_{-1}, T_i, T_{-i}\}$

We know that the permutation T_g sends each element x to xg - i.e., right multiplication by g.

$$T_1 = \begin{pmatrix} 1 & -1 & i & -i \\ 1.1 & (-1)1 & (i)1 & (-i)1 \end{pmatrix}$$

$$T_1 = \begin{pmatrix} 1 & -1 & i & -i \\ 1 & -1 & i & -i \end{pmatrix}$$

$$T_{-1} = \begin{pmatrix} 1 & -1 & -i & i \\ -1 & 1 & i & -i \end{pmatrix}$$

$$T_i = \begin{pmatrix} 1 & -1 & i & -i \\ i & -i & -1 & 1 \end{pmatrix}$$

$$T_{-i} = \begin{pmatrix} 1 & -1 & i & -i \\ -i & i & 1 & -1 \end{pmatrix}$$

EXERCISES

1. If $A = \begin{pmatrix} 1 & 2 & 3 \\ 3 & 1 & 2 \end{pmatrix}$, $B = \begin{pmatrix} 1 & 2 & 3 \\ 2 & 3 & 1 \end{pmatrix}$ and $C = \begin{pmatrix} 1 & 2 & 3 \\ 3 & 2 & 1 \end{pmatrix}$ are permutations on the set {1, 2, 3}, then

   (i) Find *AB, BA, AC, CA, BC* and *CB*. **(ii)** Verify the associative Axiom $A(BC) = (AB)C$.

   (iii) Show that $A^{-1} = B$, $B^{-1} = A$, $C^{-1} = A$. **(iv)** Verify $(AC)^{-1} = C^{-1}A^{-1}$.

(v) Show that $(AB)^{-1} = B^{-1} A^{-1} = I$.

(vi) Do the above permutations form a group under permutation multiplication?

2. If $f = \begin{pmatrix} 1 & 2 & 3 \\ 3 & 1 & 2 \end{pmatrix}$ and $g = \begin{pmatrix} 1 & 2 & 3 \\ 1 & 3 & 2 \end{pmatrix}$ and I denote the identity permutation.

(i) find fg, gf, f^{-1}, g^{-1}. Is $fg = gf$? **(ii)** Verify $ff^{-1} = gg^{-1} = 1$.

3. If $\alpha = \begin{pmatrix} 1 & 2 & 3 & 4 & 5 \\ 2 & 3 & 4 & 1 & 5 \end{pmatrix}$, $\beta = \begin{pmatrix} 1 & 2 & 3 & 4 & 5 \\ 3 & 1 & 4 & 5 & 2 \end{pmatrix}$. Find $\alpha\beta$, $\beta\alpha$, α^{-1}, β^{-1}, $\alpha\beta^2$.

4. Find all the elements of S_1 and S_2 and prepare their multiplication table.

5. Find the right regular representation of the multiplication group $G = \{1, \omega, \omega^2\}$ of cube root of unity.

Rings, Integral Domains and Fields

2

2.1 Introduction

The readers are familiar with the algebraic structure called group and some of the results related to it. A group is an algebraic structure with one binary operation. We shall study in this chapter an algebraic structure with two appropriately restricted and related binary operations. This algebraic structure is called a ring. We shall see some special types of rings namely integral domain and field and the results connected to these.

2.2 Rings

Definition : An algebraic structure $(R, +, \cdot)$ consisting of a non empty set R with two binary operations called addition and multiplication, is called a ring if,

1. $(R, +)$ is an abelian group.
2. The binary operation '$\cdot$' is associative in R. That is,
$$\forall\, a, b, c \in R,\ a \cdot (b \cdot c) = (a \cdot b) \cdot c$$
3. The binary operation '$\cdot$' is both left as well as right distributive over the operation '+' in R.

   (i) $a \cdot (b + c) = a \cdot b + a \cdot c$ **(left distributive law)**

   (ii) $(b + c) \cdot a = b \cdot a + c \cdot a$ **(right distributive law)**

Note :

1. In a ring $(R, +, \cdot)$ the identify element w.r.t. addition is denoted by 0 and is called the **zero** of the ring or the zero element of the ring. Thus,
$$\forall\, a \in R,\ a + 0 = 0 + a = a$$
2. In a ring $(R, +, \cdot)$ the inverse of an element a w.r.t. addition is denoted by $-a$ and is called **additive inverse or negative** of a in R. Thus,
$$\forall\, a \in R, a + (-a) = (-a) + a = 0$$

Further we define,
$$\forall\, a, b \in R,\ a - b = a + (-b)$$

2.3 Types of Rings

A ring $(R, +, \cdot)$ is called

1. **a commutative ring** if the binary operation multiplication is commutative. That is,
$$\forall\, a, b \in R,\ a \cdot b = b \cdot a$$
2. **a ring with unity,** if there exists identity element w.r.t multiplication. This identity element is denoted by 1 and called **unity** of the ring R. That is,
$$1 \cdot a = a \cdot 1 = a,\ \ \forall\, a \in R$$

3. **a ring with zero divisors** if for $a, b \in R, a \neq b, b \neq 0 \Rightarrow a \cdot b = 0$. The elements a and b are respectively called as **left zero divisor** and **right zero divisor**.
4. **a ring without zero divisors** if for
$$a, b \in R,\ a \neq 0 \text{ and } b \neq 0 \Rightarrow a \cdot b \neq 0$$
5. **an integral domain** if it is a commutative ring with unit element and is without zero divisors.

Definition : Let $(R, +, \cdot)$ be a ring with unit element. An element $a \in R$ is said to be invertible, if there exists an element $b \in R$ such that, $a \cdot b = b \cdot a = 1$.

In this case the element b is called a multiplicative inverse of the element a.

Note :

(i) In a ring R with unit element, the element 0 is not invertible for all $a \in R$, $a \cdot 0 = 0$. Further the unity is invertible, for $1 \cdot 1 = 1$.

(ii) If an element $a \in R$, the ring with unit element, is invertible then it is unique and is denoted by a^{-1}.

6. **Division ring :** A ring with at least two elements is said to be **division ring or a skew field** if it has unity and such that every non zero element in invertible i.e., every non zero element has multiplicative inverse.
7. **Field :** A commutative division ring is called a field.

That is, a field is a commutative ring with at least two elements, having unit element and is such that every non zero element is inversible.

Note : In a field set of all non zero elements is a commutative group, w.r.t. multiplication.

In what follows we use the following symbols to represent the sets of numbers, mentioned below :

R = Set of all real numbers
Q = Set of all rational numbers
R^* = Set of all non zero real numbers
Q^* = Set of all non zero rational numbers
N = Set of all natural numbers
Z(or I) = Set of all integers
Z^* = Set of all positive integers
Z^- = Set of all negative integers
C = Set of all complex numbers
C^* = Set of all non-zero complex numbers

2.4 Examples of Rings

1. $(Z, +, \cdot)$ where + and $\cdot$ are usual addition and multiplication in **Z** is a commutative ring with unity and without zero divisors. It is neither a skew field nor a field.
2. Each one of the following,
$$(R, +, \cdot),\ (Q, +, \cdot),\ (C, +, \cdot)$$
is a ring with unit element without zero divisors. Further each one of them is a field.
3. For any non zero fixed integer $k \neq 1$, $(kZ, +, \cdot)$ where $kZ = \{kn \mid n \in \mathbf{Z}\}$ is a commutative ring without unity and without zero divisors.

4. Let $M_{(n \times n)}$ be the set of all matrices of order $n \times n$ matrices over the real number set R. Then $(M_{(n, n)}, +, \cdot)$, where + and $\cdot$ are usual addition and multiplication of matrices, is a non commutative ring with unit element, identity element being I_n. This ring has divisors. It is not an integral domain and not a field.

2.5 Properties of Rings

Few basic properties of rings are given in the form of theorems.

Theorem 1 : In a ring $(R, +, \cdot)$

(i) $\forall\, a \in R,\ a \cdot 0 = 0 \cdot a = 0$ where 0 being the additive identity.

(ii) $\forall\, a, b \in R,\ a(-b) = (-a)b = -(ab)$.

(iii) $\forall\, a, b \in R,\ (-a)(-b) = ab$.

Proof : (i) We have,

$$a \cdot 0 = a \cdot (0 + 0)$$

$$\Rightarrow \quad a \cdot 0 = a \cdot 0 + a \cdot 0 \qquad \text{(by left distributive law)}$$

Also, $$a \cdot 0 = a \cdot 0 + 0$$

Thus, $$a \cdot 0 + a \cdot 0 = a \cdot 0 + 0 \Rightarrow a \cdot 0 = 0 \qquad \text{(by left cancellation law)}$$

Similarly, $$0 \cdot a = (0 + 0) \cdot a = 0 \cdot a + 0 \cdot a. \text{ Also, } 0 \cdot a = 0 \cdot a = 0$$

$$\Rightarrow \quad 0 \cdot a + 0 = 0 \cdot a + 0 \cdot a \Rightarrow 0 \cdot a = 0$$

Hence, $a \cdot 0 = 0 \cdot a = 0$.

(ii) Consider, $$ab + a(-b) = a \cdot \{b + (-b)\} = a \cdot 0 = 0$$

Also, $$a(-b) + ab = a \cdot [(-b) + b] = a \cdot 0 = 0$$

Thus, $a(-b)$ is the additive inverse of $a \cdot b$.

But the inverse of $a \cdot b$ is denoted by, $-(ab) \Rightarrow a \cdot (-b) = -(ab)$

Similarly, $(-a) \cdot b = -(ab)$ can be proved.

Hence, $a \cdot (-b) = (-a) \cdot b = -(ab)$.

(iii) We note that, $$(-a) \cdot (-b) = -[a(-b)] \qquad \text{(by (ii))}$$

$$= -[-(ab)] \qquad \text{(again by (ii))}$$

$$= ab \qquad (\because \text{ inverse of } -ab \text{ is } ab \text{ itself})$$

Hence, $(-a)(-b) = a \cdot b$.

Definition : Cancellation laws in a ring.

In a ring R, we say that the cancellation laws holds, if

(i) $a \neq 0,\ ab = ac \Rightarrow b = c$ (left cancellation law)

(ii) $a \neq 0,\ ba = ca \Rightarrow b = c$ (right cancellation law)

Theorem 2 : A ring is without zero divisors if and only if the cancellation laws hold in it.

Proof : Let R be a ring without zero divisors. Let, $a, b, c \in R$ such that, $a \neq 0$ and $ab = ac$.

$$\Rightarrow \quad a \neq 0 \quad \text{and} \quad ab - ac = 0$$

$$\Rightarrow \quad a \neq 0 \quad \text{and} \quad a(b - c) = 0$$

$$\Rightarrow \quad b - c = 0 \qquad (\because a \neq 0 \text{ and } R \text{ being without zero divisors})$$

$$\Rightarrow \quad b = c$$

Thus $a \neq 0,\ ab = ac \Rightarrow b = c$

Hence the left cancellation law holds. Similarly, right cancellation law holds.

Conversely, let R be a ring in which the cancellation laws hold. We shall show that R has no zero divisors. To this end we have to show that

$$a \cdot b = 0 \Rightarrow a = 0 \text{ or } b = 0$$

Supposing, $ab = 0$, with $a \neq 0$

$\Rightarrow$ $a \neq 0$, $ab = a \cdot 0$ $(\because a \cdot 0 = 0)$

$\Rightarrow$ $b = 0$ (by left cancellation law)

Again, supposing, $ab = 0$, with $b \neq 0$

$\Rightarrow$ $b \neq 0, ab = a \cdot b$

$\Rightarrow$ $a = 0$ (by right cancellation law)

Thus, $a \cdot b = 0 \Rightarrow$ either $a = 0$ or $b = 0$. Hence R is a ring without zero divisors.

Corollary : The cancellation laws hold in an integral domain.

Proof : An integral domain is a commutative ring with unity and having no zero divisors. Hence by theorem 2, it follows that the cancellation laws hold in an integral domain.

Theorem 3 : Every field is an integral domain.

Proof : Let F be a filed. Then F is a commutative ring with unity such that every non zero element has multiplicative inverse. In order to show that F is an integral domain, we shall show that F has no zero divisor.

Let, $a, b \in R$ such that, $a \cdot b = 0$ with $a \neq 0$.

Since, $a \neq 0$ and F in field, a^{-1} exists in F.

Now, $a \neq 0,\ ab = 0 \Rightarrow a^{-1}(ab) = a^{-1} \cdot 0$

$\Rightarrow (a^{-1}a)b = 0$

$\Rightarrow 1 \cdot b = 0 \Rightarrow b = 0$ $(\because 1 \cdot b = b)$

Similarly for $a, b \in R$ such that $ab = 0$ and $b \neq 0 \Rightarrow a = 0$

Thus for $a, b \in F$, $ab = 0 \Rightarrow a = 0$ or $b = 0$.

Hence F is without zero divisors and therefore F is an integral domain.

Note : The converse of the above theorem is not true. That is every integral domain is not a field.

For example, the ring of integers $(\mathbf{Z}, +, \cdot)$ is an integral domain but not a field. However we have the following theorem which asserts every finite integral domain is a field.

Theorem 4 : Every finite integral domain is a field.

Proof : Let D be a finite integral domain with n elements.

Since every integral domain is a commutative ring with unity, it is sufficient if we show that every non zero element of D has multiplicative inverse - i.e., every non zero element of D is invertible.

Let, $D = \{0, 1, a_1, a_2, ..., a_{n-2}\}$ with a_j are distinct. Let, $a_i (\neq 0) \in D$ be arbitrary.

Consider the set,

$$B = \{a_i \cdot 0,\ a_i \cdot 1,\ a_2 \cdot a_1, ..., a_i \cdot a_{n-2}\}$$

Clearly each element of B is an element of D. Further,

$$a_i \cdot a_k = a_i \cdot a_j \Rightarrow a_k = a_j \quad \text{(by cancellation law)}$$

This shows that all the elements of B are distinct. Thus, $D = B$.

Thus, for every element $a_k \in R$ there exists an element a_j such that

$$a_k = a_i \cdot a_j$$

In particular as $1 \in D$, $\exists$ an element a_j such that, $1 = a_i \cdot a_j$.

Further as D is commutative, we have $1 = a_j \cdot a_i$.

Thus the element a_j acts as inverse of $a_i \Rightarrow a_i$ is invertible. Thus, every non zero element of D is invertible. Hence D is a field.

2.6 Ring of Integers Modulo n

Consider the set, $Z_n = \{0, 1, 2, ..., (n-1)\}$ of positive integers from 0 to $n-1$.

This set Z_n, together with the binary operations $+_n$ addition modulo n and $\times_n$ multiplication modulo n is a commutative ring with unity element as 1.

Note : $(Z_n, +_n, \times_n)$ of integers modulo n, may or may not have zero divisors.

For example consider $(Z_6, +_6, \times_6)$. This ring has zero divisors - for example $2 \neq 0, 3 \neq 0$ but $2 \times_6 3 = 0$. That is 2 and 3 are zero divisors in Z_6.

But the ring $(Z_7, +_7, \times_7)$ has no zero divisors. The following theorem gives us a necessary and sufficient condition for the ring Z_n of integers modulo n to be without zero divisors.

Theorem 1. The ring $(Z_n, +_n, \times_n)$ is an integral domain and hence a field if and only if n is a prime integer.

Proof : We shall show first, that, Z_n is an integral domain $\Rightarrow$ n is a prime

To this end we shall prove

n is not prime $\Rightarrow$ Z_n is not an integral domain

That is, n is not prime $\Rightarrow$ Z_n has zero divisors

Now, n is not prime $\Rightarrow$ n is composite

$\Rightarrow$ $\exists$ two integers m and l such that $n = m \cdot l$, with $1 < m < n$, $1 < l < n$.

$\Rightarrow$ $m \neq 0$, $n \neq 0 \in Z_n$ with $m \times_n l = 0$ $(\because m \cdot l = 1 \cdot n + 0)$

$\Rightarrow$ m is a zero divisor in Z_n

$\Rightarrow$ Z_n is not an integral domain.

Hence if Z_n is an integral domain, n must be prime.

Conversely we shall show that

n is a prime $\Rightarrow$ Z_n is an integral domain

Since $(Z_n, +_n, \times_n)$ is a commutative ring with unity, it is sufficient if we show that,

n is a prime $\Rightarrow$ Z_n has no zero divisors

Let $a \times_n b = 0$ for some $a, b \in Z_n$

$\Rightarrow$ $a \cdot b = nk$ for some integer k

$\Rightarrow$ $n \mid a \cdot b$

$\Rightarrow$ $n \mid a$ or $n \mid b$ $(\because n$ is a prime$)$

$\Rightarrow$ $a = 0$ or $b = 0$ $(\because a, b < n)$

Thus, $a \times_n b = 0 \Rightarrow a = 0$ or $b = 0$

Hence Z_n has no zero divisors and therefore it is an integral domain.

Further, as Z_n is a finite set, if n is a prime then it is a finite integral domain and therefore it is a field.

Theorem 2 : Let R be a commutative ring with unity. Then R is an integral domain if and only if the set R^* of all non zero elements of R is closed under multiplication.

Proof : Let R be an integral domain and $a, b \in R^*$ be arbitrary.

$$\Rightarrow \quad a \neq 0 \text{ and } b \neq 0$$

$$\Rightarrow \quad a \cdot b \neq 0 \qquad (\because R \text{ is an integral domain})$$

$$\Rightarrow \quad a \cdot b \neq R^*$$

Thus R^* is closed under multiplication.

Conversely let R^* be closed under multiplication.

$$\Rightarrow \quad \forall\ a,\ b \in R^*,\ a \cdot b \in R^*$$

$$\Rightarrow \quad a \neq 0,\ b \neq 0 \Rightarrow a \cdot b \neq 0$$

Thus, R has no zero divisors and hence is an integral domain.

Theorem 3 : In an integral domain the non-zero elements form a semigroup w.r.t. multiplication.

Proof : Let R be an integral domain and $a, b \in R^*$ be arbitrary.

$$\Rightarrow \quad a \neq 0 \text{ and } b \neq 0 \Rightarrow a \cdot b \neq 0 \qquad (\because R \text{ is an integral domain})$$

That is, R^* is closed under multiplication.

Further as the multiplication is associative in R, so is associative in R^*.

Hence R^* is a semigroup, w.r.t. multiplication.

Example 1. Show that an integral domain with six elements does not exist.

Solution : Let D be an integral domain with 6 elements. Therefore $(D, +)$ is an abelian group of order 6. Thus there exists two non-zero elements a and b in D such that,

$$O(a) = 2, \quad O(b) = 3$$

Thus, 2 and 3 are least positive integers such that, $2a = 0$, $3b = 0$ respectively.

Now, $\quad 2a \cdot b = 0 \cdot b = 0 \Rightarrow a \cdot (2b) = 0$ where $a \neq 0$ and $2b \neq 0$

This implies that D has zero divisors and hence D is not an integral domain.

Hence there exists no integral domain having six elements.

Example 2. An element a of ring R is said to be idempotent if $a^2 = a$. Show that the only idempotent elements of an integral domain are 0 and 1.

Solution : Let R be an integral domain and $x \in R$ such that $x^2 = x$.

Supposing, $x \neq 0$, then

$$x^2 = x \Rightarrow x \cdot x = x \cdot 1 \Rightarrow x = 1 \qquad \text{(by cancellation law)}$$

Again, if, $x \neq 1$

$$x^2 = x \Rightarrow x \cdot x - x = 0 \Rightarrow x(x-1) = 0 \Rightarrow x = 0$$

$$(\because R \text{ is an integral domain})$$

Hence, $\quad x^2 = x$ in $R \Rightarrow x = 1$ or $x = 0$.

Example 3 : Show that the set of all matrices of the form $\begin{pmatrix} 0 & x \\ 0 & y \end{pmatrix}$, $x, y \neq Q$ is a non-commutative ring without unity, the binary operation being addition and multiplication of matrices.

Solution : Let, $R = \left\{ \begin{pmatrix} 0 & x \\ 0 & y \end{pmatrix} \mid x, y \in Q \right\}$, $R \neq \phi$ as $\begin{pmatrix} 0 & 0 \\ 0 & 0 \end{pmatrix} \in R$.

Let, $A = \begin{pmatrix} 0 & a \\ 0 & b \end{pmatrix}$ and $B = \begin{pmatrix} 0 & c \\ 0 & d \end{pmatrix} \in R$ be arbitrary.

Then, $A + B = \begin{pmatrix} 0 & a+c \\ 0 & b+d \end{pmatrix} \in R$ and $A \cdot B = \begin{pmatrix} 0 & ad \\ 0 & bd \end{pmatrix} \in R.$

That is, R is closed under matrix addition and multiplication. Matrix addition is associative and commutative. The zero matrix $O = \begin{pmatrix} 0 & 0 \\ 0 & 0 \end{pmatrix}$ acts as additive identity in R. For every $A = \begin{pmatrix} 0 & a \\ 0 & b \end{pmatrix}$ the matrix $-A = \begin{pmatrix} 0 & -a \\ 0 & -b \end{pmatrix}$ acts a the additive inverse of A.

Hence, $(R, +)$ is an abelian group.

The matrix multiplication is associative and matrix multiplication is distributive over addition.

Hence, $(R, +, \cdot)$ is a ring.

Now, let $A = \begin{pmatrix} 0 & 2 \\ 0 & 3 \end{pmatrix}$ and $B = \begin{pmatrix} 0 & 5 \\ 0 & 1 \end{pmatrix}$, then $AB = \begin{pmatrix} 0 & 2 \\ 0 & 3 \end{pmatrix}$ and $BA = \begin{pmatrix} 0 & 15 \\ 0 & 3 \end{pmatrix} \Rightarrow AB \neq BA.$

Thus ***R* is non-commutative**.

Supposing there exists an element say $I = \begin{pmatrix} 0 & g \\ 0 & h \end{pmatrix}$ of R such that for every ,

$$A = \begin{pmatrix} 0 & a \\ 0 & b \end{pmatrix} \neq 0,\ AI = IA = A.$$

Then, $AI = A \Rightarrow \begin{pmatrix} 0 & ah \\ 0 & bh \end{pmatrix} = \begin{pmatrix} 0 & a \\ 0 & b \end{pmatrix} \Rightarrow ah = a$ and $bh = b \Rightarrow h = 1$

Again, $IA = A \Rightarrow \begin{pmatrix} 0 & gb \\ 0 & hb \end{pmatrix} = \begin{pmatrix} 0 & a \\ 0 & b \end{pmatrix} \Rightarrow gb = a$ and $hb = b$

But there is no unique g which satisfies $gb = a$ for all rationals a and b.

Hence, ***R* is a ring without unity.**

Example 4. Consider the set, $Z[i] = \{a + ib \mid a, b \in Z\}$. The elements of $Z[i]$ are called Gaussian integers. Show that the set Z[i] is a ring under usual addition and multiplication of complex numbers. Also show that the only elements which are invertible are 1, –1, *i*, –*i*.

Solution : $Z[i]$ is a commutative ring-proof left as an exercise.

This ring $Z[i]$ is called the ring of Gaussian integers. The element $1 = 1 + i0$ in $Z[i]$ is unity element in the ring. We shall show that the only elements which are invertible are 1, –1, *i*, –*i*.

Supposing, $a + ib \in Z[i]$ is invertible and let $c + id \in Z[i]$ be its multiplicative inverse.

$\Rightarrow \quad (a + ib)(c + id) = 1$

$\Rightarrow \quad |(a + ib)(c + id)| = 1$

$\Rightarrow \quad (a^2 + b^2)(c^2 + d^2) = 1$

$\Rightarrow \quad a^2 + b^2 = 1$ (because *a*, *b*, *c*, d are integers)

$\Rightarrow \quad a^2 = 1,\ b^2 = 0$ or $a^2 = 0,\ b^2 = 1$

$\Rightarrow \quad a \pm 1,\ b = 0$ or $a = 0,\ b = \pm 1$

$\Rightarrow \quad a + ib = \pm 1, \pm i.$ Hence, only invertible elements are 1, –1, *i*, –*i*.

Example 5. State all invertible elements in the ring $M_2(Z)$.

Solution : In the ring $M_2(\mathbf{Z})$ an element $A = \begin{pmatrix} a & b \\ c & d \end{pmatrix}$ is invertible if and only if there exists an element $B \in M_2(\mathbf{Z})$ such that

$$AB = BA = \begin{pmatrix} 1 & 0 \\ 0 & 1 \end{pmatrix}$$

$$\Leftrightarrow \quad |\text{AB}| = \begin{vmatrix} 1 & 0 \\ 0 & 1 \end{vmatrix} = 1$$

$$\Leftrightarrow \quad |A| \cdot |B| = 1 \quad \Leftrightarrow \quad |A| = \pm 1 \Leftrightarrow ad - bc = \pm 1$$

Thus an element A is invertible if and only if $|A| = \pm 1$.

Example 6. Given an example of

(i) a commutative ring with unity

(ii) a commutative ring without unity

(iii) a commutative ring without zero divisors

(iv) a commutative ring with zero divisors

(v) a finite commutative ring with unity which is not an integral domain

(vi) a finite non commutative ring

(vii) an infinite non commutative ring

(viii) an infinite ring without unity

Solution :

(i) Ring of integers $(Z, +, \cdot)$ (ii) The ring $(2Z, +, \cdot)$

(iii) Ring of integers $(Z, +, \cdot)$ (iv) Ring $(Z_6, +_6, \times_6)$

(v) Ring $(Z_6, +_6, \times_6)$

(vi) See the ring given in the example 7 below.

(vii) Ring $M_2(Z)$ of all 2×2 matrices over set of integers.

(viii) Ring $(2Z, +, \cdot)$

Example 7. Show that the set $T = \{a, b, c, d\}$ is a non commutative ring where addition and multiplication are defined by the following tables.

+	a	b	c	d
a	a	b	c	d
b	b	a	d	c
c	c	d	a	b
d	d	c	b	a

·	a	b	c	d
a	a	a	a	a
b	a	b	c	d
c	a	a	a	a
d	a	b	c	d

Solution : Verification for T to be a ring is left as an exercise. The ring T is non commutative - for from the multiplication table we have $bc = c$ and $cb = a$. Further this ring has no unity.

Example 8. If the additive group of a ring R is cyclic, then show that R is commutative.

Solution : By data $(R, +)$ is cyclic. Let $g \in R$ be a generator of R. Thus,

$$\forall\, x \in R, \exists \text{ an integer } n \text{ such that } x = ng.$$

Let, $x, y \in R$ be arbitrary. Then there exist two integers, say n and m such that,

$$x = ng,\ y = mg$$

Consider, $xy = (ng)(mg)$

$= (n \cdot m)(gg)$ because $(na)(mb) = (mn)(ab)$ (prove)

$= (m \cdot n)(gg) = (mg)(ng) = yx$

Hence R is commutative.

Example 9. If $(R, +, \cdot)$ is a system satisfying all the conditions for a ring with unity 1, except $a + b = b + a$, show that the condition $a + b = b + a$ also hold in R and hence R is a ring.

Solution : Let $a, b \in R$ be arbitrary.

Consider, $(a + b)(1 + 1) = (a + b) \cdot 1 + (a + b) \cdot 1$

$= (a + b) + (a + b)$ .... (1)

Again, $(a + b)(1 + 1) = a(1 + 1) + b(1 + 1)$

$= a \cdot 1 + a \cdot 1 + b \cdot 1 + b \cdot 1 = (a + a) + (b + b)$ .... (2)

From (1) and (2) we have,

$$(a + b) + (a + b) = (a + a) + (b + b)$$

$\Rightarrow$ $a + (b + a) + b = a + (a + b) + b$

$\Rightarrow$ $b + a = a + b$ (by cancellation laws in $(R, +)$)

Hence $b + a = a + b$ and therefore R is a ring.

Example 10. If R is a ring such that $a^2 = a,\ \forall\ a \in R$, then show that

(i) the additive inverse of each element of R is itself

(ii) $a + b = 0 \Rightarrow a = b$

(iii) R is a commutative ring.

Solution : (i) To show that the additive inverse of any element $a \in R$ is itself, we shall show that $a + a = 0$, where 0 is the additive identity.

By data, $a = a^2\ \forall\ a \in R$

$\Rightarrow$ $(a + a) = (a + a)^2$

$= (a + a)(a + a)$

$= a^2 + a^2 + a^2 + a^2$

$= a + a + a + a$ $(\because a^2 = a)$

$\Rightarrow$ $(a + a) + 0 = (a + a) + (a + a)$

$\Rightarrow$ $a + a = 0$ (by cancellation law in R)

(ii) Let $a + b = 0$ and by (i) $a + a = 0\ \forall\ a \in R$. Thus,

$$a + b = a + a \Rightarrow b = a.$$

(iii) Let, $a, b \in R$. By data,

$a + b = (a + b)^2 = (a + b)(a + b)$

$= a^2 + ba + ab + b^2$

$= a + ba + ab + b$

Thus, $(a + b) + 0 = (a + b) + (ab + ba)$ $(\because R$, is an abelian group)

$\Rightarrow$ $ab + ba = 0 \Rightarrow ab = ba$ (by (ii))

Hence, R is commutative.

Example 11. Show that in the ring Z_n the set of non invertible elements (the elements which has no multiplicative inverse) consists of zero and the zero divisors.

Solution : Let, $a \in Z_n$ is not invertible and $a \neq 0$, then

$$1 < a < n \text{ with } g.c.d\ \{a, n\} \neq 1$$

Thus, $\exists$ k such that $k | a$, $k | n$, where $1 < k < n$.

$\Rightarrow$ $a = mk$, $n = qk$; $1 < m < n$ and $1 < q < n$

Now, $aq = mkq \Rightarrow aq = mn \Rightarrow aq \equiv 0 \pmod n$

Thus, $aq = 0$ in Z_n $\Rightarrow$ a is a zero divisor.

EXERCISES

1. Show that the set R of numbers of the form $a + b\sqrt{2}$, where a and b are integers, is a ring with respect to usual addition and multiplication of numbers.
2. Show the set of all matrices of order 2 × 2 forms a ring under matrix addition and multiplication. Is this a commutative ring?
3. Show that the set of all 2 × 2 complex matrices of the form $\begin{pmatrix} a & -b \\ \overline{b} & \overline{a} \end{pmatrix}$, where $\overline{a}$ denotes the complex conjugate of a, is a division ring under addition and multiplication of matrices. Is it a field?
4. Show that the set of all matrices of the form $\begin{pmatrix} a & b \\ -b & a \end{pmatrix}$ $a, b \in \mathbf{R}$ is a field under matrix addition and multiplication.
5. Let $\mathbf{R}^+$ be the set of all positive real numbers and q is a fixed positive real number not equal to 1. If $a, b \in \mathbf{R}^+$ define addition + and multiplication * on $\mathbf{R}^+$ by $a + b = a \cdot b$ (usual multiplication) and $a * b = a^{\log(qb)}$. Show that $\mathbf{R}^+$ is a field.
6. Let $(G, +)$ be an abelian group whose identity element is denoted by O. Define the binary operation * in $G a * b = O \ \forall \ a, b \in G$. Show that $(G, +, *)$ is a ring.
7. If $P(S)$ is the power set, then show that $(P(S), \Delta, \cap)$ is a ring, where Δ is the symmetric differences of two sets and $\cap$ is the inter-section of two sets.
8. Consider the set $R = \{f | f : S \to \mathbf{Z}\}$ where S is any non empty set. For any $f, g \in R$ define $f + g$ and $f \cdot g$ by
$$(f + g)(x) = f(x) + g(x)$$
$$(f \cdot g)(x) = f(x) \cdot g(x) \ \forall \ x \in S$$
Show that $(R, +, \cdot)$ is a ring.
9. Let C[0, 1] be the set of all real valued continuous functions defined on the closed interval [0, 1]. For any $f, g \in$ C[0, 1] define $f + g$ and $f \cdot g$ as in the problem 8. Show that C[0, 1] is a commutative ring with unity. Is it an integral domain?
10. Show that $(\mathbf{Z}, \oplus, \cdot)$, where $\oplus$ and $\cdot$ are defined by
$$a \oplus b = a + b - 1 \ \forall \ a, b \in \mathbf{Z}$$
$$a \cdot b = a + b - ab \ \forall \ a, b \in \mathbf{Z}$$
is a commutative ring with identity.
11. Let $(G, +)$ be any abelian group and Hom (G) be the set of all homomorphisms of G into itself. For any $f, g \in$ Hom (G) define $f + g$ and $f \cdot g$ by $(f + g)(x) = f(x) + g(x) \ \forall$ $x \in$ G and $f \cdot g = fog$ (composition of functions). Show that (HOM(G), +, ·) is a non-commutative ring.

12. Is the set **Z** of all integers is a ring w.r.t the compositions ⊕ and ⊕ defined by

$$a \oplus b = ab$$

and $$a \cdot b = a + b \quad \forall\ a, b \in \mathbf{Z}$$

13. Let R be a ring with unity 1. Show that the set of all invertible elements of R, is a group under multiplication.
14. R is an integral domain and $a, b \in R$. If $a^m = b^m$, $a^n = b^n$ $(m, n) = {}_1$, show that $a = b$.
15. Find all the zero divisors of the ring $(z_6, +_6, \times_6)$ and the ring $(z_8, +_8, \times_8)$.
16. If D is an integral domain prove that the equation $x^2 = 1$ has only two solutions in D.
17. Determine (giving a proof or counter example) which of the following statements are true.

(a) The invertible elements of an integral domain form a cyclic group under multiplication.

(b) If R is a ring with an identity and the non-zero elements of R form a cyclic group under multiplication, then R is a field.

[**(a)** false for, $r^n = 1 \Rightarrow r = \pm 1$ which is impossible, **(b)** true]

2.7 Sub Rings

Definition : A non empty subset of S of ring $(R, +, \cdot)$ is called a sub ring of R, if S itself is a ring by itself under the operations induced by + and · in S.

If S is a subring of a ring R, then clearly S must be closed under addition and multiplication.

Every ring $(R, +, \cdot)$ is a subring of itself. Also the set $\{0\}$, containing the zero of the ring R, is a subring of the ring R. These two subrings are called *trivial* or *improper* subring of the ring R. A subring other than these is called *proper* subring of R.

Examples of subrings

1. The ring Z of integers is a subring of the ring Q of rationals.
2. The ring Q of rationals is a subring of the ring R of real integers.
3. The ring R of reals is a subring of the ring C of complex numbers.
4. Let n be an arbitrary but fixed non zero integer. Then the set $|nx|\ x \in Z|$ is a subring of the ring Z of integers.
5. The set D of all $n \times n$ diagonal matrices over the set of reals is a subring of the ring of all $n \times n$ matrices over the set of reals..

Note :

1. If R is a ring without zero divisors, then any subring S of R must also be without zero divisors.
2. If a ring R is commutative, then any subring S of R must also be commutative.
3. If R is a ring with unity, then it may have a subring without unity.

Theorem 1. A non empty subset S of a ring is a subring of R if and only if

$$a \in S, b \in S \Rightarrow a - b \in S \text{ and } a \in S, b \in S \Rightarrow ab \in S$$

Proof : Let S be a subring of ring R. Thus S is a ring by itself.

Now, $a \in S,\ b \in S \quad \Rightarrow \quad a \in S,\ -b \in S$

$\Rightarrow \quad a + (-b) \in S$

$\Rightarrow \quad a - b \in S$

Further, $a \in S,\ b \in S \quad \Rightarrow \quad ab \in S$

Thus the conditions are necessary.

Conversely, let S be a non empty subset of R such that

$$a \in S,\ b \in S \Rightarrow a - b \in S \text{ and } a \in S,\ b \in S \Rightarrow ab \in S$$

Since $\forall\, a \in S,\ b \in S \Rightarrow a + (-b) \in S$, the non empty subset S or R, is a subgroup of R.

Thus, S is a group by itself. Moreover it is an abelian as R is abelian.

The second condition implies, that S is closed under multiplication.

As S is a subset of R, and further as the associative law for multiplication and distributive laws hold in R, they are also true for the elements of S. Hence S is a ring by itself and hence is a subring of R.

Theorem 2. A non empty subset S of a ring R is a subring of R if and only if

$$\textbf{(i) } S + (-S) = S \textbf{ and (ii) } SS \subseteq S$$

where - S is the set of all negatives of all elements of S.

Proof : Let S be a subring of the ring R. Thus S is a ring by itself.

S is an abelian a group and hence a subgroup of the group $R \Rightarrow S + (-S) = S$.

Further as S is a ring by itself, we have,

$x \in SS \Rightarrow \exists\, a \in S,\ b \in S$ such that $x = ab$

$\Rightarrow \quad ab \in S \quad$ ($\because$ S is closed under multiplication)

$\Rightarrow \quad SS \subseteq S$

Conversely, let S be a non empty subset of R such that (i) and (ii) are true.

The conditions (i) $\Rightarrow$ S is a sub group of R

$\Rightarrow$ S is a group by itself, further it is abelian.

By condition (ii), we have,

$$a \in S,\ b \in S \Rightarrow ab \in SS \Rightarrow ab \in S \qquad (\because SS \subseteq S)$$

Thus, S is closed under multiplication.

The remaining axioms of the ring are satisfied for the elements of S as they are true in R.

Hence S is a subring of R.

Theorem 3. The intersection of any two subrings of a ring R is again a subring of R.

Proof : Let S_1 and S_2 be any two subrings of the ring R.

Let, $\quad a, b \in S_1 \cap S_2$ be arbitrary

$\Rightarrow \quad a, b \in S_1$ and $a, b \in S_2$

$\Rightarrow \quad (a - b \in S_1,\ ab \in S_1)$ and $(a - b \in S_2,\ ab \in S_2)$ $\quad$ ($\because$ S_1 and S_2 are subrings)

$\Rightarrow \quad (a - b) \in S_1 \cap S_2$ and $ab \in S_1 \cap S_2$

$\Rightarrow \quad S_1 \cap S_2$ is a subring of R.

Note :

1. The result of the above theorem can be extended to any arbitrary collection of subrings of a ring R. That is if $\{S_\alpha : \alpha \in \Lambda\}$ be any arbitrary collection of subrings of a ring, then $\bigcap_\alpha \{S_\alpha : \alpha \in \Lambda\}$ is also a subring of R.

2. The union of two subrings of a ring need not be a subring. For example, consider the subrings 2Z and 3Z of the ring (Z, +, ·) of integers. Now, $2, 3 \in 2Z \cup 3Z$, but $2 + 3 = 5 \notin 2Z \cup 3Z$. That is the union of 2Z and 3Z is not closed under addition. Hence $2Z \cup 3Z$ is not a subring of **Z**.

Example 1. Show that the set $S = \left\{\begin{pmatrix} a & 0 \\ b & c \end{pmatrix} : a, b, c \in Z\right\}$ **is a subring of the ring M_2(Z) for all 2 × 2 matrices over the set of integers.**

Solution : Let, $A = \begin{pmatrix} a_1 & 0 \\ b_1 & c_1 \end{pmatrix}$ and $B = \begin{pmatrix} a_2 & 0 \\ b_2 & c_2 \end{pmatrix}$ be any two elements of S.

Then, $$A - B = \begin{pmatrix} a_1 - a_2 & 0 \\ b_1 - b_2 & c_1 - c_2 \end{pmatrix} \in S$$

Also, $$AB = \begin{pmatrix} a_1a_2 & 0 \\ a_2b_1 - c_1b_2 & c_1c_2 \end{pmatrix} \in S$$

Hence S is a subring of $M_2(\mathbf{Z})$.

Example 2. Show that every sub group of (Z, +) is a subring of the ring (Z, +, ·) of integers.

Solution : Let S be a subgroup of (**Z**, +, ·). Then, $S = n\mathbf{Z}$ for some integer n

Now to show that S is a subring **Z**, we only have to show that S is closed under multiplication.

Let, $a, b \in S$ be arbitrary $\Rightarrow$ $a = nx$, $b = ny$ for some integers x, y

Now, $a \cdot b = (nx)(ny) \Rightarrow ab = n(nxy) \in n\mathbf{Z} = S$

Thus, $S = n\mathbf{Z}$ is a sub ring of (**Z**, +, ·)

Example 3. Can a subring of a non commutative ring be commutative. Justify the answer.

Solution : A subring of a non commutative ring may be commutative. For example consider the subset.

$$S = \left\{\begin{pmatrix} a & 0 \\ 0 & b \end{pmatrix} : a, b, \in Z\right\}$$

of the non commutative ring $M_2(\mathbf{Z})$ of all 2 × 2 matrices over the set of integers.

Since the zero matrix of order 2 × 2 belongs to S, the set S is non empty

Let, $A = \begin{pmatrix} a_1 & 0 \\ 0 & b_1 \end{pmatrix}$ and $B = \begin{pmatrix} a_2 & 0 \\ 0 & b_2 \end{pmatrix}$ be any two arbitrary elements of S.

Then, $A - B = \begin{pmatrix} a_1 - a_2 & 0 \\ 0 & b_1 - b_2 \end{pmatrix} \in S$. Also, $A \cdot B = \begin{pmatrix} a_1a_2 & 0 \\ 0 & b_1b_2 \end{pmatrix} \in S$.

Thus, S is a sub ring of $M_2(Z)$.

Further, $AB = BA$, $\forall A, B \in S$. Hence, S is commutative.

Example 4. Show by example that a ring R without unity, may have a subring S with unity.

Solution : Consider the set, $$R = \left\{\begin{pmatrix} a & b \\ 0 & 0 \end{pmatrix} : a, b, \in \mathbf{Z}\right\}$$

Then R is a ring under matrix addition and multiplication. Further it has no unit element.

Let, $$R = \left\{\begin{pmatrix} a & b \\ 0 & 0 \end{pmatrix} : a,\ b,\ \in \mathbf{Z}\right\}$$

Clearly S is a subset of R. Further it can be shown that S is a subring of R.

Now, $$\begin{pmatrix} a & 0 \\ 0 & 0 \end{pmatrix}\begin{pmatrix} 1 & 0 \\ 0 & 0 \end{pmatrix} = \begin{pmatrix} 1 & 0 \\ 0 & 0 \end{pmatrix}\begin{pmatrix} a & 0 \\ 0 & 0 \end{pmatrix} = \begin{pmatrix} a & 0 \\ 0 & 0 \end{pmatrix},\ \forall\, a \in \mathbf{Z}$$

Hence, S is a subring of R with unity $\begin{pmatrix} 1 & 0 \\ 0 & 0 \end{pmatrix}$.

Example 5. Show that the unity element of a subring S of a ring R (with unit element), may be different from the unity R.

Solution : Consider the ring $M_2(Z)$ of all 2×2 matrices over Z. This is a ring with unit element as I_2.

Let, $$S = \left\{\begin{pmatrix} a & 0 \\ 0 & 0 \end{pmatrix} : a,\ b,\ \in \mathbf{Z}\right\}$$

Clearly S is a subring of $M_2(Z)$. Further the element $\begin{pmatrix} 1 & 0 \\ 0 & 0 \end{pmatrix}$ of S, acts as unit element S. Observe that

$$\begin{pmatrix} 1 & 0 \\ 0 & 1 \end{pmatrix} \neq \begin{pmatrix} 1 & 0 \\ 0 & 0 \end{pmatrix}$$

Example 6. Let R be a ring and a be a fixed element of R. Show that $R_a = \{x \in R\ ;\ ax = 0\}$ is a subring of R.

Solution : Since $a \cdot 0 = 0,\ 0 \in R_a$. Thus, $R_a \neq \phi$.

Let, $x, y \in R_a$ be arbitrary $\Rightarrow\ ax = 0$ and $ay = 0$.

Now, $a \cdot (x - y) = ax - ay = 0 \ \Rightarrow\ x - y \in R$

Further, $a \cdot (xy) = (ax)y = 0 \cdot y = 0 \ \Rightarrow\ xy \in R$

Hence R_a is subring of R.

Example 7. Let R be a ring and let $C = \{x \in R : xy = yx\}\ \forall\, y \in R$, show that C, called the centre of R, is a subring of R.

Solution : We havr for every $y \in R,\ 0 \cdot y = y \cdot 0 = 0$. Thus, $0 \in C \ \Rightarrow\ C \neq \phi$.

Let, $x_1, x_2 \in C$ be arbitrary $\Rightarrow\ x_1y = yx_1$ and $x_2y = yx_2\ \ \forall\, y \in R$.

Now, $(x_1 - x_2)y = x_1y - x_2y = yx_1 - yx_2 = y(x_1 - x_2),\ \forall\, y \in R \ \Rightarrow\ x_1 - x_2 \in C$

Again, $(x_1x_2)y = x_1 \cdot (x_2y)$
$$\begin{aligned} &\Rightarrow & (x_1x_2)y &= x_1(yx_2) \\ &\Rightarrow & (x_1x_2)y &= (x_1y)x_2 \\ &\Rightarrow & (x_1x_2)y &= (yx_1)x_2 \\ &\Rightarrow & (x_1x_2)y &= y(x_1x_2) \ \Rightarrow\ x_1x_2 \in C \end{aligned}$$

Hence C is a subring.

Example 8. Show that the set K of matrices of the form $\begin{pmatrix} a & -b \\ b & a \end{pmatrix}$, where $a, b \in Z_3$ form a subring of $M_2(Z_3)$ of all 2×2 matrices over Z_3. Show also (1) K is a field, (2) the non zero

elements of K form a cyclic group of order S under multiplication (3) find the generator of this cyclic group.

Solution : K is a sub ring proof left as an exercise. K is commutative (verify!)

Clearly $\begin{pmatrix} 1 & 0 \\ 0 & 1 \end{pmatrix}$ is the identity of K.

0, 1 are the only squares in Z_3. Thus for $a, b \in Z_3$ we have $a^2 + b^2 = 0$ only if $a = b = 0$. Hence non zero element $\begin{pmatrix} a & -b \\ b & a \end{pmatrix}$ of K has inverse $\begin{pmatrix} at & bt \\ -bt & at \end{pmatrix}$ in K where $t = (a^2 + b^2)^{-1}$. Thus K is a field.

There are three possibilities for each of a and b, so there are $3 \times 3 = 9$ elements in K. Hence K contains 8 non zero elements.

One can show that each non zero element of K is a power of $\begin{pmatrix} 1 & 2 \\ 2 & 1 \end{pmatrix}$. Thus non zero element of K form a cyclic group with $\begin{pmatrix} 1 & 2 \\ 2 & 1 \end{pmatrix}$ as a generator.

EXERCISES

1. Show that the ring of Gaussian integers is a subring of ring $(\mathbf{C}, +, \cdot)$ of complex numbers.
2. Show that the sub sets S and T defined below are the subrings of ring $M_2(\mathbf{Z})$.

$$S = \left\{\begin{pmatrix} a & 0 \\ b & 0 \end{pmatrix} \mid a, b, \in \mathbf{Z}\right\}, \quad T = \left\{\begin{pmatrix} a & 0 \\ b & c \end{pmatrix} \mid a, b, c, \in \mathbf{Z}\right\}.$$

3. Show that the union of two subrings of a ring R is a subring of R if and only if one is contained in the other.
4. Let S be a subring of a finite ring $(R, +, \cdot)$. Show that the order of S divides the order of R.
5. Show that every subring with unity, of a field is an integral domain.
6. An element a of a ring R is nilpotent if $a^n = 0$ for some positive integer n. Show that the collection of all nilpotent elements of a ring forms of subring.
7. R is a ring and $a \in R$. Let $C(a) = \{r \in R \mid ar = ra\}$. Show that $C(a)$ is a sub ring of R.
8. Let λ be a fixed real number. Show that the set $R(\lambda)$ of matrices of the form $\begin{pmatrix} x+y & y \\ \lambda y & x \end{pmatrix}$ where $x, y \in \mathbf{R}$, is a subring of the ring $M_2(\mathbf{R})$ of 2×2 matrices. Show that $R(\lambda)$ has an identity for every λ.
9. Show that the set F of matrices of the form $\begin{pmatrix} x & x \\ x & x \end{pmatrix}$ where $x \in \mathbf{R}$, a subring of $M_2(\mathbf{R})$.

   Show that F has an identity element different from the identity element of $M_2(\mathbf{R})$. Also show that every non zero element of F has a multiplicative inverse.

   [**Hint :** Identity element in F is $\begin{pmatrix} 1/2 & 1/2 \\ 1/2 & 1/2 \end{pmatrix}$ and inverse of $\begin{pmatrix} x & x \\ x & x \end{pmatrix}$, $x \neq 0$, is $\begin{pmatrix} 1/4x & 1/4x \\ 1/4x & 1/4x \end{pmatrix}$].

10. Show that the set H of matrices of the form $\begin{pmatrix} a+ib & c+id \\ -c+di & a-ib \end{pmatrix}$ where $a, b, c, d \in \mathbf{R}$ is a subring of $M_2(\mathbf{C})$.

11. Show that the set R of matrices of the form $\begin{pmatrix} a & b\sqrt{5} \\ -b\sqrt{5} & a \end{pmatrix}$ where $a, b \in \mathbf{Z}$ forms a subring of the ring $M_2(\mathbf{R})$. Show that R is commutative and has an identity.

12. Let R be the set of rationals $\frac{a}{b}$ (in lowest terms) such that b is not a multiple of 3. Show that R is a subring of Q. (Hint use the fact that 3 is a prime).

2.8 Ideals

In this section we shall consider a class of subrings which are closed under multiplication by an arbitrary element of the ring.

Definition 1. A non empty subset I of a ring $(R, +, \cdot)$ is said to be a left ideal of R, if

(i) $\forall\, a, b \in I,\ a - b \in I$ **(ii)** $\forall\, a \in I$, **and** $\forall\, r \in R,\ r \cdot a \in I$

Definition 2. A non empty subset I of a ring $(R, +, \cdot)$ is said to be a right ideal of R, if

(i) $\forall\, a, b \in I,\ a - b \in I$ **(ii)** $\forall\, a \in I$, **and** $\forall\, r \in R,\ a \cdot r \in I$

Definition 3. A non empty subset I of a ring $(R, +, \cdot)$ is said to be a two sided ideal (or simply an ideal) of R if

(i) $\forall\, a, b \in I,\ a - b \in I$ **(ii)** $\forall\, a \in I$, **and** $\forall\, r \in R,\ a \cdot r$ **and** $r \cdot a \in I$

Note : In a commutative ring, every left ideal is also a right ideal. Thus there is no distinction between the left and right ideals in a commutative ring.

For any ring R, the sets $\{0\}$ and R are ideals and these referred as ***'trivial'*** or ***'improper'*** ideals. The other ideals of R are called ***'proper ideals'***.

A ring has no proper ideals is called a **simple ring.**

Theorem 1. Every left (right or two sided) ideal of a ring R is a subring of R.

Proof : Let I be a left ideal of R.

By definition, of I we have $\forall\, a, b \in I, a - b \in I, \forall\, a \in I$ and $\forall\, r \in R,\ r \cdot a \in I$.

In particular $\forall\, a \in I$ and $\forall\, b \in I\ (\subset R),\ b \cdot a \in I$. Thus I is a sub ring of R.

The converse of the above theorem is not true. That is every subring of a ring is not an ideal of R - for example $\mathbf{Z}$ is a subring of Q, but $\mathbf{Z}$ is not an ideal of Q, since $1 \in \mathbf{Z}$, $\frac{1}{4} \in Q$ but $1 \cdot \frac{1}{4} = \frac{1}{4} \notin \mathbf{Z}$.

Theorem 2. The intersection of any two ideals of a ring is again an ideal of the ring.

Proof : Let I_1 and I_2 be any two left ideal of a ring R.

Since, $0 \in I_1$ and I_2, it follows $0 \in I_1 \cap I_2 \therefore I_1 \cap I_2 \neq \phi$.

Let, $a, b \in I_1 \cap I_2 \Rightarrow a, b \in I_1$ and $a, b \in I_2$

$\Rightarrow \quad a - b \in I_1$ and $a - b \in I_2$ $\quad$ ($\because I_1$ and I_2 are ideals)

$\Rightarrow \quad a - b \in I_1 \cap I_2$

Now, let $a \in I_1 \cap I_2$ and $r \in R$

$\Rightarrow \quad a \in I_1$ and $a \in I_2$ and $r \in R$

$\Rightarrow \quad r \cdot a \in I_1$ and $r \cdot a \in I_2$ $\quad (\because I_1$ and I_2 are ideals)

$\Rightarrow \quad r \cdot a \in I_1 \cap I_2$

Hence $I_1 \cap I_2$ is a left ideal of R.

Similarly one can show that the intersection of two right ideals is also a right ideal of the ring.

Note : The union of two ideals of a ring R need not be an ideal of R. For example, consider the two ideals 2Z and 3Z of the ring Z of integers. Now $3, 2 \in 2Z \cup 3Z$, but $3 - 1 = 2 \notin 2Z \cup 3Z$. Thus $2Z \cup 3Z$ is not an ideal of Z.

Theorem 3. Let a be an element of a commutative ring R. Then $aR = \{ar \mid r \in R\}$ is an ideal of R.

Proof : Since R is a ring it contains at least one element namely 0 of the ring.

Thus, $\quad 0 = 0 \cdot a \in aR \Rightarrow ar \neq \phi$

Let, $ar, as \in aR$ be arbitrary $\Rightarrow ar - as = a(r - s) \in aR \quad (\because r - s \in R)$

Again let, $ar \in aR$ and $s \in R$

$\Rightarrow \quad (ar)s = a(rs) \in aR \quad (\because rs \in R)$

Further, $\quad s(ar) = (ar) \cdot s \quad (\because R$ is commutative)

$\Rightarrow \quad s(ar) = a(rs) \in R$

Hence aR is an ideal of R.

Note : The element a need not be an element of aR. For example in the ring 2Z, the ideal 4(2Z) is

$$(\ldots\ldots\ldots\ldots -16, -8, 0, 8, 16, \ldots\ldots\ldots\ldots)$$

Clearly 4 is not an element of this ideal. However if R is a commutative ring with unity then $a = a \cdot 1 \in aR$.

Example 1. Show that 2Z is an ideal of the ring (Z, +, ·).

Solution : Let $a, b \in 2Z$ be arbitrary. Then $\exists$ two integers m and n such that $a = 2m$ and $b = 2n$.

Consider, $\quad a - b = 2m - 2n = 2(m - n) \in 2Z \quad (\because m - n \in Z)$

Again consider, $\quad a \cdot r = (2m)r$ (where $r \in$ Z be arbitrary) $= 2(mr) \in$ **2Z**

Further, $\quad a \cdot r = r \cdot a = 2(mr) \in$ **2Z. Hence 2Z is an ideal in Z.**

Example 2. Let R be the ring of all matrices of 2 × 2 over Z and $S = \left\{\begin{pmatrix} a & 0 \\ b & 0 \end{pmatrix} \mid a, b \in Z\right\}$ **Show that S is a left ideal in R but not a right ideal.**

Solution : Clearly $S \neq \phi$, as $\begin{pmatrix} 0 & 0 \\ 0 & 0 \end{pmatrix} \in S$

Let, $A = \begin{pmatrix} a & 0 \\ b & 0 \end{pmatrix}$ and $B = \begin{pmatrix} c & 0 \\ d & 0 \end{pmatrix}$ be any two elements of S

Then, $A - B = \begin{pmatrix} a - c & 0 \\ b - d & 0 \end{pmatrix} \in S$

Let, $T = \begin{pmatrix} p & q \\ r & s \end{pmatrix} \in R$ be arbitrary. Consider, $T \cdot A = \begin{pmatrix} p & q \\ r & s \end{pmatrix}\begin{pmatrix} a & 0 \\ b & 0 \end{pmatrix} = \begin{pmatrix} pa + qb & 0 \\ ra + sb & 0 \end{pmatrix} \in S.$

Hence, **S is a left ideal of R.**

To show that S is not a right ideal, consider,

$$A = \begin{pmatrix} 2 & 0 \\ 1 & 0 \end{pmatrix} \in S \text{ and } T = \begin{pmatrix} 3 & 1 \\ 2 & 4 \end{pmatrix} \in R$$

Now, $$A \cdot T = \begin{pmatrix} 2 & 0 \\ 1 & 0 \end{pmatrix} \begin{pmatrix} 3 & 1 \\ 2 & 4 \end{pmatrix} = \begin{pmatrix} 6 & 2 \\ 3 & 1 \end{pmatrix} \notin S$$

Hence, S is not a right ideal of R.

Example 3. Let R be a commutative ring with unity whose only ideals are $\{0\}$ and $\{R\}$ itself. Then prove that R is a field (i.e., if R is a simple commutative ring with unity then it is a field).

Solution : Since R is a commutative ring with unity, to show that R is a field it is sufficient if we establish that each non zero element R of has multiplicative inverse.

Let $a \neq 0$ be arbitrary element R.

Then the set $aR = \{ar \mid r \in R\}$ is an ideal in R (by theorem 3).

Since $a \in R$, we have $a \cdot 1 \in aR \Rightarrow a \in aR$, where $a \neq 0$. Thus, the ideal $Ra \neq \{0\}$.

As the ring R has no proper ideals, the only possibility is that $Ra = R$.

$\Rightarrow \forall\, x \in R.\ \exists\, y \subseteq R$ such that $x = ya$.

In particular, as $1 \in R, \exists\, b \in R.$ such that $b \cdot a = 1 \Rightarrow a^{-1} = b$.

Thus, each non zero element of R has multiplicative inverse. **Hence R is a field.**

Example 4. if I is an ideal of a ring R with unity, and $1 \in I$ then $I = R$.

Solution : Let $r \in R$ be arbitrary. Now,

$I \in I$ and $r \in R \Rightarrow 1 \cdot r \in I$ ($\because$ I is an ideal)

$\Rightarrow r \in I$

Thus every element of R is in I. That is, $R \subseteq I$. Further $I \subseteq R$. **Hence $I = R$.**

Example 5. Show that a field F has no proper ideals. That is the only idals of a field F are $\{0\}$ and F.

Solution : Let I be any ideal of F and $I \neq \{0\}$.

Let, $a \in I \Rightarrow a \neq 0$ as $I \neq \{0\}$

$\Rightarrow$ a has multiplication inverse $a^{-1} \in F$ ($\because$ f is a field)

Now, $a \in I$ and $a^{-1} \in F \Rightarrow a \cdot a^{-1} \in I$

$\Rightarrow 1 \in I$

$\Rightarrow I = F$ (by example 4)

Thus, If $I \neq \{0\}$ then $I = F$.

Example 6. Let R be the set of rationals $\frac{a}{b}$ (in lowest terms) such that b is not a multiple of 3. Show that R is a subring of Q. Let I be the subset of R consisting of those elements whose numerators are divisible by 3. Show that I is an ideal of R.

Solution : $R = \left\{\frac{a}{b} \mid a, b \in \mathbf{Z}, b \notin 3\mathbf{Z}\right\}$.

Now, let, $\frac{a}{b}, \frac{c}{d} \in R$ then $\frac{a}{b} - \frac{c}{d} = \frac{ad - bc}{bd}$ and $\frac{a}{b} \cdot \frac{c}{d} = \frac{ac}{bd}$

Since 3 is prime, $3 \mid bd \Rightarrow 3 \mid b$ or $3 \mid d$. But $3 \nmid b$ and $3 \nmid d$. Thus, $3 \nmid bd$.

$\Rightarrow \quad \frac{a}{b} - \frac{c}{d}$ and $\frac{a}{b} \cdot \frac{c}{d} \in R$. Hence is **R subring**.

Let, $\quad I = \left\{\frac{a}{b} \in R \mid 3 \text{ divides } a\right\}$

Now, $\quad \frac{a}{b}, \frac{c}{d} \in I \Rightarrow 3 \mid a,\ 3 \mid c \Rightarrow 3 \mid ad,\ 3 \mid bc \Rightarrow 3 \mid ad - bc$

$\therefore \quad \frac{a}{b} - \frac{c}{d} = \frac{ad - bc}{bd} \in I$

Similarly, $\quad \frac{e}{f} \in R, \frac{a}{b} \in I \Rightarrow \frac{e}{f} \cdot \frac{a}{b} \Rightarrow \frac{ea}{fb} \in I$. Thus, ***I* is an ideal.**

Example 7. Given subsets *A* and *B* of a ring *R*, define $\{A + B\} = \{a + b \mid a \in A, b \in B\}$. Are the following true? Justify.

(a) If *A* and *B* are sub rings then so is *A* + *B*

(b) If *A* and *B* are ideals so is *A* + *B*

(c) If *A* is a left ideal and *B* is right ideal then *A* + *B* is a sub ring

(d) If *A* is a two sided ideal and *B* is left ideal then *A* + *B* is a left ideal.

Solution : (a) and (c) are false. For consider,

A = Set of all matrices of the form $\begin{pmatrix} a & 0 \\ b & 0 \end{pmatrix}$ where $a, b \in \mathbf{R}$

B = Set of all matrices of the form $\begin{pmatrix} a & b \\ 0 & 0 \end{pmatrix}$ where $a, b \in \mathbf{R}$

A is a left ideal and B is a right ideal of the ring $M_2(\mathbf{R})$.

$\Rightarrow \quad A$ and B are sub rings $M_2(\mathbf{R})$.

But $A + B$ is not a sub ring of $M_2(R)$ - for consider

$$\begin{pmatrix} 0 & 0 \\ 1 & 0 \end{pmatrix}\begin{pmatrix} 0 & 1 \\ 0 & 0 \end{pmatrix} = \begin{pmatrix} 0 & 0 \\ 0 & 1 \end{pmatrix} \notin A + B$$

($\because A + B$ consists of the matrices of the form $\begin{pmatrix} a & b \\ c & 0 \end{pmatrix}$, $a, b, c \in \mathbf{R}$)

(b) and (d) are true - for let A and B be left ideals.

Let, $a_1, a_2 \in A$ and $b_1, b_2 \in B$. Then, $(a_1 + b_2) - (a_2 + b_2) = (a_1 - a_2) + (b_1 - b_2) \in A + B$

Now for any $r \in R$, $\quad r(a + b) = ra + rb \in A + B$. Thus, ***A* + *B* is an ideal.**

Theorem 4. A subset *I* of the ring *Z* is an ideal of *Z* if and only *I* = *nZ* for some non-negative integer *n*-That is the ideals of *Z* are precisely the sub-groups of (*Z*, +).

Proof : Let *I* be an ideal of **Z**. Then $(I, +)$ is a sub group of $(\mathbf{Z}, +)$.

We know that any sub-group of **Z**, is the form, $n\mathbf{Z}$, for some non negative integer n.

Thus, $I = n\mathbf{Z}$, for some non negative integer.

Conversely, let $I = n\mathbf{Z}$, for some non negative integer. Then,

$$I = \{nr \mid r \in \mathbf{Z}\}$$

Clearly, $n\mathbf{Z} = I$ is an ideal of **Z**, by theorem 3.

Example 8. If A and B are ideals of a ring R such that $A \cap B = \{0\}$, show that for all $a \in A$, $b \in B$, $ab = 0$.

Solution : $a \in A$ and A is an ideal. Thus, $\forall\, r \in R,\ ar \in A$. In particular, for $b \in R,\ ab \in$ A.

Again, $b \in B$ and B is an ideal. Thus, $\forall\ r \in R,\ br \in A$. In particular, for $a \in R,\ ab \in B$.

Thus, $ab \in A \cap B \;\Rightarrow\; \mathbf{ab = 0}$ as $A \cap B = \{0\}$.

Example 9. R be a commutative ring and I is an ideal of R. Let $\sqrt{I} = \{x \in R \mid x^n \in I\}$ for some positive integer n}. Show that $\sqrt{I}$ is an ideal containing I.

Solution : Let $a, b \in \sqrt{I}$. Then there existstwo positive integers m, n such that $a^m \in I$ and $b^m \in I$.

Consider, $(a-b)^{m+n} = a^{m+n} - (m+n)\,C_1\, a^{m+n-1} \cdot b + ... + (-1)^{m+n} \cdot b^{m+n}$

Since I is an ideal each term of the right side belongs to I. Hence $(a-b)^{m+n} \in I$.

Thus $a - b \in \sqrt{I}$.

Now, $\quad a \in \sqrt{I}$ and $r \in R \quad \Rightarrow \quad a^m \in I, r^m \in R$

$\Rightarrow \quad r^m \cdot a^m \in I$

$\Rightarrow \quad (ra)^m \in I$

$\Rightarrow \quad r \cdot a \in \sqrt{I}$

Hence $\sqrt{I}$ is an ideal.

Further $\forall\ a \in I$, we have, $a^{+1} \in I \;\Rightarrow\; a \in \sqrt{I} \;\Rightarrow\; I \subset \sqrt{I}$. Hence $\sqrt{I}$ **is an ideal containing I.**

EXERCISES

1. Show that the union of two ideals of a ring R is an ideal of R if and only if one is contained in the other.
2. Let I and J be ideals of a ring R. Then show that $I + J = \{x + y \mid x \in I, y \in J\}$ is also an ideal in R.
3. If U is an ideal of a ring r, let, $r(U) = \{x \in R \mid xu = 0,\ \forall\ u \in U\}$. Show that $r(U)$ is an ideal in R.
4. R is any commutative ring and a is a fixed element of R. Show that the set I of all elements of the form $na + ra$ where $n \in Z$ and $r \in R$ is an ideal of R. ($na = a + a + ...$ $+ a$, n times).
5. If I is an ideal of a ring R, show that $I + I = I$, where $I + I = \{a + b \mid a, b \in I\}$.
6. Show that the subset, $S = \left\{\begin{pmatrix} a & b \\ 0 & 0 \end{pmatrix} \mid a, b \in Z\right\}$ is a right ideal but not a left ideal of $M_2(Z)$.
7. R is a ring and I is a left ideal of R. Let, $\lambda(I) = \{x \in R \mid xa = 0,\ \forall\ a \in I\}$. Show that $\lambda(I)$ is a two sided ideal of R.
8. R is a ring and $a \in R$. If $r(a) = \{x \in R \mid ax = 0\}$, show that $r(a)$ is right ideal.
9. I is a right ideal and J is a left ideal of a ring R such that $I \cap J = \{0\}$. Show that $ab = 0$ for all $a \in I,\ b \in J$.
10. If S is an ideal of a ring R with unit element 1 andif $1 \in S$ then show that $S = R$.
11. An element of a ring R is said to be nilpotent if $a^n = 0$ for some positive integer n. Show that the set of all nilpotent elements of a commutative rng is an ideal of R.

Line and Multiple Integrals

3

3.1 Introduction

In this chapter we consider the concepts of line, double and triple integrals, and differentiation under the integral sign-Leibnitz rule.

3.2 Line Integral Over a Plane Curve and Space curves

The concept of line integral is a natural generalization of the concept of definite integral.

$$\int_a^b f(x)\, dx$$

In the case of definite integral, we evaluate along the x-axis between the limits a and b, further $f(x)$ is defined at every point between a and b.

In the case of a line integral, we integrate along a curve in the plane (or in space) and the integrand is a function defined at the points of the curve under consideration.

A plane curve is a set of points (x, y) for which

$$x = f(t),\ y = g(t)\ ;\ a \le t \le b$$

where f and g are functions defined over $[a, b]$.

The points on the curve corresponding to $t = a$ and $t = b$ are called the end points of the curve and if $f(a) = f(b)$ and $g(a) = g(b)$, then the curve is said to be *closed* - i.e., the end points coincide.

We consider a curve C defined by

$$x = f(t),\ y = g(t),\ a \le t \le b$$

which is continuous and has a continuous first derivative (which is different from zero for all t). This implies that the curve C has unique tangent at each of its points. The curve satisfying these conditions is called a **smooth curve.**

A curve C is said to be **piecewise smooth curve** in $[a, b]$ if it is composed of finite number of smooth curves.

Let C be a piecewise smooth curve defined by

$$x = f(t),\ y = g(t),\ a \le t \le b$$

Let P and Q be two continuous functions of x and y defined at every point on the curve C, then the integral defined by

$$\int_C P\, dx + Q\, dy$$

is called the **line integral** of $P\,dx + Q\,dy$ along the curve. Further its value is given by

$$\int_a^b \{P[f(t), g(t)]\, f'(t) + Q[f(t), g(t)]\, g'(t)\}\, dt$$

A curve C in space is defined by

$$x = f(t),\quad y = g(t),\quad z = h(t),\quad a \le t \le b$$

where $f(t)$, $g(t)$ and $h(t)$ are defined on $[a, b]$.

If C is a piecewise smooth curve in space and P, Q and R are continuous functions of x, y and z defined at every point on the curve C, then the integral

$$\int_C P\,dx + Q\,dy + R\,dz$$

is called the **line integral** of $P\,dx + Q\,dy + R\,dz$ along the curve C. Its value is given by

$$\int_a^b \{P[f(t), g(t), h(t)]\, f'(t) + Q[f(t), g(t), h(t)]\, g'(t) + R[f(t), g(t), h(t)]\, h'(t)\}\, dt$$

The following properties of line integrals can be easily verified.

1. If k is any scalar, then,
$$\int_C kP\,dx + kQ\,dy + kR\,dz = k\int_C P\,dx + Q\,dy + R\,dz$$

2. $$\int_C (P_1 + P_2)\,dx + (Q_1 + Q_2)\,dy + (R_1 + R_2)\,dz$$
$$= \int_C P_1\,dx + Q_1\,dy + R_1\,dz + \int_C P_2\,dx + Q_2\,dy + R_2\,dz$$

3. If C is a curve in space, one of the two directions along C is taken as positively oriented direction and opposite direction along C is negatively oriented direction and it is denoted by –C. Now we have,
$$\int_{-C} P\,dx + Q\,dy + R\,dz = -\int_C P_2\,dx + Q_2\,dy + R_2\,dz$$

4. $$\int_{C_1 + C_2} P\,dx + Q\,dy + R\,dz = \int_{C_1} P\,dx + Q\,dy + R\,dz + \int_{C_2} P\,dx + Q\,dy + R\,dz$$

This property can be extended to any finite number of arcs.

Example 1. Evaluate $\int_C (3x + y)\,dx + (2y - x)\,dy$

(a) along the curve $y = x^2 + 1$ from (0, 1) and (3, 10).

(b) along the line joining (0, 1) and (3, 10).

Solution : Now, $\int_C P\,dx + Q\,dy = \int_C (3x + y)\,dx + (2y - x)\,dy$

(a) The curve is $y = x^2 + 1 \;\Rightarrow\; dy = 2x\,dx$. Also x varies from $x = 0$ to $x = 3$.

$$\Rightarrow \quad \int_C P\,dx + Q\,dy = \int_0^3 \left(3x + x^2 + 1\right) dx + \left(2x^2 + 2 - x\right) 2x\,dx$$

$$= \int_0^3 \left(4x^3 - x^2 + 7x + 1\right) dx$$

$$= \left[x^4 - \frac{x^3}{3} + \frac{7x^2}{2} + x\right]_0^3 = 81 - 9 + \frac{63}{2} + 3 = \mathbf{\frac{213}{3}}$$

(b) The equation of the line joining the points (0, 1) and (3, 10) is given by

$$\frac{y-10}{x-3} = \frac{10-1}{3-0} \Rightarrow 3y - 30 = 9x - 27 \Rightarrow y = 3x + 1$$

Now, $y = 3x + 1 \Rightarrow dy = 3dx$. Further x varies from $x = 0$ to $x = 3$.

Now, $$\int_C P\,dx + Q\,dy = \int_0^3 (3x + 3x + 1)\,dx + (6x + 2 - x)\,3\,dx$$

$$= \int_0^3 (21x + 7)\,dx = \left[\frac{21x^2}{2} + 7x\right]_0^3 = \frac{189}{2} + 21 = \frac{231}{2}$$

Example 2. $\int_C \left(x^2 - y^2\right) dx + x^3 y$, **where C is the semi circle with centre at (0, 4) and radius 2 units and lying in the first quadrant.**

Solution : The parametric equations of the circle with centre at (x_1, y_1) and radius r is given by

$$x = x_1 + r\cos\theta, \quad y = y_1 + r\sin\theta$$

In the given problem parametric equations of the circle are

$$x = 2\cos\theta, \quad y = 4 + 2\sin\theta.$$

Here θ is such that $-\frac{\pi}{2} \le \theta \le \frac{\pi}{2}$.

Now on C, we have,

$$P = x^2 - y^2 = 4\cos^2\theta - (4 + 2\sin\theta)^2 \Rightarrow P = 4\cos^2\theta - 4\sin^2\theta - 16\sin\theta - 16$$

$$Q = x^3 y = 8\cos^3\theta(4 + 2\sin\theta) \Rightarrow Q = 32\cos^3\theta + 16\cos^3\theta\sin\theta$$

Also, $x = 2\cos\theta \Rightarrow dx = -2\sin\theta\,d\theta$ and $y = 4 + 2\sin\theta \Rightarrow dy = 2\cos\theta\,d\theta$

Thus, $$\int P\,dx + Q\,dy = \int_{-\pi/2}^{\pi/2} \left(4\cos^2\theta - 4\sin^2\theta - 16\sin\theta - 16\right)(-2\sin\theta)\,d\theta + 8\cos^3\theta\,(4 + 2\sin\theta)\,2\cos\theta\,d\theta$$

$$= \int_{-\pi/2}^{\pi/2} \left[-8\cos^2\theta\sin\theta + 8\sin^3\theta + 32\sin^2\theta + 32\sin\theta + 64\cos^2\theta + 32\cos^4\theta\sin\theta\right] d\theta$$

We have from the properties of definite integral,

$$\int_{-a}^{a} f(x)\,dx = \begin{cases} 2\int_0^a f(x)\,dx & \text{if } f(-x) = f(x) \\ 0 & \text{if } f(-x) = -f(x) \end{cases}$$

Thus we have,

$$\int_C P\,dx + Q\,dy = \int_0^{\pi/2} 32\sin^2\theta\,d\theta + 2\int_0^{\pi/2} 64\cos^4\theta\,d\theta$$

$$= 64\cdot\frac{\pi}{2} + 128\cdot\frac{3\cdot 1}{4\cdot 2}\cdot\frac{\pi}{2} = \mathbf{56\pi}$$

Example 3. Evaluate $\int_C \left[\left(x^2 - 2xy\right)dx + \left(x^2y + 3\right)dy\right]$ **around the boundary of region defined by** $y^2 = 8x$ **and** $x = 2$.

Solution : The curve C is made up of two curves C_1 and C_2, where,

$$C_1 : y^2 = 8x \quad \text{.... (1)}$$

$$C_2 : x = 2 \quad \text{.... (2)}$$

Solving (1) and (2), we have the points of intersections of the curves

$$A = (2, 4), \quad B = (2, -4)$$

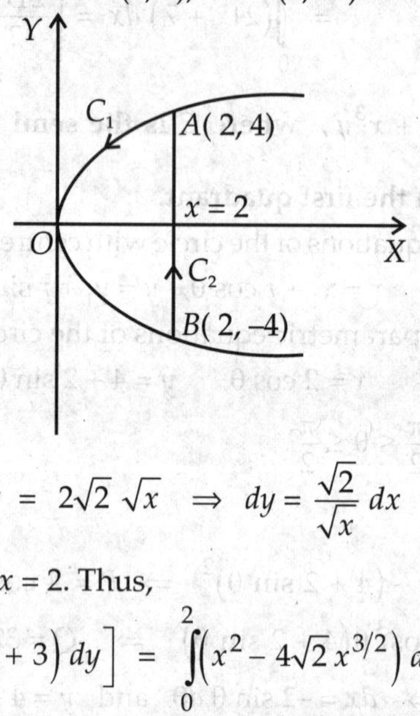

Now along C_1 we have, $y = 2\sqrt{2}\sqrt{x} \Rightarrow dy = \frac{\sqrt{2}}{\sqrt{x}}dx$

Also x varies from $x = 0$ to $x = 2$. Thus,

$$\int_{C_1}\left[\left(x^2 - 2xy\right)dx + \left(x^2y + 3\right)dy\right] = \int_0^2 \left(x^2 - 4\sqrt{2}\,x^{3/2}\right)dx + \left(2\sqrt{2}\,x^{5/2} + 3\right)\frac{\sqrt{2}}{\sqrt{x}}\,dx$$

$$= \int_0^2 \left[5x^2 - 4\sqrt{2}\,x^{3/2} + 3\sqrt{2}\,\frac{1}{\sqrt{x}}\right]dx$$

$$= \left[5\frac{x^3}{3} - \frac{8\sqrt{2}}{5}x^{5/2} + 6\sqrt{2}\sqrt{x}\right]_0^2$$

$$= \frac{40}{3} - \frac{64}{5} + 12 = \mathbf{\frac{188}{15}}$$

Now along C_2 we have, $x = 2 \Rightarrow dx = 0$ and y varies from -4 to 4. Thus,

$$\int_{C_2}\left[\left(x^2 - 2xy\right)dx + \left(x^2y + 3\right)dy\right] = \int_{-4}^{4} (4 - 4y)\cdot 0 + (4y + 3)\,dy$$

$$= \int_{-4}^{4} (4y+3)\,dy = \left[2y^2 + 3y\right]_{-4}^{4} = 24$$

Now,
$$\int_C P\,dx + Q\,dy = \int_{C_1} P\,dx + Q\,dy + \int_{C_2} P\,dx + Q\,dy$$

$$= \frac{188}{15} + 24 = \mathbf{\frac{548}{15}}$$

Example 4. Evaluate $\int_C \left[(x+2y)\,dx + (4-2x)\,dy\right]$ around the ellipse $\frac{x^2}{16} + \frac{y^2}{9} = 1$ in the counter clockwise.

Solution : The parametric equations of the ellipse $\frac{x^2}{16} + \frac{y^2}{9} = 1$ are, $x = 4\cos\theta$, $y = 3\sin\theta$.

Now, $x = 4\cos\theta \Rightarrow dx = -4\sin\theta\,d\theta$ and $y = 3\sin\theta \Rightarrow dy = 3\cos\theta\,d\theta$

The parameter θ is such that, $0 \le \theta \le 2\pi$. Thus we have,

$$\int_C \left[(x+2y)\,dx + (4-2x)\,dy\right]$$

$$= \int_0^{2\pi} \left[(4\cos\theta + 6\sin\theta)(-4\sin\theta\,d\theta) + (4 - 8\cos\theta)\,3\cos\theta\,d\theta\right]$$

$$= \int_0^{2\pi} (12\cos\theta - 16\sin\theta\cos\theta - 24)\,d\theta = \left[12\sin\theta - 8\sin^2\theta - 24\theta\right]_0^{2\pi} = \mathbf{-48\pi}$$

Example 5. Evaluate $\int_C \left[(3x-2y)\,dx + (y+2z)\,dy - x^2\,dz\right]$ where C is the curve $x = t$, $y = 2t^2$, $z = 3t^3$ and $0 \le t \le 1$.

Solution : We have,

$x = t \Rightarrow dx = dt$, $y = 2t^2 \Rightarrow dy = 4t\,dt$, $z = 3t^3 \Rightarrow dz = 9t^2\,dt$

By data $0 \le t \le 1$. Thus we have,

$$\int_C \left[(3x-2y)\,dx + (y+2z)\,dy - x^2\,dz\right]$$

$$= \int_0^1 \left[\left(3t - 4t^2\right)dt + \left(2t^2 + 6t^3\right)4t\,dt - t^2\left(9t^2\,dt\right)\right]$$

$$= \int_0^1 \left(15t^4 + 8t^3 - 4t^2 + 3t\right)dt = \left[3t^5 + 2t^4 - \frac{4}{3}t^3 + \frac{3}{2}t^2\right]_0^1 = \mathbf{\frac{31}{6}}$$

Example 6. Evaluate $\int_C \left[xy\,dx + x^2z\,dy + xyz\,dz\right]$ where C is given by $x = e^t$, $y = e^{-t}$, $z = t^2$ and $0 \le t \le 1$.

Solution : We have,

$$x = e^t \Rightarrow dx = e^t\, dt, \quad y = e^{-t} \Rightarrow dy = -e^{-t}\, dt, \quad z = t^2 \Rightarrow dz = 2t\, dt$$

By data $0 \le t \le 1$. Thus we have,

$$\int_C \left[xy\, dx + x^2 z\, dy + xyz\, dz\right] = \int_0^1 \left[e^t + t^2 e^{2t}\left(-e^{-t}\, dt\right) + t^2\, 2t\, dt\right]$$

$$= \int_0^1 \left[e^t - t^2 e^t + 2t^3\right] dt = \left[e^t - t^2 e^t + 2te^t - 2e^t + \frac{t^2}{2}\right]_0^1$$

$$= \left(e - e + 2e - 2e + \frac{1}{2}\right) - (1 - 2)$$

$$= \mathbf{\frac{3}{2}}$$

Example 7. Evaluate $\int x^2 y^2\, ds$ around the circle $x^2 + y^2 = 1$.

Solution : The parametric equations of the circle $x^2 + y^2 = 1$, are $x = \cos\theta,\ y = \sin\theta$.

Now if s is the arc length then we have,

$$\left(\frac{ds}{d\theta}\right)^2 = \left(\frac{dx}{d\theta}\right)^2 + \left(\frac{dy}{d\theta}\right)^2 \Rightarrow \left(\frac{ds}{d\theta}\right)^2 = \sin^2\theta + \cos^2\theta = 1 \Rightarrow \frac{ds}{d\theta} = 1$$

Now the parameter θ varies from $\theta = 0$ to $\theta = 2\pi$. Thus we have,

$$\int_C x^2 y^2\, ds = \int_0^{2\pi} \cos^2\theta \sin^2\theta \frac{ds}{d\theta} \cdot d\theta = \int_0^{2\pi} \frac{1}{4} \sin^2 2\theta\, d\theta$$

$$= \frac{1}{8}\int_0^{2\pi} (1 - \cos 4\theta)\, d\theta = \frac{1}{8}\left[\theta - \frac{\sin 4\theta}{4}\right]_0^{2\pi} = \frac{2\pi}{8} = \mathbf{\frac{\pi}{4}}$$

Example 8. Evaluate $\int_C (x + y + z)\, ds$ where C is the line joining the points (1, 2, 3) and (4, 5, 6).

Solution : Equations of the line joining the points $P(1, 2, 3)$ and $Q(4, 5, 6)$ are

$$\frac{x-1}{3} = \frac{y-2}{3} = \frac{z-3}{3}$$

$$\Rightarrow \quad \frac{x-1}{3} = \frac{y-2}{3} = \frac{z-3}{3} = t \text{ (say)}$$

$$\Rightarrow \quad x = t + 1, \quad y = 2 + t, \quad z = 3 + t$$

Now, $\left(\frac{ds}{dt}\right)^2 = \left(\frac{dx}{dt}\right)^2 + \left(\frac{dy}{dt}\right)^2 + \left(\frac{dz}{dt}\right)^2 \Rightarrow \left(\frac{ds}{dt}\right)^2 = 1 + 1 + 1 \Rightarrow \frac{ds}{dt} = \sqrt{3}$

Now, when $x = 1,\ y = 2,\ z = 3$, we have $t = 0$ and when $x = 4, y = 5, z = 6$, we have $t = 3$

Thus we have,

$$\int_C (x + y + z)\, ds = \int_0^3 (t + 1 + 2 + t + 3 + t) \frac{ds}{dt} \cdot dt$$

$$= \int_0^3 (3t + 6)\sqrt{3}\, dt = \sqrt{3}\left[\frac{3t^2}{2} + 6t\right]_0^3 = \sqrt{3}\left[\frac{27}{2} + 18\right] = \mathbf{\frac{63\sqrt{3}}{2}}$$

EXERCISES

1. Evaluate $\int\limits_C \left[\left(2y + x^2\right) dx + \left(3x - y\right) dy\right]$ along the curve $x = 2t$, $y = t^2 + 3$, where $0 \le t \le 1$.

2. Evaluate $\int\limits_C \left[\left(xy + y^2\right) dx + x^2 dy\right]$ where C is a closed curve of the region bounded by $y = x$, $y = x^2$.

3. Evaluate $\int\limits_C \left[(x + y)\, dx + (y - x)\, dy\right]$

(a) along the parabola $y^2 = x$ from (1, 1) to (4, 2)

(b) along the line joining (1, 1) and (4, 2)

(c) along the curve $x = 2t^2 + t + 1$, $y = t^2 + 1$.

4. Evaluate $\int\limits_C \left[(2x - y + 4)\, dx + (5y + 3x - 6)\, dy\right]$ around the triangle whose vertices are (0, 0), (3, 0), (3, 2) traversed in the a counter clockwise.

5. If C is an ellipse with centre at the origin, major axis of length $2a$ along the x-axis, minor axis of length $2b$ along y-axis, traversed once in the counter clockwise direction, evaluate $\int\limits_C \left[(x + y)\, dx + (x - y)\, dy\right]$.

6. Evaluate $\int\limits_C \left[\left(x^2 - y\right) dx + \left(y^2 + x\right) dy\right]$ where C is the curve given by $x = t$, $y = t^2 + 1$, $0 \le t \le 1$.

7. Evaluate $\int\limits_C \left[(2xy - 1)\, dx + \left(x^2 + 1\right) dy\right]$

(a) along the curve $y = x + 1$ from (0, 1) to (2, 3)

(b) along the curve $y = \dfrac{x^2}{2} + 1$ from (0, 1) to (2, 3)

8. Compute $\int\limits_C \left[(x\, dy - y\, dx)\right]$

(a) along the line $y = x$ from (0, 0) to (1, 1)

(b) along the parabola $y^2 = x$ from (0, 0) to (1, 1)

(c) along the circle $x = a \cos t$, $y = 1 + \sin t$, $-\dfrac{\pi}{2} \le t \le 0$.

(d) around the square (0, 0), (1, 0), (1, 1), (0, 1)

9. Show that $\int\limits_C \left[\dfrac{x}{x^2 + y^2}\, dy - \dfrac{y}{x^2 + y^2}\, dx\right] = 2\pi$ where C is the circle $x^2 + y^2 = 1$ on the xy plane in the counter clockwise direction.

10. Evaluate $\int_C (2x + y)\, ds$ where C is the circle $x^2 + y^2 = 25$ on the xy plane from the point (3, 4) to (4, 3).

11. Evaluate $\int_C x\, ds$,

(a) where C is the curve $y = x$ from (0, 0) to (1, 1)

(b) where C is the curve $y = x^2$ from (0, 0) to (1, 1)

(c) where C is the curve $y = x^{3/2}$ from (0, 0) to (1, 1).

12. Evaluate $\int_C [xy\, dx + yz\, dy + zx\, dz]$ where C is the curve given by $x = t$, $y = t^2$, $z = t^3$ and t varying from –1 to 1.

13. Evaluate $\int_C \left[\frac{yz}{x}\, dx + e^y\, dy + \sin z\, dz\right]$ where C is the curve given by $x = t^3$, $y = t$, $z = t^2$, $2 \le t \le 3$.

14. Evaluate $\int_C \left[x^2\, dx + y^3\, dy + \cos z\, dz\right]$ where C is the above given by $x = t^2 + t^3$, $y = \sqrt{t^2 + 1}$, $z = e^{t^2}$, $-1 \le t \le 1$.

15. Evaluate $\int_C \left[y\, dx + x\, dy - z^2\, dz\right]$ where C is the curve given by $x = \sin t$, $y = \cos t$, $z = t^2$, $0 \le t \le 1$.

16. Evaluate $\int_C \left[3x^2\, dx + (2xz - y)\, dy + z\, dz\right]$

(a) along the line joining (0, 0, 0) and (2, 1, 3)

(b) along the curve defined by $x^2 = 4y$, $3x^3 = 8z$ from $x = 0$ to $x = 2$ (**Hint:** put $x = t$).

17. Evaluate $\int_C \left[\left(xy + z^2\right) ds\right]$ where C is the arc of the helix $x = \cos t$, $y = \sin t$, $z = t$ which joins the points (1, 0, 0) and (–1, 0, π).

18. Evaluate $\int_C \left(\frac{a^2y^2}{b^2} + \frac{b^2x^2}{a^2}\right) ds$ around the ellipse $\frac{x^2}{a^2} + \frac{y^2}{b^2} = 1$.

Answers

1. $\frac{33}{2}$ **2.** $-\frac{1}{20}$ **3. (a)** $\frac{34}{3}$ **(b)** 11 **(c)** $\frac{32}{3}$

4. 12 **5.** 0 **6.** 2 **7. (a)** $\sqrt{2}$ **(b)** $\sqrt{2}$

8. (a) 0 **(b)** $\frac{1}{3}$ **(c)** $\frac{\pi}{2} - 1$ **(d)** 2 **10.** 15

11. **(a)** $\frac{1}{\sqrt{2}}$ **(b)** $\frac{1}{12}\left(5^{3/2}-1\right)$ **(c)** $\frac{32}{81}\left[\left(\frac{13}{4}\right)^{3/2}\frac{19}{60}+\frac{2}{15}\right]$

12. πa^3 **13.** 2 **14.** $\frac{3}{8}$ **15.** $\frac{2}{3}$

16. **(a)** 16 **(b)** 16 **17.** $\frac{\sqrt{2}}{3}\pi^3$ **18.** $\pi\left(a^2+b^2\right)$

3.3 Independent of Path of Integration

The value of a line integral depends on the path along which it is evaluated. That is the value of a line integral not only depends on the end points A and B of the path C but also on the path C. However under certain conditions the value of the line integral will be independent of the path C obtained by joining the end points A and B.

Let P, Q and R be functions of x, y and z defined and continuous in a domain D. The integral

$$\int_C P\,dx + Q\,dy + R\,dz \qquad \text{.... (1)}$$

is said to be **independent of the path** in D, if for every pair of points A and B in D the value of the integral is same for the all paths C in D, joining A and B.

We shall state a condition for the line integral (1) to be independent of path.

The line integral $\int_C P\,dx + Q\,dy + R\,dz$ is independent of the path in a domain D, if and only if the expression $P\,dx + Q\,dy + R\,dz$ is exact in the domain D.

The expression $P\,dx + Q\,dy + R\,dz$ is said to be exact if there exists a single valued function $\varphi(x, y, z)$ such that

$$d\varphi = P\,dx + Q\,dy + R\,dz$$

In other words if there exists a function $\varphi(x, y, z)$ such that

$$\frac{\partial\varphi}{\partial x} = P, \qquad \frac{\partial\varphi}{\partial y} = Q, \qquad \frac{\partial\varphi}{\partial z} = R$$

Further the conditions for the expression $P\,dx + Q\,dy + R\,dz$ to be exact are

$$\frac{\partial P}{\partial y} = \frac{\partial Q}{\partial x}, \qquad \frac{\partial Q}{\partial z} = \frac{\partial R}{\partial y}, \qquad \frac{\partial R}{\partial x} = \frac{\partial P}{\partial z}$$

Similarly the line integral $\int_C P\,dx + Q\,dy$ will be independent of the path if and only if the expression $P\,dx + Q\,dy$ is exact. The condition for $P\,dx + Q\,dy$ to be exact is $\frac{\partial P}{\partial y} = \frac{\partial Q}{\partial y}$.

Note : If the line integral

$$\int_C P\,dx + Q\,dy + R\,dz \quad \text{or} \quad \int_C P\,dx + Q\,dy$$

is independent of the path then its value is zero for every simple closed path C in D and conversely.

Example 1. Show that $\int_C \left[y^2\, dx + 2xy\, dy\right]$ is independent of the path joining (0, 1) and (1, 3) and hence evaluate.

Solution : Let $P = y^2$ and $Q = 2xy$. Now, $\frac{\partial P}{\partial y} = 2y$ and $\frac{\partial Q}{\partial x} = 2y \Rightarrow \frac{\partial P}{\partial y} = \frac{\partial Q}{\partial x}$.

Thus, $P\, dx + Q\, dy$ is exact and hence the given integral is independent of the path.

Also, $$P\, dx + Q\, dy = y^2 dx + 2xy\, dy \Rightarrow P\, dx + Q\, dy = d\left(xy^2\right)$$

Now, $$\int_C P\, dx + Q\, dy = \int_{(0,1)}^{(1,3)} d\left(xy^2\right) = \left[xy^2\right]_{(0,1)}^{(1,3)} = \mathbf{9}$$

Example 2. Let C be any path leading from the origin to the point (1, 1, 1). Show that
$$\int_C \left[2xy\, dx + \left(x^2 + 2yz\right) dy + \left(y^2 + 1\right) dz\right] = 3.$$

Solution : Let, $P = 2xy, \quad Q = x^2 + 2yz, \quad R = y^2 + 1.$
$$\Rightarrow \frac{\partial P}{\partial y} = 2x,\ \frac{\partial P}{\partial z} = 0,\ \frac{\partial Q}{\partial x} = 2x,\ \frac{\partial Q}{\partial z} = 2y,\ \frac{\partial R}{\partial x} = 0,\ \frac{\partial R}{\partial y} = 2y$$

Clearly, $\frac{\partial P}{\partial y} = \frac{\partial Q}{\partial x},\ \frac{\partial Q}{\partial z} = \frac{\partial R}{\partial y},\ \frac{\partial R}{\partial x} = \frac{\partial P}{\partial z}.$

Thus $P\, dx + Q\, dy + R\, dz$ is exact and hence the given integral is independent of the path.

Now,
$$\begin{aligned} P\, dx + Q\, dy + R\, dz &= 2xy\, dx + (x^2 + 2yz)dy + (y^2 + 1)dz \\ &= (2xy\, dx + x^2)dy + (2yz\, dy + y^2 dz) + dz \\ &= d(x^2 y) + d(y^2 z) + dz = d[x^2 y + y^2 z + z] \end{aligned}$$

Thus, $$\int_C P\, dx + Q\, dy + R\, dz = \int_{(0,0,0)}^{(1,1,1)} d\left[x^2 y + y^2 z + z\right] = \left[x^2 y + y^2 z + z\right]_{(0,0,0)}^{(1,1,1)} = \mathbf{3}$$

Example 3. Show that $\int_C \left[(x + y)\, dx + (x - y)\, dy\right] = 0$ where C is the ellipse $\frac{x^2}{a^2} + \frac{y^2}{b^2} = 1$.

Solution : Let $P = x + y$ and $Q = x - y \Rightarrow \frac{\partial P}{\partial y} = 1$ and $\frac{\partial Q}{\partial x} = 1 \Rightarrow \frac{\partial P}{\partial y} = \frac{\partial Q}{\partial x}$

Thus the given integral exact.

Further C is the ellipse $\frac{x^2}{a^2} + \frac{y^2}{b^2} = 1$ is a closed curve. Thus,

$$\int_C \left[(x + y)\, dx + (x - y)\, dy\right] = \mathbf{0}$$

EXERCISES

1. If C is any curve leading from (3, 2, –1) and (–1, 2, 3) show that

$$\int_C \left[\left(3x^2 - 3yz + 2xz\right) dx + \left(3y^2 - 3xz + z^2\right) dy + \left(3z^2 - 3xy + x^2 + 2yz\right) dz\right] = 28.$$

2. Evaluate $\int_C \left[\left(4xy - 3x^2z^2\right) dx + 2x^2 dy - 2x^3 z\, dz\right]$ along the curve C from (3, –1, 1) to (0, 1, 2).

3. Evaluate $\int_C \left[y^2 dx + 2xy\, dy\right]$ along C from (0, 1) to (1, 3).

4. If C is a curve leading from (0, 0, 1) to $\left(1, \frac{\pi}{2}, 2\right)$, evaluate

$$\int_C \left[3xyz^2 + \left(x^2z^2 + z \cos yz\right) dz + \left(2x^2yz + y \cos yz\right) dz\right].$$

5. If C si the curve leading from (–1, 2, 4) to (6, 0, –2) compute

$$\int_C \left[\cos x\, dx + e^{-y} dy + z^2 dz\right].$$

Answers

2. 45 **3.** 9 **4.** $\pi + 1$ **5.** $\sin 6 + \sin 1 - e - e^{-2} - \frac{72}{3}$

3.4 Double Integral

The students are familiar about the concept of integration as the limit of a sum. This concept can be extended to the case of continuous functions defined over the domain in a plane or a region in space. This fact leads to the concept of double and triple integrals respectively. In this section we shall consider the concept of double integral and explain the method of evaluating it.

Let $f(x, y)$ be a continuous function defined in the closed domain D in the plane. Let the domain D be subdivided into n rectangles by drawing lines parallel to x and y axes. Let $P_i = (x_i, y_i)$ be a point in the ith rectangle (i = 1, 2, 3, ..., n). Its area will be $\partial x_i \cdot \partial y_i$. Consider the sum

$$S_n = \sum_{i=1}^{n} f(x_i, y_i)\, \partial x_i \cdot \partial y_i$$

This sum is called the integral sum of the function $f(x, y)$ over the domain D corresponding to the partition of D under consideration.

Let the partition of D be done in any manner but the length of maximum diagonal of the rectangles must tend to zero as n tends to ∞.

The double integral of the function $f(x, y)$ over the domain D is defined as

$$\lim_{n \to \infty} S_n = \lim_{n \to \infty} \sum_{i=1}^{n} f(x_i, y_i)\, \partial x_i \cdot \partial y_i$$

This is denoted by

$$\iint_D f(x, y)\, dx\, dy$$

where D is called the domain of integration.

3.5 Evaluation of Double Integral

Let the domain D under consideration be such that any line parallel to one of the coordinate axes, say y-axis and passing through an interior point of the domain cuts the boundary of the domain at two points P_1 and P_2.

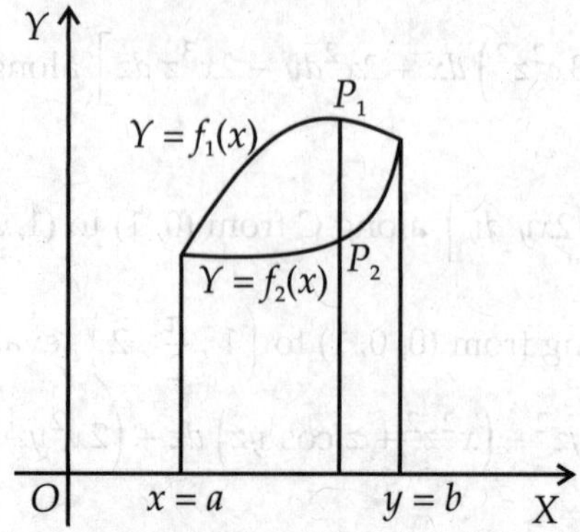

Let us assume that the domain D is bounded by the lines $x = a$, $x = b$ and the curves $y = f_1(x)$ and $y = f_2(x)$. Further let $a < b$ and $f_1(x) < f_2(x)$ with f_1 and f_2 to be continuous.

The expression

$$I = \int_a^b \left[\int_{f_1(x)}^{f_2(x)} f(x, y)\, dy \right] dx$$

is called a **two fold iterated integral** of $f(x, y)$ over D.

The above expression implies that, first we calculate

$$\int_{f_1(x)}^{f_2(x)} f(x, y)\, dy$$

treating x as constant. Clearly the resulting value will be a function of x say $g(x)$. Then we calculate

$$\int_a^b g(x)\, dx$$

The final result which will be a constant is taken as the value of the iterated integral.

Similarly one can define the iterated integral

$$\int_c^d \left[\int_{\varphi_1(y)}^{\varphi_2(y)} f(x, y)\, dx \right] dy$$

Now we shall state the method of evaluating a double integral without proof.

The double integral of a continuous function $f(x, y)$ over the domain is equal to the value of a two - fold iterated integral of $f(x, y)$ over D. That is

$$\iint_D f(x, y)\, dx\, dy = \int_a^b \left[\int_{f_1(x)}^{f_2(x)} f(x, y)\, dy \right] dx$$

or

$$\iint_D f(x, y)\, dx\, dy = \int_c^d \left[\int_{\varphi_1(y)}^{\varphi_2(y)} f(x, y)\, dx \right] dy$$

Note :

1. The brackets in the above iterated integrals can be dropped and can be written as

$$\int_a^b \int_{f_1(x)}^{f_2(x)} f(x, y)\, dy\, dx \quad \text{and} \quad \int_c^d \int_{\varphi_1(y)}^{\varphi_2(y)} f(x, y)\, dx\, dy$$

2. If the limits of integration are constants then the order in which the integration is done is immaterial. That is it is immaterial whether we first integrate w.r.t. x or w.r.t. y and vice versa. However if the limits of integration are variable the order of integration must be maintained.

Example 1. Evaluate $\int_0^1 \int_0^2 (x + y)\, dy\, dx$ **.**

Solution :

$$\int_0^1 \left[\int_0^2 (x + y)\, dy\right] dx = \int_0^1 \left[xy + \frac{y^2}{2}\right]_0^2 dx = \int_0^1 (2x + 2)\, dx = \left[x^2 + 2x\right]_0^1 = 3$$

Example 2. Evaluate $\int_0^1 \int_0^1 \frac{dx\, dy}{\sqrt{(1 - x^2)(1 - y^2)}}$ **.**

Solution :

$$\int_0^1 \int_0^1 \frac{dx\, dy}{\sqrt{(1 - x^2)(1 - y^2)}} = \int_0^1 \left[\int_0^1 \left(\frac{1}{\sqrt{(1 - y^2)}}\right) \frac{1}{\sqrt{(1 - x^2)}}\, dx\right] dy$$

$$= \left[\int_0^1 \frac{1}{\sqrt{(1 - y^2)}} \cdot \sin^{-1} x\, dx\right]_0^1 dy$$

$$= \int_0^1 \frac{1}{\sqrt{(1 - y^2)}} \cdot \left(\frac{\pi}{2}\right) dy = \frac{\pi}{2}\left[\sin^{-1} y\right]_0^1 = \frac{\pi}{2} \cdot \frac{\pi}{2} = \boldsymbol{\frac{\pi^2}{4}}$$

Example 3. Evaluate $\int_0^2 \int_0^2 x\left(x^2 + y^2\right) dy\, dx$ **.**

Solution :

$$\int_0^2 \int_0^2 x\left(x^2 + y^2\right) dy\, dx = \int_0^2 \left[\int_0^2 x\left(x^2 + y^2\right) dy\right] dx$$

$$= \int_0^2 \left[x^3 y + \frac{xy^3}{3}\right]_0^{x^2} dx$$

$$= \int_0^2 \left[x^5 + \frac{x^7}{3}\right] dx = \left[\frac{x^6}{6} + \frac{x^8}{24}\right]_0^2 = \frac{64}{6} + \frac{256}{24} = \boldsymbol{\frac{64}{3}}$$

Example 4. Evaluate $\iint\limits_R \frac{dx\,dy}{x+y+1}$ over the square $R: 0 \le x \le 1, 0 \le y \le 1, 0 \le x, y \le 1$.

Solution : The given integral is $\int_0^1\int_0^1 \frac{dx\,dy}{x+y+1}$.

$$\int_0^1\int_0^1 \frac{dx\,dy}{x+y+1} = \int_0^1\left[\int_0^1 \frac{dx}{x+y+1}\right]dy = \int_0^1 \left[\log(x+y+1)\right]_0^1 dy$$

$$\int_0^1 \left[\log(y+2) - \log(y+1)\right] dy$$

$$= \left[(y+2)\log(y+2) - (y+2) - (y+1)\log(y+1) + (y+1)\right]_0^1$$

$$= 3\log 3 - 3 - 2\log 2 + 2 - 2\log 2 + 2 + \log 1 - 1$$

$$= 3\log 3 - 4\log 2 = \log 27 - \log 16 = \mathbf{\log \frac{27}{16}}$$

Example 5. Evaluate $\int_0^1 \int_0^{\sqrt{1+x^2}} \frac{dy\,dx}{1+x^2+y^2}$.

Solution : $\int_0^1 \int_0^{\sqrt{1+x^2}} \frac{dy\,dx}{1+x^2+y^2} = \int_0^1 \left[\int_0^{\sqrt{1+x^2}} \frac{dy}{(1+x^2)+y^2}\right] dx$

$$= \int_0^1 \left[\int_0^{\sqrt{1+x^2}} \frac{dy}{\left(\sqrt{1+x^2}\right)^2 + y^2}\right] dx$$

The integral is of the form $\int \frac{1}{a^2+x^2}\,dx$

$$= \int_0^1 \left[\frac{1}{\sqrt{1+x^2}} \cdot \tan^{-1}\frac{y}{\sqrt{1+x^2}}\right]_0^{\sqrt{1+x^2}} dx$$

$$= \int_0^1 \frac{1}{\sqrt{1+x^2}} \cdot \tan^{-1}(1)\,dx$$

$$= \frac{\pi}{4}\left[\log\left(x + \sqrt{x^2+1}\right)\right]_0^1 = \mathbf{\frac{\pi}{4}\left[\log\left(1+\sqrt{2}\right)\right]}$$

Example 6. Evaluate $\iint\limits_R xy\,dx\,dy$ where R is the quadrant of the circle $x^2 + y^2 = a^2$ where $x \ge 0, y \ge 0$.

Solution : The region of integration is region of the circle $x^2 + y^2 = a^2$ which is in the first quadrant. That is the region OAB in the figure.

In this region, x varies from 0 to 1. Now for each x, the variable y varies from a point P on $y = 0$ (x – axis) to the point Q on the circle $x^2 + y^2 = a^2$. Thus as x varies from 0 to a, y varies from $y = 0$ to $y = \sqrt{a^2 - x^2}$.

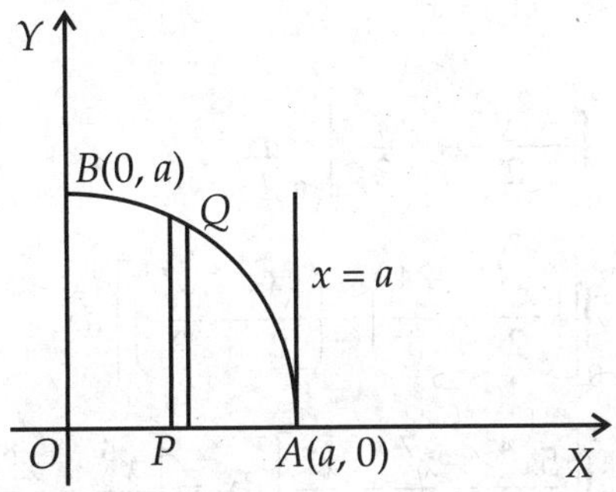

Thus, $$\iint_R xy\, dx\, dy = \int_0^a \int_0^{\sqrt{a^2-x^2}} xy\, dy\, dx$$

$$= \int_0^a \left[\int_0^{\sqrt{a^2-x^2}} xy\, dy \right] dx$$

$$= \int_0^a \left[\frac{xy^2}{2} \right]_0^{\sqrt{a^2-x^2}} dx$$

$$= \frac{1}{2} \int_0^a x\left(a^2 - x^2\right) dx = \frac{1}{2}\left[\frac{a^2x^2}{2} - \frac{x^4}{4} \right]_0^a = \frac{1}{2}\left[\frac{a^4}{2} - \frac{a^4}{4} \right] = \frac{a^4}{8}$$

Example 7. Evaluate $\iint_R xy(x + y)\, dx\, dy$ over the domain D between $y = x^2$ and $y = x$.

Solution : The domain D is the domain between $y = x^2$ and the line $y = x$, as shown in the figure.

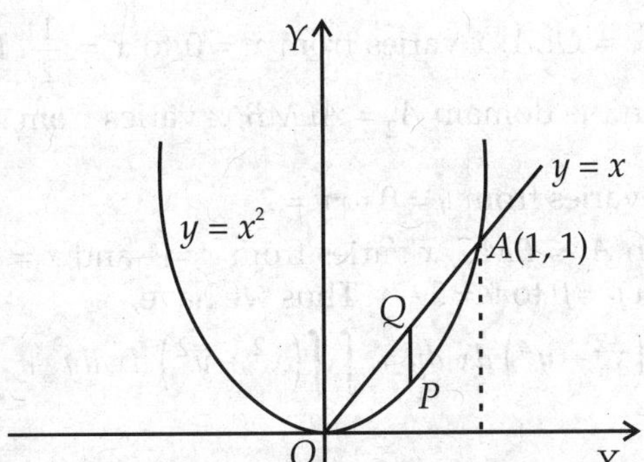

Solving the equations we get $O = (0, 0)$ and $A = (1, 1)$.

Now in the domain D, x varies from $x = 0$ to $x = 1$. For each value of x, the variable y varies from the point P on the curve $y = x^2$ and to the point Q on the line $y = x$.

Thus y varies from $y = x^2$ to $y = x$.

$$\iint_R xy(x+y)\,dx\,dy = \int_{x=0}^{1}\left[\int_{y=x^2}^{x} xy(x+y)\,dy\right]dx$$

$$= \int_0^1 \left[\frac{x^2y^2}{2} + \frac{xy^3}{3}\right]_{x^2}^{x} dx$$

$$= \int_0^1 \left[\left(\frac{x^4}{2} + \frac{x^4}{3}\right) - \left(\frac{x^6}{2} + \frac{x^7}{3}\right)\right] dx$$

$$= \int_0^1 \left[\frac{5x^4}{6} - \frac{x^7}{3} - \frac{x^6}{2}\right] dx = \left[\frac{x^5}{6} - \frac{x^8}{24} - \frac{x^7}{14}\right]_0^1 = \frac{1}{6} - \frac{1}{24} - \frac{1}{14} = \frac{3}{56}$$

Example 8. Evaluate $\iint_A (x^2 + y^2)\,dx\,dy$ where A is the area enclosed by the lines $y = 4x$, $x + y = 3$, $y = 0$ and $y = 2$.

Solution : The region of integration is the area of the trapizium $OABC$.

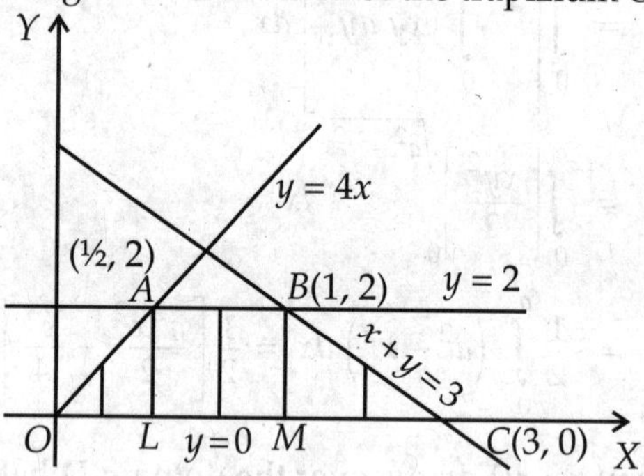

The point of intersection of $y = 4x$ and $y = 2$ is $A\left(\frac{1}{2}, 2\right)$ and the point of intersection of $y = 2$ and $x + y = 3$ is $B(1, 2)$. (Figure is not to the scale).

Now the domain $A_1 = OLA$, x varies from $x = 0$ to $x = \frac{1}{2}$. For each value of x, the variable $y = 0$ to $y = 4x$. In the domain $A_2 = ALMB$, x varies from $x = \frac{1}{2}$ to $x = 1$. For each value of x, the variable y varies from $y = 0$ to $y = 2$.

Again in the domain $A_3 = BMC$, x varies from $x = 1$ and $x = 3$. For these values of x, the variable y varies from $y = 0$ to $y = 3 - x$. Thus we have,

$$\iint_A (x^2+y^2)\,dx\,dy = \iint_{A_1} (x^2+y^2)\,dx\,dy + \iint_{A_2} (x^2+y^2)\,dx\,dy + \iint_{A_3} (x^2+y^2)\,dx\,dy$$

$$= \int_0^{1/2}\left[\int_0^{4x} (x^2+y^2)\,dy\right]dx + \int_{1/2}^{1}\left[\int_0^{2} (x^2+y^2)\,dy\right]dx + \int_1^{3}\left[\int_0^{3-x} (x^2+y^2)\,dy\right]dx$$

$$\text{Now, } I_1 = \int_0^{1/2}\left[\int_0^{4x}\left(x^2+y^2\right)dy\right]dx = \int_0^{1/2}\left[x^2y+\frac{y^3}{3}\right]_0^{4x}dx$$

$$= \int_0^{1/2}\left[4x^3+\frac{64x^3}{3}\right]dx$$

$$= \frac{76}{3}\int_0^{1/2}x^3\,dx = \frac{76}{3}\left[\frac{x^4}{4}\right]_0^{1/2} = \mathbf{\frac{19}{48}}$$

$$I_2 = \int_{1/2}^{1}\left[\int_0^{2}\left(x^2+y^2\right)dy\right]dx = \int_{1/2}^{1}\left[x^2y+\frac{y^3}{3}\right]_0^{2}dx$$

$$= \int_{1/2}^{1}\left[2x^2+\frac{8}{3}\right]dx = \left[\frac{2x^3}{3}+\frac{8}{3}x\right]_{1/2}^{1} = \mathbf{\frac{23}{12}}$$

$$I_3 = \int_1^{3}\left[\int_0^{3-x}\left(x^2+y^2\right)dy\right]dx = \int_1^{3}\left[x^2y+\frac{y^3}{3}\right]_0^{3-x}dx$$

$$= \int_1^{3}\left[x^2(3-x)+\frac{(3-x)^3}{3}\right]dx$$

$$= \left[x^3-\frac{x^4}{4}-\frac{(3-x)^4}{12}\right]_1^{3} = \mathbf{\frac{22}{3}}$$

$$\text{Thus,}\qquad \iint_A\left(x^2+y^2\right)dx\,dy = \frac{19}{48}+\frac{23}{12}+\frac{22}{3}$$

$$= \mathbf{\frac{463}{48}}$$

Example 9. Evaluate $\iint_A \sqrt{4x^2-y^2}\;dx\,dy$ where A is the area bounded by the lines $y = 0$, $y = x$ and $x = 1$.

Solution : The area of integration is the area of the triangle OBA, bounded by the lines $y = 0, y = x$ and $x = 1$.

The point of intersection of $y = x$ and $x = 1$ is $A(1, 1)$.

Now in the region OBA, x varies from $x = 0$ to $x = 1$.

Now for each value of x, the variable y varies from the point Q on $y = 0$ to P on $y = x$. That is y varies from $y = 0$ to $y = x$.

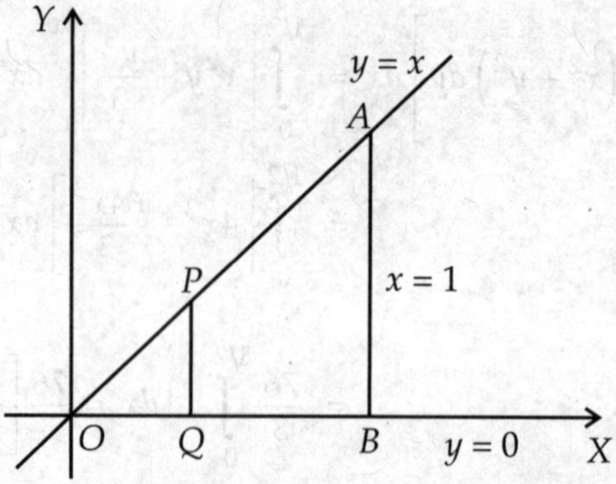

Thus,

$$\iint_A \sqrt{4x^2 - y^2}\, dx\, dy = \int_0^1 \left[\int_0^x \sqrt{4x^2 - y^2}\, dy\right] dx$$

$$= \int_0^1 \left[\frac{y}{2}\sqrt{4x^2 - y^2} + \frac{4x^2}{2}\sin^{-1}\left(\frac{y}{2x}\right)\right]_0^x dx$$

$$\left\{\because \int \sqrt{a^2 - x^2}\, dx = \frac{x}{2}\sqrt{a^2 - x^2} + \frac{a^2}{2}\sin^{-1}\frac{x}{a}\right\}$$

$$= \int_0^1 \left[\frac{x}{2}\sqrt{3x^2} + \frac{4x^2}{2}\sin^{-1}\left(\frac{1}{2}\right)\right] dx$$

$$= \int_0^1 \left[\frac{\sqrt{3}}{2}x^2 + \frac{\pi}{3}x^2\right] dx$$

$$= \left[\frac{\sqrt{3}}{2} + \frac{\pi}{3}\right]\int_0^1 x^2\, dx = \left(\frac{\sqrt{3}}{2} + \frac{\pi}{3}\right)\left[\frac{x^3}{3}\right]_0^1 = \mathbf{\frac{1}{3}\left(\frac{\sqrt{3}}{2} + \frac{\pi}{3}\right)}$$

Example 10. Evaluate $\iint_D x^2\, dx\, dy$ where D is the region in the first quadrant bounded by the hyperbola $xy = 16$ and the lines $y = x$, $y = 0$ and $x = 8$.

Solution : The region D of integration is the area of the triangle $OABC$.

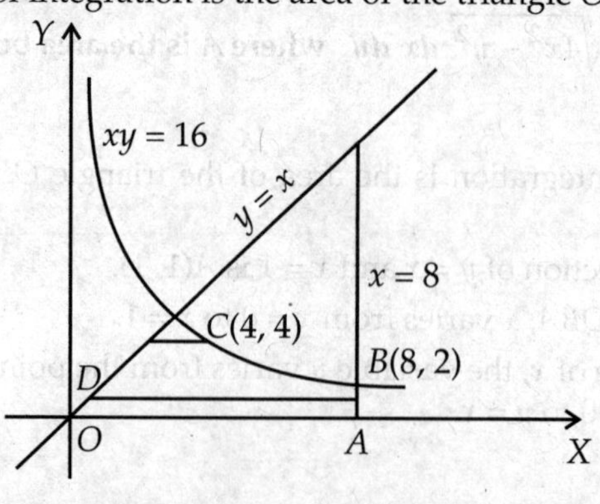

The region D is divided into two regions $D_1 : DBC$ and $D_2 : OABD$. In the region D_1, y varies from $y = 2$ to $y = 4$. For each value of y, the variable x varies from $x = y$ to $x = \frac{16}{y}$.

Again in the region D_2, y varies from $y = 0$ to $y = 2$. For each value of y, the variable x varies from $x = y$ to $x = 8$.

Thus,
$$\iint_D x^2\,dx\,dy = \iint_{D_1} x^2\,dx\,dy + \iint_{D_2} x^2\,dx\,dy$$
$$= \int_2^4 \left[\int_y^{16/y} x^2\,dx\right] dy + \int_0^2 \left[\int_y^8 x^2\,dx\right] dy$$
$$= \int_2^4 \left[\frac{x^3}{3}\right]_y^{16/y} dy + \int_0^2 \left[\frac{x^3}{3}\right]_y^8 dy$$
$$= \frac{1}{3}\int_2^4 \left[\frac{16^3}{y^3} - y^3\right] dy + \frac{1}{3}\int_0^2 \left[8^3 - y^3\right] dy$$
$$= \frac{1}{3}\left[-\frac{16^3}{2y^2} - \frac{y^4}{4}\right]_2^4 + \frac{1}{3}\left[8^3 y - \frac{y^4}{4}\right]_0^2 = \mathbf{448}$$

Example 11. Evaluate $\iint_A e^{y^2}\,dx\,dy$, where A is the region formed by the points (0, 0), (1, 1) and (–1, 1).

Solution : The region of integration is the triangle OAB, where $A = (-1, 1)$ and $B = (1, 1)$.

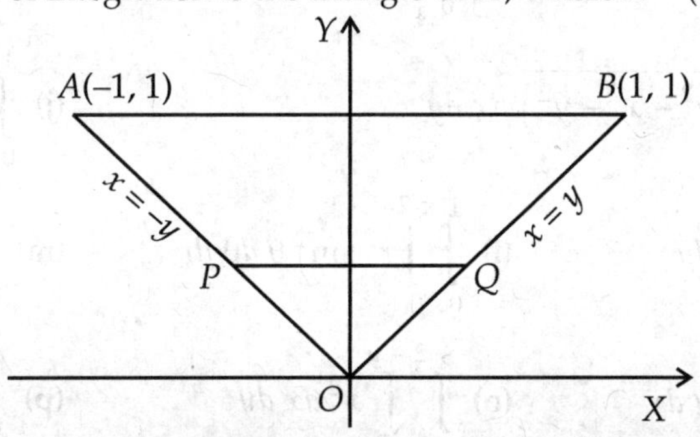

In the region OAB, y varies from $y = 0$ to $y = 1$. Now for each y, the variable varies from the point P on OA to the point Q on the line OB.

Thus as y varies from $y = 0$ to $y = 1$, the variable x varies from $x = -y$ to $x = y$.

Thus,
$$\iint_A e^{y^2}\,dx\,dy = \int_0^1 \left[\int_{x=-y}^{y} e^{y^2}\,dx\right] dy = \int_0^1 \left[x e^{y^2}\,dx\right]_{-y}^{y} dy$$
$$= \int_0^1 [y - (-y)] e^{y^2}\,dy$$
$$= \int_0^1 2y e^{y^2}\,dy = \int_0^1 e^{y^2} \cdot 2y\,dy = \left[e^{y^2}\right]_0^1 = \mathbf{e - 1}$$

EXERCISES

1. Evaluate the following integrals

(a) $\int_3^4\int_1^2 \frac{dy\,dx}{(x+y)^2}$ (b) $\int_0^1\int_1^2 \left(x^2+y^2\right) dy\,dx$ (c) $\int_0^1\int_0^1 \left(x^2+xy\right) dy\,dx$

(d) $\int_0^3\int_1^2 \left(x^2+3y^2\right) dy\,dx$ (e) $\int_1^2\int_3^4 \left(xy+e^y\right) dy\,dx$ (f) $\int_b^{2b}\int_0^a (a-y)x^2\,dy\,dx$

(g) $\int_0^4\int_0^3 y^2(x+y)^2\,dy\,dx$ (h) $\int_0^3\int_1^2 xy(1+x+y)\,dy\,dx$ (i) $\int_0^1\int_0^1 \frac{dx\,dy}{\sqrt{\left(1-x^2\right)\left(1-y^2\right)}}$

(j) $\int_0^a\int_0^b \left(x^2+y^2\right) dy\,dx$ (k) $\int_0^{\pi/2}\int_0^{\pi/6} \sin x \cos y\,dx\,dy$

2. Evaluate the following integration

(a) $\int_0^1\int_1^{1-x} xy\,dy\,dx$ (b) $\int_0^1\int_0^{x^2} \left(x^2+y^2\right) dy\,dx$ (c) $\int_0^1\int_1^{y^2} e^{x/y}\,dx\,dy$

(d) $\int_0^1\int_0^{1-x} (x+y)^2\,dy\,dx$ (e) $\int_1^2\int_0^x \frac{dy\,dx}{y^2+x^2}$ (f) $\int_0^1\int_0^x e^{x/y}\,dy\,dx$

(g) $\int_0^1\int_0^{x^2} e^{x/y}\,dy\,dx$ (h) $\int_0^1\int_{x^2}^{x} \left(x^2+3y+2\right) dy\,dx$

(i) $\int_0^a\int_0^{\sqrt{a^2-y^2}} \sqrt{\left(a^2-x^2-y^2\right)}\,dx\,dy$ (j) $\int_0^1\int_0^{\sqrt{x}} \left(x^2+y^2\right) dy\,dx$

(k) $\int_0^{4a}\int_{y^2/4a}^{2\sqrt{ay}} dx\,dy$ (l) $\int_0^1\int_0^{\pi/2} r^3\sin^2\theta\,d\theta\,dr$ (m) $\int_{-\pi/2}^{\pi/2}\int_0^{2\cos\theta} r\,dr\,d\theta$

(n) $\int_0^a\int_{x-a}^{2x} xy\,dy\,dx$ (o) $\int_0^3\int_{\sqrt{y}}^{2-y} x^2\,dx\,dy$ (p) $\int_0^a\int_0^{\sqrt{a^2-x^2}} y^3\,dy\,dx$

3. Evaluate

(a) $\iint_R (2x+3y)^2\,dy\,dx$ where R is the rectangle $0 \le x \le 1, -1 \le y \le 1$.

(b) $\iint_R xye^x\,dy\,dx$ where R is region, $0 \le x \le 1, 2 \le y \le 3$.

(c) $\iint_R x^2y^2\,dx\,dy$ where S is the region in the first quadrant bounded by the circle $x^2+y^2=a^2$.

(d) $\iint_A \sqrt{4x^2-y^2}\,dx\,dy$ where A is the region bounded by the lines $y=0, y=x, x=1$.

(e) $\iint_S \sqrt{xy - y^2}\, dx\, dy$ where S is the triangle with vertices (0, 0), (10, 1), (1, 1).

(f) $\iint \frac{xy}{\sqrt{1 - y^2}}\, dx\, dy$ over the positive quadrant of the circle $x^2 + y^2 = 1$.

(g) $\iint_R y\, dx\, dy$ where R is the region bounded by the parabolas $y^2 = 4x$ and $x^2 = 4y$.

(h) $\iint_R xy(x + y)\, dx\, dy$ where R is the region bounded by the parabolas $y^2 = x$ and $x^2 = y$.

(i) $\iint_R x^2y^2\, dx\, dy$ where R is the triangular region with vertices (0, 0), (2, 0), (2, 3).

(j) $\iint_R (x + y)^2\, dx\, dy$ where R is the region bounded by the ellipse $\frac{x^2}{a^2} + \frac{y^2}{b^2} = 1$.

Answers

1. (a) $\log \frac{25}{24}$ (b) $\frac{8}{3}$ (c) $\frac{7}{12}$ (d) 30

(e) $\frac{21}{4} + e^4 - e^2$ (f) $\frac{7}{6}a^2b^2$ (g) $\frac{3552}{5}$ (h) $\frac{123}{4}$

(i) $\frac{\pi^2}{4}$ (j) $\frac{ab}{3}(a^2 + b^2)$ (k) $1 - \frac{\sqrt{3}}{2}$

2. (a) $\frac{1}{24}$ (b) $\frac{26}{105}$ (c) $\frac{1}{2}$ (d) $\frac{1}{4}$

(e) $\frac{\pi}{4}\log 2$ (f) $\frac{1}{2}(e - 1)$ (g) $\frac{1}{2}$ (h) $\frac{7}{2}$

(i) $\frac{\pi a^3}{6}$ (j) $\frac{3}{35}$ (k) $\frac{16}{3}a^2$ (l) $\frac{\pi}{16}$

(m) $\frac{\pi}{4}$ (n) $\frac{11}{24}a^4$ (o) $\frac{1}{20}\left[25 - 24\sqrt{3}\right]$ (p) $\frac{2}{15}a^2$

3. (a) $\frac{26}{3}$ (b) $\frac{5}{2}$ (c) $\frac{\pi a^6}{96}$ (d) $\frac{1}{18}(2\pi + 3\sqrt{3})$

(e) 6 (f) $\frac{1}{6}$ (g) $\frac{48}{5}$ (h) $\frac{3}{28}$

(i) 12 (j) $\frac{\pi}{4}ab(a^2 + b^2)$

3.6 Change of Order of Integration

Consider the iterated integral

$$\int_a^b \left[\int_{f_1(x)}^{f_2(x)} f(x, y)\, dy\right] dx$$

In this iterated integral, the integration is performed first w.r.t. y and then w.r.t. x. Further the limits of integration for y are functions of x, whereas those of x are constants. Also the domain of integration is defined by the lines $x = a$, $x = b$ and the curves $y = f_1(x)$ and $y = f_2(x)$.

Sometimes it will be helpful if we specify the same domain of integration, keeping the limits of y as constants and those of x as functions of y. In this case the above iterated integral takes the form

$$\int_{y=c}^{d}\left[\int_{\varphi_1(x)}^{\varphi_2(x)} f(x, y)\, dx\right] dy$$

This process is called the **change of order** of integration. It should be noted that in order to change the order of integration, the new limits are to be obtained by examining the domain of integration. The following examples will illustrate the method.

Example 1. Change the order of integration in the integral $\int_0^{4a}\int_{x^2/4a}^{2\sqrt{ax}} dy\, dx$ and hence evaluate it.

Solution : The region of integration is bounded by $x = 0$, $x = 4a$ and the curves $y = \frac{x^2}{4a}$ and $y = 2\sqrt{ax}$.

That is $x = 0$, $x = 4a$, the curves $x^2 = 4ay$ and $y^2 = 4ax$.

Thus the region of integration is the area bounded between the curves $x^2 = 4ay$ and $y^2 = 4ax$, between $x = 0$, $x = 4ay$.

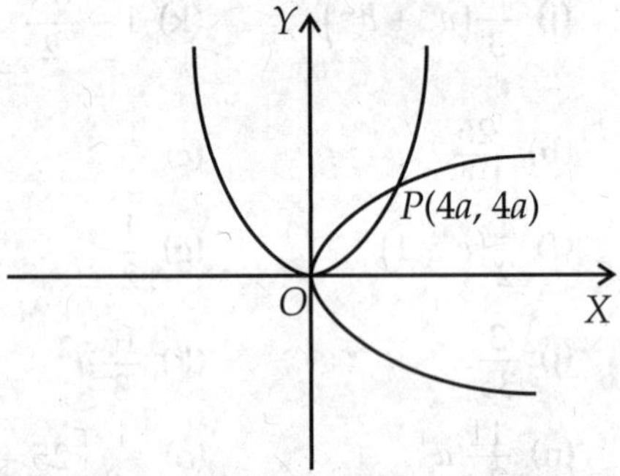

To change the order of integration we have to make limits of y independent of x.

Now, when $x = 0$, $x^2 = 4ay$ and $y^2 = 4ax$ $\Rightarrow$ $y = 0$

when $x = 4a$, $x^2 = 4ay$ and $y^2 = 4ax$ $\Rightarrow$ $y = 4a$

When y varies from $y = 0$ and $y = 4a$ then x varies from $\frac{y^2}{4a}$ to $x = 2\sqrt{ay}$. Thus we have,

$$\int_0^{4a}\int_{x^2/4a}^{2\sqrt{ax}} dy\, dx = \int_{y=0}^{4a}\int_{y^2/4a}^{2\sqrt{ay}} dx\, dy = \int_{y=0}^{4a}\left[\, x \,\right]_{y^2/4a}^{2\sqrt{ay}} dx\, dy$$

$$= \int_0^{4a}\left(2\sqrt{ay} - \frac{y^2}{4a}\right) dy$$

$$= \left[\frac{4}{3}\sqrt{a}\,y^{3/2} - \frac{y^3}{12a}\right]_0^{4a}$$

$$= \frac{4\sqrt{a}}{3}(4a)^{3/2} - \frac{64a^3}{12a} = \frac{32}{3}a^2 - \frac{16}{3}a^2 = \mathbf{\frac{16a^2}{3}}$$

Example 2. Evaluate $\int_0^1\int_x^1 (x^2+y^2)\,dy\,dx$ by changing the order of integration.

Solution : The region of integration is bounded by $x = 0$, $x = 1$, $y = x$ and $y = 1$. Thus the region of integration is the triangle OAB.

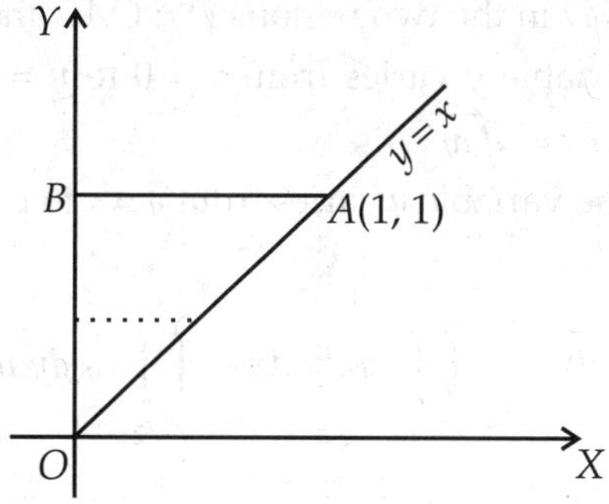

Thus, $$\int_0^1\int_x^1 (x^2+y^2)\,dy\,dx = \int_{y=0}^{1}\left[\int_{x=0}^{x=y}(x^2+y^2)\,dx\right]dy$$

$$= \int_{y=0}^{1}\left[\frac{x^3}{3} + y^2x\right]_0^y dy$$

$$= \int_0^1\left(\frac{y^3}{3} + y^3\right)dy = \int_0^1 \frac{4}{3}y^3\,dy = \left[\frac{y^4}{3}\right]_0^1 = \mathbf{\frac{1}{3}}$$

Example 3. Evaluate $\int_0^a\int_{x^2/4a}^{2a-x} xy\,dy\,dx$ by changing the order of integration.

Solution : The region of integration is bounded by the lines $x = 0$, $x = a$ and the curves $y = \frac{x^2}{a}$ and $y = 2a - x$.

Now, $y = \frac{x^2}{a} \Rightarrow x^2 = ay$.

Thus the region D of integration is the area bounded by $x^2 = ay$ and the line $x + y = 2a$. That is the region $OABCO$.

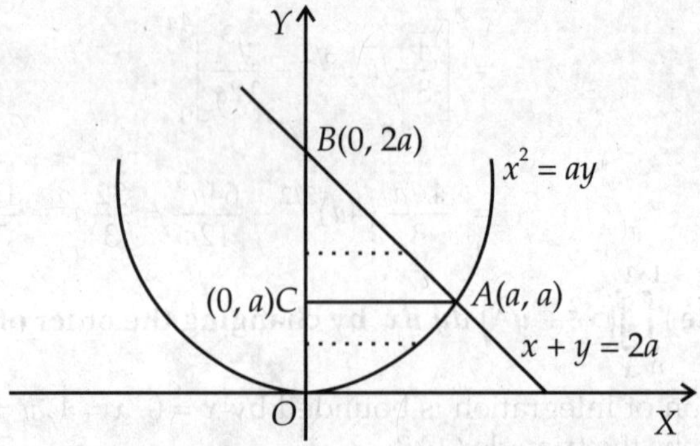

Let us divide the region D in the two regions $D_1 : OAC$ and the region $D_2 : CAB$.

In the region D_1 the variable y varies from $y = 0$ to $y = a$. For each value of y, the variable x varies from $x = 0$ to $x = \sqrt{ay}$.

Now, in the region D_2 the variable y varies from $y = a$ to $y = 2a$, whereas x varies from $x = 0$ to $x = 2a - y$.

Thus, $$\int_0^a \int_{x^2/4a}^{2a-x} xy\, dy\, dx = \iint_{D_1} xy\, dy\, dx + \iint_{D_2} xy\, dy\, dx$$

$$= \int_{y=0}^{a} \left[\int_{x=0}^{\sqrt{ay}} xy\, dx \right] dy + \int_{y=a}^{2a} \left[\int_{x=0}^{2a-y} xy\, dx \right] dy$$

$$= \int_{y=0}^{a} \left[\frac{x^2 y}{2} \right]_0^{\sqrt{ay}} dy + \int_{y=a}^{2a} \left[\frac{x^2 y}{2} \right]_0^{2a-y}$$

$$= \int_0^a \frac{ay^2}{2}\, dy + \int_a^{2a} \frac{y\,(2a-y)^2}{2}\, dy$$

$$= \left[\frac{ay^3}{6} \right]_0^a + \frac{1}{2} \int_a^{2a} \left(4a^2 y - 4ay^2 + y^3 \right) dy$$

$$= \frac{a^4}{6} + \frac{1}{2} \left[2a^2 y^2 - \frac{4ay^3}{3} + \frac{y^4}{4} \right]_a^{2a}$$

$$= \frac{a^4}{6} + \frac{1}{2} \left[8a^4 - \frac{32}{3} a^4 + 4a^4 - 2a^4 + \frac{4}{3} a^4 - \frac{a^4}{4} \right]$$

$$= \frac{a^4}{6} + \frac{1}{2} \left[10a^4 - \frac{28}{3} a^4 - \frac{a^4}{4} \right] = \frac{a^4}{6} + \frac{5}{24} a^4 = \frac{3}{8} a^4$$

Example 4. Show that $\int_0^\infty \int_0^x xe^{-x^2/y}\, dy\, dx = \frac{1}{2}$ **by changing the order of integration.**

Solution : The region of integration is bounded by the lines $x = 0$ to $x = \infty$, $y = 0$ and $y = x$.

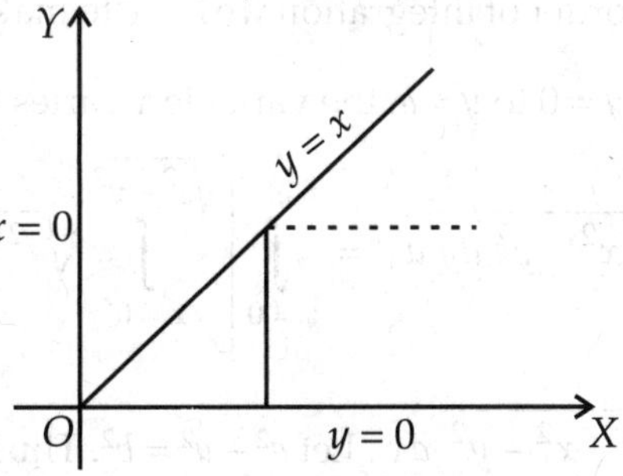

Thus the region of integration is as shown in the figure.

From the given order of integration we make out that as x varies from 0 to ∞, the variable varies from $y = 0$ to $y = x$.

Now as y varies from $y = 0$ to $y = \infty$ then x varies from $x = y$ to $x = \infty$. Thus,

$$\int_{x=0}^{\infty}\int_{y=0}^{x} xe^{-x^2/y}\,dy\,dx = \int_{y=0}^{\infty}\left[\int_{x=y}^{x=\infty} xe^{-x^2/y}\,dx\right]dy$$

$$= \int_{0}^{\infty}\left[-\frac{1}{2}\int_{y}^{\infty} e^{-\frac{1}{y}x^2}(-2x\,dx)\right]dy$$

$$= \int_{0}^{\infty} -\frac{1}{2}\left[ye^{-\left(\frac{1}{y}\right)x^2}\right]_y^{\infty} dy = \int_{0}^{\infty} -\frac{y}{2}\left[0 - e^{-y}\right]dy \quad (\because e^{-x} \to 0 \text{ as } x \to \infty)$$

$$= \frac{1}{2}\int_{0}^{\infty} y e^{-y}\,dy$$

$$= \frac{1}{2}\left[y\left(-e^{-y}\right) - \left(e^{-y}\right)\right]_0^{\infty} = \frac{1}{2}[0+1] = \frac{1}{2}$$

Example 5. Evaluate $\int_0^a \int_0^{\sqrt{a^2-x^2}} \sqrt{a^2-x^2-y^2}\,dy\,dx$ **by changing the order of integration.**

Solution : The region of integration is bounded by $x = 0$, $x = a$, $y = 0$ and the curve $y = \sqrt{a^2 - x^2}$.

Thus the region of integration is bounded by $x = 0$, $x = a$, $y = 0$ and the circle $x^2 + y^2 = a^2$.

In other words the region of integration is the circle bounded in the first quadrant (see figure).

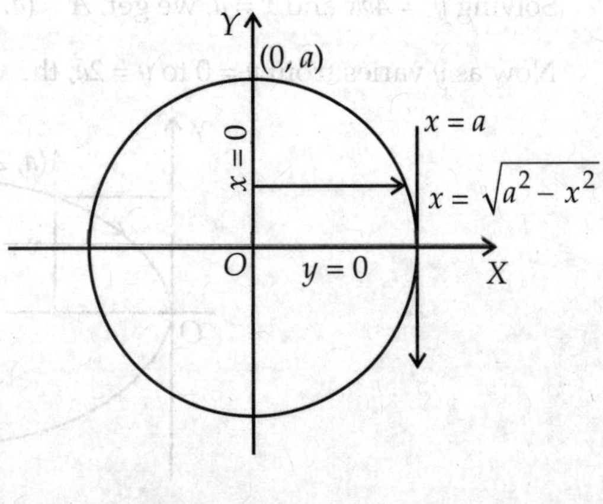

In order to change the order of integration we have to make the limits of y independent of x.

Now as y varies from $y = 0$ to $y = a$, the variable x varies from $x = 0$ to $x = \sqrt{a^2 - y^2}$.

$$\int_0^a \int_0^{\sqrt{a^2-x^2}} \sqrt{a^2 - x^2 - y^2}\, dy\, dx = \int_{y=0}^{a} \left[\int_{x=0}^{\sqrt{a^2-y^2}} \sqrt{a^2 - x^2 - y^2}\, dx \right] dy$$

Consider, $\int_{x=0}^{\sqrt{a^2-y^2}} \sqrt{a^2 - x^2 - y^2}\, dx$. Let $a^2 - y^2 = b^2$. Thus,

$$\int_{x=0}^{\sqrt{a^2-y^2}} \sqrt{a^2 - x^2 - y^2}\, dx = \int_{x=0}^{b} \sqrt{b^2 - x^2}\, dx$$

$$= \left[\frac{x}{2}\sqrt{b^2 - x^2} + \frac{b^2}{2}\sin^{-1}\frac{x}{b} \right]_0^b$$

$$= \frac{b}{2}\sqrt{b^2 - b^2} + \frac{b^2}{2}\sin^{-1}(1) = \frac{a^2 - y^2}{2}\frac{\pi}{2} = \left(a^2 - y^2\right)\frac{\pi}{4}$$

Now, $$\int_0^a \int_0^{\sqrt{a^2-x^2}} \sqrt{a^2 - x^2 - y^2}\, dy\, dx = \frac{\pi}{4}\int_{y=0}^{a} \left(a^2 - y^2\right) dy$$

$$= \frac{\pi}{4}\left[a^2 y - \frac{y^3}{3} \right]_0^a = \frac{\pi}{4}\left[a^3 - \frac{a^3}{3} \right] = \frac{\pi a^3}{6}$$

Example 6. Change the order of integration and evaluate $\int_0^a \int_0^{2\sqrt{ax}} x^2\, dy\, dx$.

Solution : The region of integration is bounded by $x = 0$, $x = a$, $y = 0$ and the curve $y = 2\sqrt{ax}$, i.e., $y^2 = 4ax$.

The region of integration AOB is as shown in the figure.

Now to change the order of integration, we have to make the limits of y independent of x.

Solving $y^2 = 4ax$ and $x = a$, we get, $A = (a, 2a)$.

Now as y varies from $y = 0$ to $y = 2a$, the variable x varies from $x = \frac{y^2}{4a}$ to $x = a$.

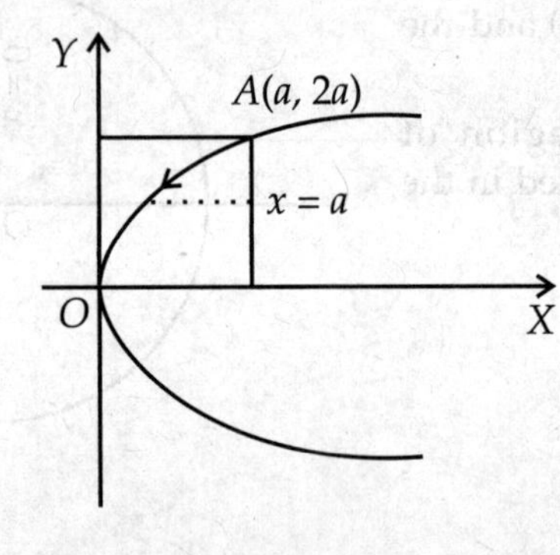

Thus we have,

$$\int_0^a \int_0^{2\sqrt{ax}} x^2\,dy\,dx = \int_{y=0}^{2a} \int_{x=\frac{y^2}{4a}}^{a} x^2\,dy\,dx = \int_{y=0}^{2a} \left[\frac{x^3}{3}\right]_{\frac{y^2}{4a}}^{a} dy$$

$$= \int_0^{2a} \left(\frac{a^3}{3} - \frac{y^6}{192a^3}\right) dy$$

$$= \left[\frac{a^3}{3}y - \frac{y^7}{7(192)a^3}\right]_0^{2a} = \frac{2}{3}a^4 - \frac{2}{21}a^4 = \frac{12}{21}a^4 = \frac{4}{7}a^4$$

Example 7. Evaluate $\int_0^1 \int_{y^2}^{\sqrt{y}} \frac{y}{x} e^x\,dx\,dy$ by changing the order of integration.

Solution : The region of integration is bounded by $y = 0$, $y = 1$ and the curves $x = y^2$ and $x = \sqrt{y}$. That is, $y = 0$, $y = 1$ and the curves $x = y^2$ and $x^2 = y$.

The region of integration is *OABCO* as shown in the figure.

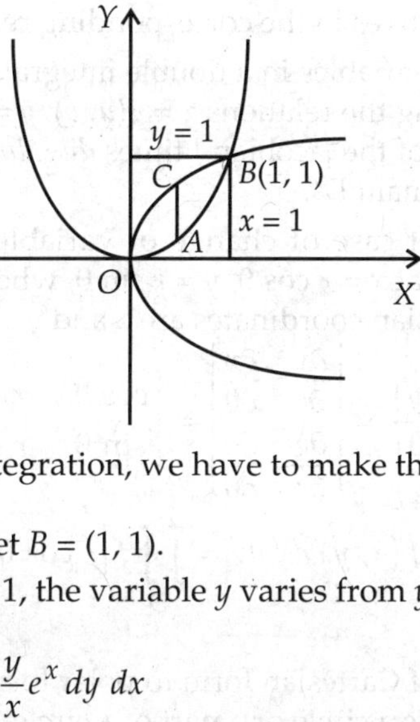

Now to change the order of integration, we have to make the limits of x independent of y.

Solving $x = y^2$ and $x^2 = y$, we get $B = (1, 1)$.

Now x varies from $x = 0$ to $x = 1$, the variable y varies from $y = x^2$ and $y = \sqrt{x}$. Thus,

$$\int_0^1 \int_{y^2}^{\sqrt{y}} \frac{y}{x} e^x\,dx\,dy = \int_{x=0}^{1} \int_{y=x^2}^{\sqrt{x}} \frac{y}{x} e^x\,dy\,dx$$

$$= \int_0^1 \left[\frac{e^x}{x}\frac{y^2}{2}\right]_{x^2}^{\sqrt{x}} dx = \int_0^1 \left(\frac{e^x}{2} - \frac{x^3 e^x}{2}\right) dx$$

$$= \left[\frac{e^x}{2} - \left\{x^3\frac{e^x}{2} - 3x^2\frac{e^x}{2} + 6x\frac{e^x}{2} - 6\frac{x^3}{2}\right\}\right]_0^1$$

$$= \left[\left(\frac{e}{2} - \frac{e}{2} + \frac{3e}{2} - \frac{6e}{2} + \frac{6e}{2}\right) - \left(\frac{1}{2} + 3\right)\right]$$

$$= \frac{3}{2}e - \frac{7}{2} = \mathbf{\frac{1}{2}(3e - 7)}$$

3.7 Change of Variables

Sometimes it is convenient to evaluate the iterated integrals (there by a double integral) by changing the variables of integration to another set of variables.

Let the variables x and y in the double integral

$$\iint_D f(x, y)\, dx\, dy$$

be related to the variables u and v, by means of the relations $x = g(u, v)$, $y = h(u, v)$. Further let the jacobian,

$$J = \frac{\partial(x, y)}{\partial(u, v)} = \begin{vmatrix} \dfrac{\partial x}{\partial u} & \dfrac{\partial x}{\partial v} \\ \dfrac{\partial y}{\partial u} & \dfrac{\partial y}{\partial v} \end{vmatrix} \neq 0$$

Then, we have,

$$\iint_D f(x, y)\, dx\, dy = \iint_{D^*} f\{g(u, v), h(u, v)\} \cdot |J|\, du\, d\theta$$

Here D is the region over which the variables x, y vary, where D^* is the corresponding region in which the variables u and v vary.

Note : The functions $g(u, v)$ and $h(u, v)$ are assumed to be continuous and have continuous partial derivatives in the corresponding region D^*.

Thus for change of variables in a double integral we replace the variables x and y by the variables u and v using the relations $x = g(u, v)$, $y = h(u, v)$ and the element of area $dx \cdot dy$ by the absolute value of the jacobian J times $du \cdot dv$, further the region of integration D by the corresponding domain D^*.

The most important case of change of variables is from Cartesian form to polar coordinates. We know that $x = r \cos \theta$, $y = r \sin \theta$, where r and θ are the polar coordinates of the point whose Cartesian coordinates are x and y.

Now, $$J = \frac{\partial(x, y)}{\partial(r, \theta)} = \begin{vmatrix} \dfrac{\partial x}{\partial r} & \dfrac{\partial x}{\partial \theta} \\ \dfrac{\partial y}{\partial r} & \dfrac{\partial y}{\partial \theta} \end{vmatrix} = \begin{vmatrix} \cos\theta & -r\sin\theta \\ \sin\theta & r\cos\theta \end{vmatrix} = r(\cos^2\theta + \sin^2\theta) = r$$

This, $$\iint_D f(x, y)\, dx\, dy = \iint_{D^*} f[r\cos\theta, r\sin\theta]\, r\, dr\, d\theta$$

Note : The change of Cartesian form to polar form is useful, most of the times, when the region of integration is a circle or a part of a circle.

Example 1. Evaluate $\iint xy\left(x^2 + y^2\right)^{3/2} dxdy$ over the positive quadrant of the circle $x^2 + y^2 = 1$.

Solution : Let us change the variables to polar coordinates. Thus,

$$\begin{aligned} \iint_R xy\left(x^2+y^2\right)^{3/2} dxdy &= \iint_{R^*} r\cos\theta\, r\sin\theta \left(r^2\right)^{3/2} r\, dr\, d\theta \\ &= \iint_{R^*} r^6 \cos\theta \sin\theta\, dr\, d\theta \end{aligned}$$

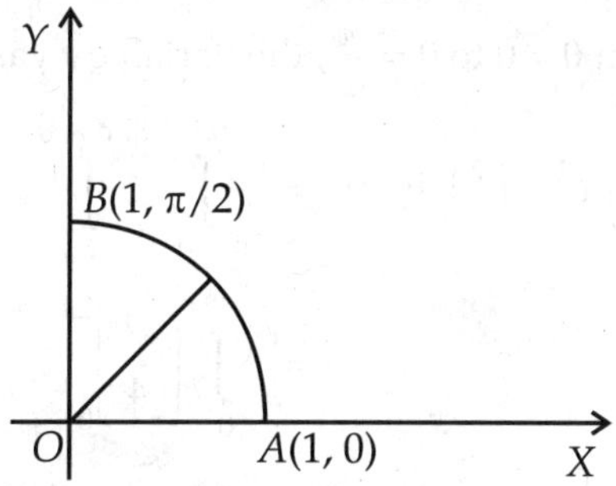

The region R of integration is the region OAB.

In this region r varies from 0 to 1, correspondingly the variable θ varies from $\theta° = 0$ to $\theta = \frac{\pi}{2}$. Thus,

$$\iint_R xy\left(x^2+y^2\right)^{3/2} dx\,dy = \int_{\theta=0}^{\pi/2}\int_{r=0}^{1} r^6 \cos\theta \sin\theta \, dr\, d\theta$$

$$= \int_0^{\pi/2}\left[\int_0^1 r^6 \sin\theta\cos\theta\, dr\right] d\theta$$

$$= \int_0^{\pi/2} \sin\theta\cos\theta \left[\frac{r^7}{7}\right]_0^1 d\theta$$

$$= \int_0^{\pi/2} \frac{1}{7}\sin\theta\cos\theta\, d\theta$$

$$= \frac{1}{14}\int_0^{\pi/2} \sin 2\theta\, d\theta = \frac{1}{14}\left[-\frac{\cos 2\theta}{2}\right]_0^{\pi/2} = -\frac{1}{28}[-1-1] = \mathbf{\frac{1}{14}}$$

Example 2. Evaluate $\int_0^{2a}\int_0^{\sqrt{2ax-x^2}} \left(x^2+y^2\right) dy\, dx$ by changing into polar coordinates.

Solution : In the given integral x varies from 0 to $2a$ and correspondingly y varies from 0 to $y = \sqrt{2ax - x^2}$, i.e., $x^2 + y^2 - 2ax = 0$, which is a circle with centre $(a, 0)$ and radius a.

Now, putting $x = r\cos\theta$, $y = r\sin\theta$ in $x^2 + y^2 - 2ax = 0$ gives us

$$r^2 - 2ar\cos\theta = 0 \;\Rightarrow\; r = 2a\cos\theta$$

The region of integration in polar coordinates is the positive quadrant of the circle $r = 2a\cos\theta$.

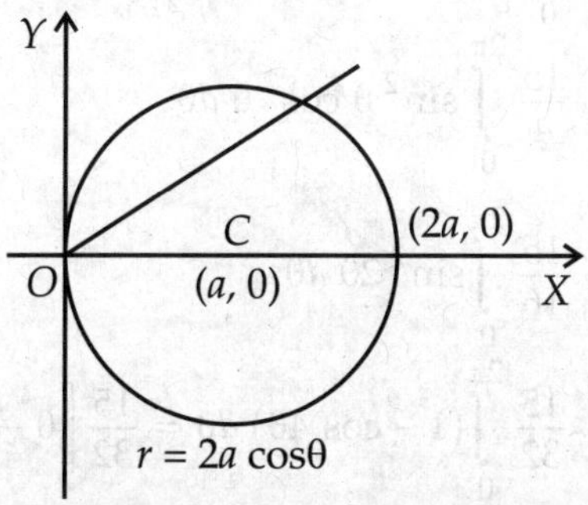

Now as θ varies from $\theta = 0$ to $\theta = \frac{\pi}{2}$, the variable r varies from $r = 0$ to $r = 2a\cos\theta$.

$$\text{Thus, } \int_0^{2a}\int_0^{\sqrt{2ax-x^2}} \left(x^2+y^2\right) dy\, dx = \int_{\theta=0}^{\pi/2}\int_{r=0}^{2a\cos\theta} r^2 \cdot r\, dr\, d\theta$$

$$= \int_0^{\pi/2} \left[\frac{r^4}{4}\right]_0^{2a\cos\theta} d\theta$$

$$= 4a^4 \int_0^{\pi/2} \cos^4\theta\, d\theta = 4a^4 \cdot \frac{3\cdot 1}{4\cdot 2}\cdot\frac{\pi}{2} = \frac{3}{4}a^4\pi$$

Example 3. Evaluate $\iint_D \frac{x^2y^2}{x^2+y^2}\, dxdy$, where D is the annular region between the circles $x^2 + y^2 = 2$ and $x^2 + y^2 = 1$.

Solution : We shall change the variables to polar coordinates.

The region of integration is as shown in the figure.

In the region of integration, θ varies from 0 to 2π. For each value of θ, the variable r varies from $r = 1$ to $r = 2$.

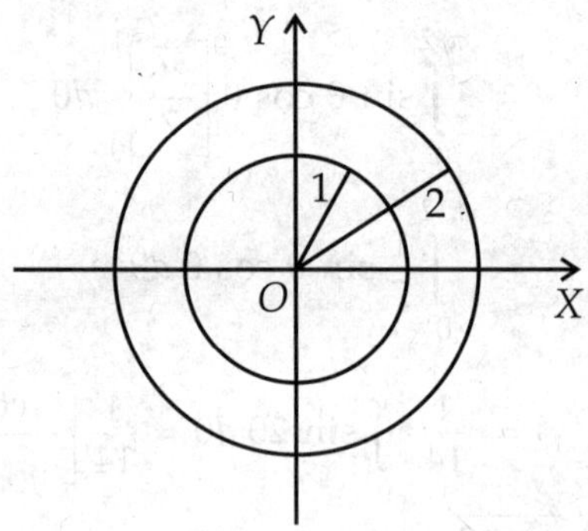

Thus we have,

$$\iint_D \frac{x^2y^2}{x^2+y^2}\, dx\, dy = \int_{\theta=0}^{2\pi}\int_{r=1}^{2} \frac{r^4\cos^2\theta\sin^2\theta}{r^2}\, r\, dr\, d\theta$$

$$= \int_0^{2\pi}\left[\int_1^2 r^3\cos^2\theta\sin^2\theta\, dr\right] d\theta$$

$$= \int_0^{2\pi}\left[\cos^2\theta\sin^2\theta\cdot\frac{r^4}{4}\right]_1^2 d\theta$$

$$= \frac{15}{4}\int_0^{2\pi}\sin^2\theta\cos^2\theta\, d\theta$$

$$= \frac{15}{16}\int_0^{2\pi}\sin^2 2\theta\, d\theta$$

$$= \frac{15}{32}\int_0^{2\pi}(1-\cos 4\theta)\, d\theta = \frac{15}{32}\left[\theta - \frac{1}{4}\sin 4\theta\right]_0^{2\pi} = \frac{15}{32}[2\pi] = \mathbf{\frac{15}{16}\pi}$$

Example 4. Evaluate $\int_0^\infty \int_0^\infty e^{-(x^2+y^2)}\,dx\,dy$ by changing into polar coordinates.

Solution : The region of integration in the given integral is whole of first quadrant.

In this region, r varies from 0 to ∞ and correspondingly θ varies from $\theta = 0$ to $\theta = \frac{\pi}{2}$.

Thus,
$$\int_0^\infty \int_0^\infty e^{-(x^2+y^2)}\,dx\,dy = \int_{\theta=0}^{\pi/2} \int_{r=0}^{\infty} e^{-r^2} r\,dr\,d\theta$$
$$= \int_0^{\pi/2} \left[\int_{r=0}^{\infty} re^{-r^2} dr \right] d\theta$$
$$= \int_0^{\pi/2} \left[-\frac{1}{2} \int_0^{\infty} e^{-r^2} (-2r)\,dr \right] d\theta$$
$$= \int_0^{\pi/2} \left[-\frac{1}{2} e^{-r^2} \right]_0^{\infty} d\theta = \int_0^{\pi/2} \frac{1}{2}\,d\theta = \frac{1}{2}[\theta]_0^{\pi/2} = \frac{\pi}{4}$$

Example 5. Evaluate $\int_0^a \int_y^a \frac{x^2}{\sqrt{x^2+y^2}}\,dxdy$, by changing to polar coordinates.

Solution : In the given integral, y varies from 0 to a and x varies from y to a.

The region R of integration is bounded by $y = 0$, $y = a$, $x = y$ and $x = a$. The region R is OAB as shown in the figure.

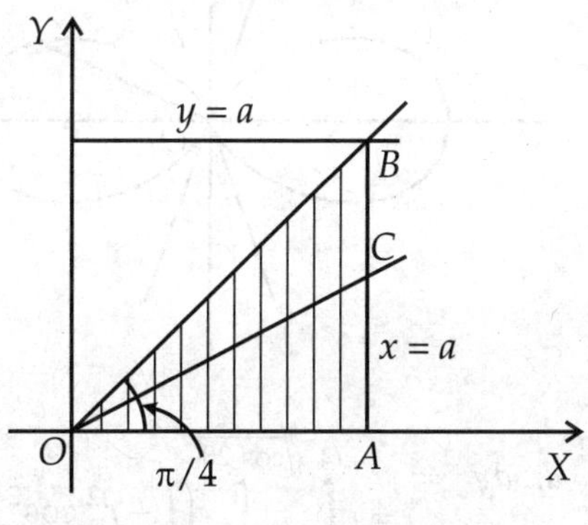

Here θ varies from 0 to $\frac{\pi}{4}$. For each value of θ, the variable r varies from the origin O to a point C on the line $x = a$. Let (r, θ) be the polar coordinates of C. Then $OC = r$ and $A\hat{O}C = \theta$.

Now, $\cos A\hat{O}C = \frac{OA}{OC} \Rightarrow \cos\theta = \frac{a}{OC} \Rightarrow OC = a\sec\theta$

Thus r varies from $r = 0$ to $r = a\sec\theta$. Thus,
$$\int_0^a \int_y^a \frac{x^2}{\sqrt{x^2+y^2}}\,dx\,dy = \int_{\theta=0}^{\pi/4} \int_{r=0}^{a\sec\theta} \frac{r^2\cos^2\theta}{r}\, r\,dr\,d\theta$$

$$= \int_0^{\pi/4} \left[\int_0^{a \sec\theta} r^2 \cos^2\theta \, dr \right] d\theta$$

$$= \int_0^{\pi/4} \cos^2\theta \left[\frac{r^3}{3} \right]_0^{a \sec\theta} d\theta$$

$$= \int_0^{\pi/4} \frac{a^3}{3} \sec\theta \, d\theta$$

$$= \frac{a^3}{3} \left[\log(\sec\theta + \tan\theta) \right]_0^{\pi/4} d\theta = \frac{a^3}{3} \log\left(\sqrt{2} + 1\right)$$

Example 6. Evaluate $\iint_D \left(1 + x^2 + y^2\right)^{-1/2} dx\,dy$, where D is the interior of lemniscate $r^2 = \cos 2\theta$.

Solution : The region of integration is as shown in the figure. The curve $r^2 = \cos 2\theta$ is symmetric about x-axis and y-axis. Thus,

$$\iint_D \left(1 + x^2 + y^2\right)^{-1/2} dx\,dy = 4 \iint_{D_1} \left(1 + x^2 + y^2\right)^{-1/2} dx\,dy$$

where D_1 is the region bounded by the curve in the first quadrant.

Clearly θ varies from 0 to $\frac{\pi}{4}$ and for each value of θ the variable r varies from $r = 0$ to $r = \sqrt{\cos 2\theta}$.

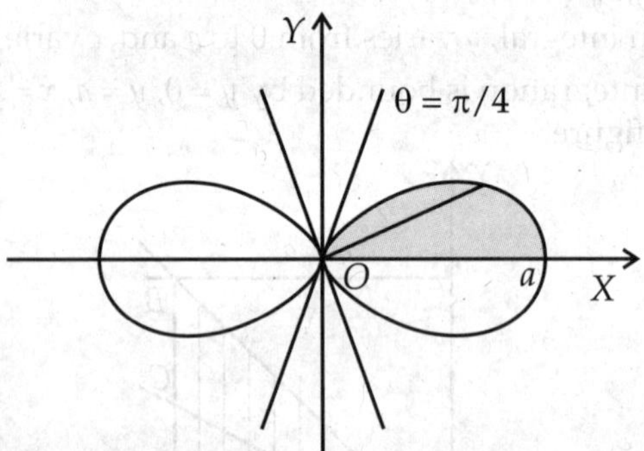

Thus,

$$\iint_D \left(1 + x^2 + y^2\right)^{-1/2} dx\,dy = 4 \int_0^{\pi/4} \int_0^{\sqrt{\cos 2\theta}} \left(1 + r^2 \cos^2\theta + r^2 \sin^2\theta\right)^{-1/2} r\,dr\,d\theta$$

$$= 4 \int_0^{\pi/4} \int_0^{\sqrt{\cos 2\theta}} \left(1 + r^2\right)^{-1/2} r\,dr\,d\theta$$

$$= 4 \int_0^{\pi/4} \left[\int_0^{\sqrt{\cos 2\theta}} \frac{1}{2\sqrt{1 + r^2}} (2r\,dr) \right] d\theta$$

$$= 4 \int_0^{\pi/4} \left[\sqrt{1 + r^2} \right]_0^{\sqrt{\cos 2\theta}} d\theta$$

$$= 4 \int_0^{\pi/4} \left[\sqrt{1 + \cos 2\theta} - 1\right] d\theta$$

$$= 4 \int_0^{\pi/4} \left[\sqrt{2} \cos \theta - 1\right] d\theta$$

$$= 4\left[\sqrt{2} \sin \theta - \theta\right]_0^{\pi/4} = 4\left[\sqrt{2} \cdot \frac{1}{\sqrt{2}} - \frac{\pi}{4}\right] = \mathbf{4\left(1 - \frac{\pi}{4}\right)}$$

Example 7. Evaluate $\iint_R x^2\, dx\, dy$, where R is the region bounded by the ellipse $\frac{x^2}{a^2} + \frac{y^2}{b^2} = 1$.

Solution : The parametric equations of the ellipse $\frac{x^2}{a^2} + \frac{y^2}{b^2} = 1$ are $x = a \cos \theta$, $y = b \sin \theta$, with $0 \le \theta \le 2\pi$.

Let us use the transformation $x = au \cos t$ and $y = bu \sin t$, with $0 \le u \le 1$, $0 \le t \le 2\pi$.

Now, $$I = \frac{\partial(x, y)}{\partial(u, v)} = \begin{vmatrix} a \cos t & -au \sin t \\ b \sin t & bu \cos t \end{vmatrix} = abu$$

$$\therefore \quad \iint_R x^2\, dx\, dy = \int_{\theta=0}^{2\pi} \int_{u=0}^{1} a^2 u^2 \cos^2 \theta \cdot abu\, du\, d\theta$$

$$= \int_0^{2\pi} \left[\int_0^1 a^3 b u^3 \cos^2 \theta\, du\right] d\theta$$

$$= \int_0^{2\pi} \left(a^3 b \cos^2 \theta\right) \left[\frac{u^4}{4}\right]_0^1 d\theta$$

$$= \int_0^{2\pi} \frac{a^3 b}{4} \cos^2 \theta\, d\theta$$

$$= \frac{a^3 b}{8} \int_0^{2\pi} (1 + \cos 2\theta)\, d\theta = \frac{a^3 b}{8}\left[\theta + \frac{1}{2} \sin 2\theta\right]_0^{2\pi} = \mathbf{\frac{a^3 b}{4} \pi}$$

EXERCISES

1. Evaluate the following integrals by changing them into polar coordinates.

(a) $\int_0^a \int_0^{\sqrt{a^2 - x^2}} \sqrt{a^2 - x^2 - y^2}\, dy\, dx$ **(b)** $\int_0^a \int_0^{\sqrt{a^2 - x^2}} x^2\, dy\, dx$ **(c)** $\int_0^a \int_0^{\sqrt{a^2 - x^2}} y^2 \sqrt{x^2 + y^2}\, dy\, dx$

(d) $\int_0^{2a} \int_0^{\sqrt{2ax - x^2}} x^2\, dy\, dx$ **(e)** $\int_0^a \int_y^a \frac{x^2}{\sqrt{x^2 + y^2}}\, dx\, dy$ **(f)** $\int_0^1 \int_0^x \sqrt{x^2 + y^2}\, dy\, dx$

(g) $\int_0^1 \int_0^x \left(x^2 + y^2\right)^{3/2} dy\, dx$ **(h)** $\int_0^\infty \int_0^\infty e^{-(x^2 + y^2)}\, dx\, dy$

(i) $\iint_{\varphi} \frac{x^2y^2}{x^2+y^2}\, dx\, dy$, outside the circle $x^2 + y^2 = a^2$ and inside the circle $x^2 + y^2 = b^2$, $b > a$.

2. Evaluate $\iint_R \sqrt{x^2+y^2}\, dx\, dy$, where R is the region bounded by the circles $x^2 + y^2 = a^2$, $x^2 + y^2 = b^2$, $a < b$.

3. Evaluate $\iint_R \frac{1-x^2-y^2}{1+x^2+y^2}\, dx\, dy$, where R is the region bounded by the circle $x^2 + y^2 = 1$.

4. Evaluate $\iint_R e^{-(x^2+y^2)}\, dx\, dy$, where R is the region bounded by the circle $x^2 + y^2 = a^2$.

5. Evaluate $\iint_R \frac{xy(x+y)^2}{x^2+y^2}\, dx\, dy$, where R is the region bounded by $y = 0$, $y = x$ and $x^2 + y^2 = a^2$.

6. Evaluate $\iint r^3\, dr\, d\theta$, the region bounded between the circles $r = 2 \sin \theta$ and $r = 4 \sin \theta$.

7. Evaluate $\iint \frac{r\, dr\, d\theta}{\sqrt{a^2+r^2}}$, over one loop of the lemniscate $r^2 = a^2 \cos 2\theta$.

8. Evaluate $\iint e^{-r^2/a^2}\, r \sin\theta \cos\theta\, dr\, d\theta$, over the upper half of the circle $r = a^2 \cos \theta$.

9. By changing the variables approximately, evaluate $\iint_R \left(1 - \frac{x^2}{a^2} - \frac{y^2}{b^2}\right)^{1/2} dx\, dy$, where R is the region bounded by the ellipse $\frac{x^2}{a^2} + \frac{y^2}{b^2} = 1$.

10. Evaluate $\iint_R (x+y)^2\, dx\, dy$, where R is the region bounded by the lines $x + y = 0$, $x + y = 2$, $3x - 2y = 0$, $3x - 2y = 3$. **[Hint :** $x + y = u$, $3x - 2y = v$**]**

11. If R is the region bounded by $x + y = 1$, $x = 0$, $y = 0$, show that,

$$\iint_R \cos\left(\frac{x-y}{x+y}\right) dx\, dy = \frac{1}{2} \sin 1.$$

Answers

1. (a) $\frac{\pi a^3}{6}$ **(b)** $\frac{\pi a^4}{16}$ **(c)** $\frac{\pi a^5}{20}$ **(d)** $\frac{5\pi a^4}{8}$

(e) $\frac{1}{3} a^2 \log\left(1+\sqrt{2}\right)$ **(f)** $\frac{1}{6}\left[\sqrt{2} + \log\left(1+\sqrt{2}\right)\right]$

(g) $\frac{1}{40}\left[7\sqrt{2} + 3\log\left(\sqrt{2}+1\right)\right]$ **(h)** $\frac{\pi}{4}$ **(i)** $\frac{1}{16}\pi\left(b^4 - a^4\right)$

2. $\frac{2}{3}\pi\left(b^3 - a^3\right)$ **3.** $\frac{\pi}{8}(\pi - 2)$ **4.** $\pi\left(1 - e^{-a^2}\right)$ **5.** $\frac{\pi + 4}{64}a^4$

6. $\frac{45}{2}\pi$ **7.** $2a\left(1 - \frac{\pi}{4}\right)$ **8.** $\frac{a^2}{16}\left(3 + e^{-4}\right)$ **9.** $\frac{2}{3}\pi ab$ **10.** $\frac{8}{5}$

3.8 Triple Integral

The triple integral is defined in a manner similar to that of double integral.

Let $f(x, y, z)$ be a continuous and single valued function defined over a region V of space. Divide the region V into subregions of volumes ΔV_i, i = 1, 2, 3, ... Let $P_i(x_i, y_i, z_i)$ be any point in the region ΔV_i, and the value of the function $f(x, y, z)$ at P_i is $f(x_i, y_i, z_i)$.

The triple integral of $f(x, y, z)$ over the region V is defined and denoted by

$$\lim_{n \to \infty} \sum_{i=1}^{n} f(x_i, y_i, z_i)\,\Delta V_i = \iiint_V f(x, y, z)\,dx\,dy\,dz$$

provided the left hand side limit exists.

As in the case of double integrals, the triple integrals are also found by evaluating the corresponding iterated integrals.

$$\iiint_V f(x, y, z)\,dx\,dy\,dz = \int_{z_1}^{z_2}\left[\int_{f_1(z)}^{f_2(z)}\left[\int_{g_1(y,z)}^{g_2(y,z)} f(x, y, z)\,dx\right]dy\right]dz$$

The limits of integration for x, y, z are suitably chosen that they define the region V which is under consideration.

When $f(x, y, z) = 1$, the corresponding integral

$$\iiint_V f(x, y, z)\,dx\,dy\,dz$$

gives us the **volume of the region V in the space.**

Example 1. Evaluate $\int_0^1\int_0^2\int_1^2 x^2yz\,dx\,dy\,dz$.

Solution :

$$\int_0^1\int_0^2\int_1^2 x^2yz\,dx\,dy\,dz = \int_0^1\int_0^2\left[\frac{x^3yz}{3}\right]_1^2 dy\,dz$$

$$= \int_0^1\left[\int_0^2\frac{1}{3}(8yz - yz)\,dy\right]dz$$

$$= \int_0^1\left[\frac{1}{3}\left(4zy^2 - \frac{zy^2}{2}\right)\right]_0^2 dz$$

$$= \int_0^1\frac{1}{3}(16z - 2z)\,dz = \frac{1}{3}\left[\left(8z^2 - z^2\right)\right]_0^1 = \frac{7}{3}$$

Example 2. Evaluate $\int_0^1\int_0^1\int_0^1 e^{x+y+z}\,dx\,dy\,dz$.

Solution : $\int_0^1\int_0^1\int_0^1 e^{x+y+z}\,dx\,dy\,dz = \int_0^1\int_0^1\left[\int_0^1 e^{x+y+z}\,dx\right]dy\,dz$

$$= \int_0^1\int_0^1\left[e^{x+y+z}\,dx\right]_0^1 dy\,dz$$

$$= \int_0^1\left[\int_0^1\left(e^{1+y+z} - e^{y+z}\right)dy\right]dz$$

$$= \int_0^1\left[e^{1+y+z} - e^{y+z}\right]_0^1 dz$$

$$= \int_0^1\left(e^{2+z} - e^{1+z} - e^{1+z} + e^{z}\right)dz$$

$$= \int_0^1\left(e^{2+z} - 2e^{1+z} + e^{z}\right)dz$$

$$= \left[e^{2+z} - 2e^{1+z} + e^{z}\right]_0^1 = e^3 - 3e^2 + 3e - 1 = \mathbf{(e-1)^3}$$

Example 3. Evaluate $\iiint_R (x - 2y + z)\,dx\,dy\,dz$**, where** R **is the region determined by** $0 \le x \le 1,\quad 0 \le y \le x^2,\quad 0 \le z \le x + y$.

Solution : $\iiint_R (x - 2y + z)\,dx\,dy\,dz = \int_0^1\int_0^{x^2}\int_0^{x+y}(x - 2y + z)\,dz\,dy\,dx$

$$= \int_0^1\int_0^{x^2}\left[\int_0^{x+y}(x-2y)z + \frac{z^2}{2}\right]_0^{x+y} dy\,dx$$

$$= \int_0^1\int_0^{x^2}\left[(x-2y)(x+y) + \frac{(x+y)^2}{2}\right]dy\,dx$$

$$= \int_0^1\int_0^{x^2}\left[x^2 - xy - 2y^2 + \frac{1}{2}x^2 + xy + \frac{y^2}{2}\right]dy\,dx$$

$$= \int_0^1\int_0^{x^2}\left[\frac{3}{2}x^2 - \frac{3}{2}y^2\right]dy\,dx$$

$$= \int_0^1\left[\frac{3}{2}x^2y - \frac{y^3}{2}\right]_0^{x^2} dx$$

$$= \int_0^1\left[\frac{3}{2}x^4 - \frac{x^6}{2}\right]dx = \left[\frac{3x^5}{10} - \frac{x^7}{14}\right]_0^1 = \frac{3}{10} - \frac{1}{14} = \mathbf{\frac{8}{35}}$$

Example 4. Evaluate $I = \iiint \frac{dx\,dy\,dz}{(x+y+z+1)^3}$ **over the region bounded by the coordinate planes and the plane** $x + y + z = 1$.

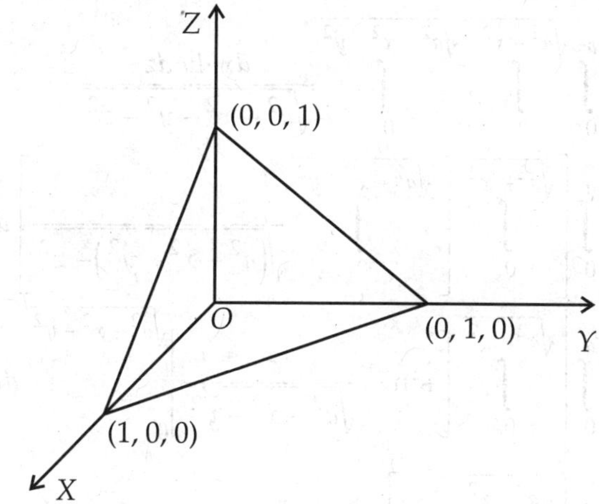

Solution : The region of integration is tetrahedron bounded by the planes $x = 0, y = 0, z = 0$ and the plane $x + y + z = 1$.

Clearly z varies from 0 to $z = 1 - x - y$. For $x = 0$, y varies from $y = 0$ to $y = 1 - x$. Now for $y = 0$ and $z = 0$, the variable x varies from $x = 0$ to $x = 1$.

$$I = \iiint \frac{dx\,dy\,dz}{(x+y+z+1)^3} = \int_0^1 \int_0^{1-x} \int_0^{1-x-y} \frac{1}{(x+y+z+1)^3}\,dz\,dy\,dx$$

$$= \int_0^1 \int_0^{1-x} \left[\frac{-1}{2(x+y+z+1)^2}\right]_0^{1-x-y} dy\,dx$$

$$= \int_0^1 \int_0^{1-x} \left(\frac{1}{2(x+y+1)^2} - \frac{1}{8}\right) dy\,dx$$

$$= \int_0^1 \left[\frac{-1}{2(x+y+1)} - \frac{y}{8}\right]_0^{1-x} dx$$

$$= \int_0^1 \left[-\frac{1}{4} - \frac{1-x}{8} + \frac{1}{2(1+x)}\right] dx$$

$$= \left[-\frac{x}{4} - \frac{1}{8}x + \frac{x^2}{16} + \frac{1}{2}\log(1+x)\right]_0^1$$

$$= -\frac{1}{4} - \frac{1}{8} + \frac{1}{16} + \frac{1}{2}\log 2 = \mathbf{\log 4 - \frac{5}{16}}$$

Example 5. Evaluate $I = \int_0^a \int_0^{\sqrt{a^2-x^2}} \int_0^{\sqrt{a^2-x^2-y^2}} \frac{dx\,dy\,dz}{\sqrt{a^2-x^2-y^2-z^2}}$.

Solution : Now, $I = \int_0^a \int_0^{\sqrt{a^2-x^2}} \int_0^{\sqrt{a^2-x^2-y^2}} \frac{dx\,dy\,dz}{\sqrt{a^2-x^2-y^2-z^2}}$

$$= \int_0^a \left[\int_0^{\sqrt{a^2-x^2}} \left[\int_0^{\sqrt{a^2-x^2-y^2}} \frac{dz}{\sqrt{(a^2-x^2-y^2)-z^2}} \right] dy \right] dx$$

$$= \int_0^a \left[\int_0^{\sqrt{a^2-x^2}} \left[\sin^{-1} \frac{z}{\sqrt{a^2-x^2-y^2}} \right]_0^{\sqrt{a^2-x^2-y^2}} dy \right] dx$$

$$= \int_0^a \left[\int_0^{\sqrt{a^2-x^2}} \frac{\pi}{2}\, dy \right] dx$$

$$= \int_0^a \left[\frac{\pi}{2} y \right]_0^{\sqrt{a^2-x^2}} dx = \frac{\pi}{2} \int_0^a \sqrt{a^2-x^2}\, dx$$

$$= \frac{\pi}{2} \left[\frac{x}{2}\sqrt{a^2-x^2} + \frac{a^2}{2} \sin^{-1}\frac{x}{a} \right]_0^a = \frac{\pi}{2}\left[\frac{a^2}{2} \cdot \frac{\pi}{2} \right] = \frac{\pi^2 a^2}{8}$$

Example 6. Find the volume of the tetrahedron formed by the planes $x = 0$, $y = 0$, $z = 0$ and $6x + 4y + 3z = 12$.

Solution : The volume of the region V in the space is given by, $\iiint_V dx\,dy\,dz$.

By data in the given region, z varies from $z = 0$ to $z = \frac{1}{3}(12 - 6x - 4y)$. Now for $x = 0$ and $z = 0$, y varies from $y = 0$ to $y = 3\left(1 - \frac{x}{2}\right)$. For $y = 0$ and $z = 0$ the variable x varies from $x = 0$ to $x = 2$.

Thus, Volume $= \int_0^2 \int_0^{3(1-x/2)} \int_0^{1/3(12-6x-4y)} dz\,dy\,dz$

$$= \int_0^2 \int_0^{3(1-x/2)} [z]_0^{1/3(12-6x-4y)}\, dy\,dx$$

$$= \int_0^2 \left[\int_0^{3(1-x/2)} \frac{1}{3}(12-6x-4y)\, dy \right] dx$$

$$= \int_0^2 \left[\frac{1}{3}\left(12y - 6xy - 2y^2\right)\right]_0^{3(1-x/2)} dx$$

$$= \int_0^2 \frac{1}{3}\left[36\left(1-\frac{x}{2}\right) - 18x\left(1-\frac{x}{2}\right) - 18\left(1-\frac{x}{2}\right)^2\right] dx$$

$$= \int_0^2 \frac{1}{3}\left[18 - 18x + \frac{9}{2}x^2\right] dx$$

$$= \int_0^2 \frac{1}{3}18\left[1 - x + \frac{x^2}{4}\right] dx$$

$$= 6\int_0^2 \left(1-\frac{x}{2}\right) dx = 6\left[\frac{-2\left(1-\frac{x}{2}\right)^3}{3}\right]_0^2 = -4(0-1) = \mathbf{4}$$

Example 7. Find the volume common to the cylinders $x^2 + y^2 = a^2$ and $x^2 + z^2 = a^2$.

Solution : Here,

z varies from $-\sqrt{a^2 - x^2}$ to $\sqrt{a^2 - x^2}$, y varies from $-\sqrt{a^2 - x^2}$ to $\sqrt{a^2 - x^2}$,
x varies from $-a$ to a.

Thus, Volume $= \int_{-a}^{a} \int_{-\sqrt{a^2-x^2}}^{\sqrt{a^2-x^2}} \int_{-\sqrt{a^2-x^2}}^{\sqrt{a^2-x^2}} dz\, dy\, dx$

$$= \int_{-a}^{a} \int_{-\sqrt{a^2-x^2}}^{\sqrt{a^2-x^2}} \left[\, z \,\right]_{-\sqrt{a^2-x^2}}^{\sqrt{a^2-x^2}} dy\, dx$$

$$= \int_{-a}^{a} \int_{-\sqrt{a^2-x^2}}^{\sqrt{a^2-x^2}} 2\sqrt{a^2-x^2}\, dy\, dx$$

$$= \int_{-a}^{a} \left[2\sqrt{a^2-x^2} \cdot y\right]_{-\sqrt{a^2-x^2}}^{\sqrt{a^2-x^2}} dy\, dx$$

$$= \int_{-a}^{a} 2\sqrt{a^2-x^2}\left[2\sqrt{a^2-x^2}\right] dx$$

$$= 4\int_{-a}^{a} \left(a^2 - x^2\right) dx$$

$$= 4\left[a^2x - \frac{x^3}{3}\right]_{-a}^{a} = 4\left[a^3 - \frac{a^3}{3} + a^3 - \frac{a^3}{3}\right] = \mathbf{\frac{16}{3}}\,a^3$$

EXERCISES

1. Evaluate the following triple integrals.

(a) $\int_0^1\int_0^2\int_1^2 x^2yz\,dz\,dy\,dx$ (b) $\int_1^2\int_1^2\int_1^2\left(\frac{x}{y}+\frac{y}{z}+\frac{z}{x}\right)dx\,dy\,dz$

(c) $\int_0^1\int_0^1\int_0^y xyz\,dx\,dy\,dz$ (d) $\int_1^2\int_0^1\int_{-1}^1\left(x^2+y^2+z^2\right)dz\,dy\,dx$

(e) $\int_0^2\int_1^3\int_1^2 xy^2z\,dx\,dy\,dz$ (f) $\int_0^1\int_{x^2}^{\sqrt{x}}\int_0^3 dx\,dy\,dz$

(g) $\int_0^1\int_0^1\int_{\sqrt{x^2+y^2}}^2 xyz\,dz\,dy\,dx$ (h) $\int_0^1\int_0^{1-x}\int_0^{1-x-y} dx\,dy\,dz$

(i) $\int_0^a\int_0^x\int_0^{x+y} e^{x+y+z}\,dx\,dy\,dz$ (j) $\int_0^1\int_0^{1-x}\int_0^{1-x-y} xyz\,dz\,dy\,dx$

(k) $\int_0^1\int_0^{1-x}\int_0^{1-y^2} z\,dz\,dy\,dx$ (l) $\int_0^1\int_0^{a-x}\int_0^{a-x-y}\left(x^2+y^2+z^2\right)dz\,dy\,dx$

(m) $\int_0^1\int_0^{1-x}\int_0^{1-x-y}\frac{dx\,dy\,dz}{(1+x+y+z)^3}$ (n) $\int_0^{\pi/2}\int_0^{a\sin\theta}\int_0^{\frac{a^2-r^2}{a}} r\,dz\,dr\,d\theta$

(o) $\int_0^1\int_0^{\sqrt{1-x^2}}\int_0^{\sqrt{a^2-x^2-y^2}} xyz\,dx\,dy\,dz$ (p) $\int_0^a\int_0^{\sqrt{a^2-x^2}}\int_0^{\sqrt{a^2-x^2-y^2}}\frac{dx\,dy\,dz}{\sqrt{a^2-x^2-y^2-z^2}}$

2. Evaluate $\iiint_D xy^2z^3\,dx\,dy\,dz$, where D is given by $0 < x < 10, 0 < y < 1, 0 < z < 1$.

3. Evaluate $\iiint_D (x+y+z)\,dx\,dy\,dz$, where D is the region bounded by the planes $x = 0, y = 0, z = 0$ and $x + y + z = 1$.

4. Evaluate $\iiint_R \frac{1}{xyz}\,dx\,dy\,dz$, where R is given by $1 < x < 2, -2 < y < -1, 2 < z < 3$.

5. If R is the region bounded by the planes $x = 0, y = 0, z = 0$ and $x + y + z = 1$ evaluate $\iiint_R z\,dx\,dy\,dz$.

6. If D is the region determined by the inequality $x > 0, y > 0, z < 4$ and $z > x^2 + y^2$ then evaluate $\iiint_D 2x\,dx\,dy\,dz$.

7. Evaluate $\iiint_D \frac{dx\,dy\,dz}{(1+x+y+z)^3}$ where D is the region bounded by $x = 0, y = 0$ and $x + y + z = 1$.

8. Evaluate $\iiint\limits_{D}$ where D is the positive octant in the sphere $x^2 + y^2 + z^2 = 1$.

9. Find the volume of the region bounded by the surfaces $x^2 + y^2 + z^2 = a^2$ and $x^2 + y^2 = ax$.

10. Find the volume inside $x^2 + y^2 = 9$ above $z = 0$ below $x + z = 4$.

11. Find the volume of the region bounded by the surface $z = x^2 + y^2$ and the plane $z = 2x$.

12. Find the volume of the tetrahedron bounded by the coordinate plane $\frac{x}{a} + \frac{y}{b} + \frac{z}{c} = 1$.

Answers

1. **(a)** 1 **(b)** $9 \log \sqrt{2}$ **(c)** $\frac{1}{16}$ **(d)** 6

(e) 26 **(f)** 1 **(g)** $\frac{3}{8}$ **(h)** $\frac{1}{6}$

(i) $\frac{1}{8}\left(e^{4a} - 6e^{2a} + 8e^{a} - 3\right)$ **(j)** $\frac{1}{720}$ **(k)** $\frac{11}{20}$

(l) $\frac{a^5}{20}$ **(m)** $\frac{1}{2}\log\left(2 - \frac{5}{8}\right)$ **(n)** $-\frac{3}{64}\pi a^3$ **(o)** $\frac{1}{48}$

(p) $\frac{\pi a^2}{8}$

2. $\frac{1}{24}$ **3.** $\frac{1}{8}$ **4.** $\log\left(\frac{3}{2}\right)\cdot(\log 2)^2$ **5.** $\frac{1}{24}$

6. $\frac{128}{15}$ **7.** $\frac{1}{2}\log 2 - \frac{5}{16}$ **8.** $\frac{1}{48}$ **9.** $\frac{2}{3}\pi a^3$

10. 36π **11.** $\frac{\pi}{2}$ **12.** $\frac{abc}{6}$

3.9 Change of Variables in the Triple Integral

A triple integral

$$\iiint\limits_{R} f(x, y, z)\, dx\, dy\, dz$$

can be evaluated by changing the variables x, y, z to some new variables u, v, w which are related to x, y, z. In this case if

$$J = \frac{\partial(x, y, z)}{\partial(u, v, w)} \neq 0$$

then it can be shown

$$\iiint\limits_{R} f(x, y, z)\, dx\, dy\, dz = \iiint\limits_{D} \varphi(u, v, w)\,|J|\, du\, dv\, dw$$

The variables x, y, z vary in the region R and the variables u, v, w vary over the corresponding region D.

We shall consider two particular transformation.

1. Cylindrical Polar Coordinates

Let (x, y, z) be related to three variables (R, φ, z) by the relations

$$x = R\cos\varphi, \quad y = R\sin\varphi, \quad z = z$$

The coordinates (R, φ, z) are called **cylindrical polar coordinates** of the point (x, y, z).

In this case

$$J = \frac{\partial(x, y, z)}{\partial(R, \varphi, z)} = \begin{vmatrix} \cos\varphi & -R\sin\varphi & 0 \\ \sin\varphi & R\cos\varphi & 0 \\ 0 & 0 & 1 \end{vmatrix} = R$$

Thus we have,

$$\iiint_R f(x, y, z)\,dx\,dy\,dz = \iiint_D \varphi(R, \varphi, z)\,R\,dR\,d\varphi\,dz$$

D is the region in which (R, φ, z) vary as (x, y, z) vary over R, where,

$$\varphi(R, \varphi, z) = f(R\cos\varphi, R\sin\varphi, z)$$

2. Spherical Polar Coordinates

Let (x, y, z) be related to three variables (ρ, θ, φ) by the relations

$$x = \rho\sin\theta\cos\varphi, \quad y = \rho\sin\theta\sin\varphi, \quad z = \rho\cos\theta$$

The coordinates (ρ, θ, φ) are called **spherical polar coordinates** of the point (x, y, z).

In this case we have,

$$J = \frac{\partial(x, y, z)}{\partial(\rho, \theta, \varphi)} = \begin{vmatrix} \sin\theta\cos\varphi & \rho\cos\theta\cos\varphi & -\rho\sin\theta\sin\varphi \\ \sin\theta\sin\varphi & \rho\cos\theta\sin\varphi & \rho\sin\theta\cos\varphi \\ \cos\theta & -\rho\sin\theta & 0 \end{vmatrix} = \rho^2\sin\theta$$

Thus we have,

$$\iiint_R f(x, y, z)\,dx\,dy\,dz = \iiint_D \varphi(\rho, \theta, \varphi)\,\rho^2\sin\theta\,d\rho\,d\theta\,d\varphi$$

D is the region in which (ρ, θ, φ) vary as (x, y, z) vary over R. Here,

$$\phi(\rho, \theta, \varphi) = f[\rho\sin\theta\cos\theta, \rho\sin\theta\sin\varphi, \rho\cos\theta]$$

Example 1. Evaluate $\iiint_R xyz\,dx\,dy\,dz$ over the positive octant of the sphere $x^2 + y^2 + z^2 = a^2$ by transforming into cylindrical polar coordinates.

Solution : In the given relation R varies from 0 to a, φ varies from 0 to $\frac{\pi}{2}$ and z varies from 0 to a, where (R, φ, z) are cylindrical polar coordinates. Thus,

$$\iiint_R xyz\,dx\,dy\,dz = \frac{1}{3}\int_0^a\int_0^{\pi/2}\int_0^a R\cos\varphi\cdot R\sin\varphi\cdot z\,R\,dR\,d\varphi\,dz$$

$$= \frac{1}{3}\int_0^a\int_0^{\pi/2}\left[\int_0^a R^3\sin\varphi\cos\varphi\,z\,dR\right]d\varphi\,dz$$

$$= \frac{1}{3}\int_0^a\int_0^{\pi/2} z\left[\frac{R^4}{4}\sin\varphi\cos\varphi\right]_0^a d\varphi\,dz$$

$$= \frac{1}{3}\int_0^a\int_0^{\pi/2} z\left[\frac{a^4}{4}\sin\varphi\cos\varphi\right]d\varphi\, dz$$

$$= \frac{1}{3}\int_0^a z\left[\frac{a^4}{4}\,\frac{\sin^2\varphi}{2}\right]_0^{\pi/2} dz$$

$$= \frac{1}{3}\int_0^a \frac{a^4}{8}\, z\, dz = \frac{1}{3}\left[\frac{a^4z^2}{16}\right]_0^a = \frac{1}{3}\frac{a^6}{16} = \frac{a^6}{48}$$

Example 2. Evaluate $\iiint_R (x+y+z)\, dx\, dy\, dz$ where R is the region in the first octant bounded by the sphere $x^2 + y^2 + z^2 = a^2$ by changing it to spherical polar coordinates.

Solution :

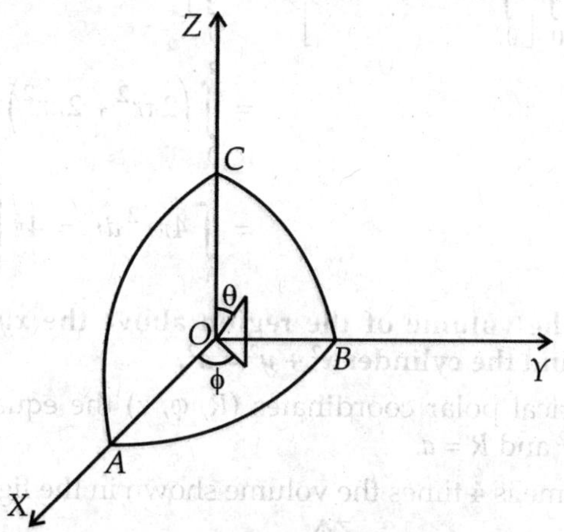

If (r, θ, φ) are the spherical coordinates, then in the given region, r varies from 0 to a, θ varies from 0 to $\frac{\pi}{2}$ and φ varies from 0 to $\frac{\pi}{2}$. Thus,

$$\iiint_R (x+y+z)\, dx\, dy\, dz$$

$$= \int_0^a\int_0^{\pi/2}\int_0^{\pi/2} (r\sin\theta\cos\varphi + r\sin\theta\sin\varphi + r\cos\theta)\cdot r^2\sin\theta\, d\varphi\, d\theta\, dr$$

$$= \int_0^a\int_0^{\pi/2}\int_0^{\pi/2} r^3\left[\sin^2\theta(\cos\varphi+\sin\varphi) + \sin\theta\cos\theta\right] d\varphi\, d\theta\, dr$$

$$= \int_0^a\int_0^{\pi/2} r^3\left[\sin^2\theta(\sin\varphi - \cos\varphi) + (\sin\theta\cos\theta)\varphi\right]_0^{\pi/2} d\theta\, dr$$

$$= \int_0^a\left[\int_0^{\pi/2} r^3\left\{2\sin^2\theta + \frac{\pi}{2}\sin\theta\cos\theta\right\} d\theta\right] dr$$

$$= \int_0^a r^3\left\{2\frac{\pi}{4} + \frac{\pi}{2}\left[\frac{\sin^2\theta}{2}\right]_0^{\pi/2}\right\} dr = \int_0^a r^3\left\{\frac{\pi}{2} + \frac{\pi}{4}\right\} dr = \frac{3}{4}\pi\left[\frac{r^4}{4}\right]_0^a = \frac{3}{16}\pi a^4$$

Example 3. Find the volume of the sphere $x^2 + y^2 + z^2 = a^2$.

Solution : Let (r, θ, φ) be the spherical coordinates of any point on the sphere.

For the sphere $x^2 + y^2 + z^2 = a^2$, the centre is $(0, 0, 0)$ and the radius is a.

Thus on the sphere, r varies from 0 to a, θ varies from 0 to π and φ varies from 0 to 2π.

Thus,

$$\iiint_R dx\, dy\, dz = \int_0^a \int_0^\pi \int_0^{2\pi} r^2 \sin\theta\, d\varphi\, d\theta\, dr$$

$$= \int_0^a \int_0^\pi \left[\left(r^2 \sin\theta\right)\varphi\right]_0^{2\pi} d\theta\, dr$$

$$= \int_0^a \left[\int_0^\pi 2\pi r^2 \sin\theta\, d\theta\right] dr = \int_0^a \left[-2\pi r^2 \cos\theta\right]_0^\pi dr$$

$$= \int_0^a \left(2\pi r^2 + 2\pi r^2\right) dr$$

$$= \int_0^a 4\pi r^2\, dr = 4\pi\left[\frac{r^3}{3}\right]_0^a = \frac{4}{3}\pi a^3$$

Example 4. Evaluate the volume of the region above the xy-plane bounded by the paraboloid $z = x^2 + y^2$ and the cylinder $x^2 + y^2 = a^2$.

Solution : In a cylindrical polar coordinates (R, φ, z) the equations of paraboloid and cylinder becomes $z = R^2$ and $R = a$.

The required volume is 4 times the volume shown in the figure.

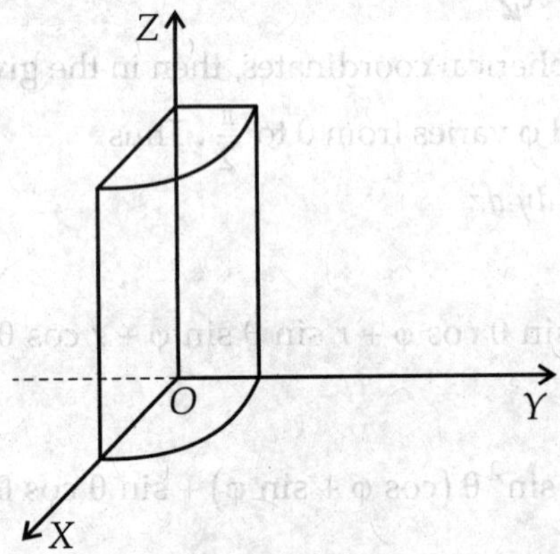

Thus,

$$4\iiint_D dx\, dy\, dz = 4\int_0^{\pi/2}\int_0^a\int_0^{x^2} r\, dz\, dr\, d\theta$$

$$= 4\int_0^{\pi/2}\int_0^a [rz]_0^{r^2}\, dr\, d\theta$$

$$= 4\int_0^{\pi/2}\left[\int_0^a r^3\, dr\right] d\theta = 4\int_0^{\pi/2}\frac{a^4}{4}\, d\theta = \frac{a^4\pi}{2}$$

EXERCISES

1. Evaluate $\int_{x=0}^{1}\int_{y=0}^{1}\int_{z=\sqrt{x^2+y^2}}^{2} xyz\, dz\, dy\, dx$ by changing into cylindrical polar coordinates.

2. Evaluate $\int_{0}^{1}\int_{0}^{\sqrt{1-x^2}}\int_{0}^{\sqrt{1-x^2-y^2}} \frac{dz\, dy\, dx}{\sqrt{1-x^2-y^2-z^2}}$ by changing into spherical polar coordinates.

3. Evaluate $\iiint_R \left(x^2+y^2+z^2\right) dx\, dy\, dz$ by changing into spherical polar coordinates, where R is the region bounded by the sphere $x^2 + y^2 + z^2 = a^2$.

4. Evaluate $\iiint_R xyz\, dx\, dy\, dz$ by changing into cylindrical polar coordinates, where R is the region bounded by the planes $x = 0$, $y = 0$, $z = 1$ and the cylinder $x^2 + y^2 = 1$.

5. Find the volume bounded by the surface $z = 4 - x^2 - y^2$ and the xy-plane by using cylindrical coordinates.

6. Find the volume bounded by the xy-plane, the cylinder $x^2 + y^2 = 1$ and the plane $x + y + z = 2$.

7. Find the volume bounded by the paraboloid $z = x^2 + y^2$ and the cone $z = \sqrt{x^2+y^2}$ by using cylindrical coordinates.

8. Find the volume of the portion of the sphere $x^2 + y^2 + z^2 = a^2$ lying inside the cylinder $x^2 + y^2 = ax$.

Answers

1. $\frac{8}{3}$	2. $\frac{\pi^2}{8}$	3. $\frac{4}{5}\pi a^5$	4. $\frac{1}{16}$
5. 8π	6. 2π	7. $\frac{\pi}{6}$	8. $\left(2-\sqrt{2}\right)\frac{\pi}{3}$

3.10 Differentiation Under the Integral Sign

Leibnitz's Rule

Let $f(x, \alpha)$ be a function, α being a parameter. In this function $f(x, \alpha)$ is integrated w.r.t. x, between the limits a and b, then resultant will be a function of α say $F(\alpha)$. That is,

$$\int_a^b f(x, \alpha)\, dx = F(\alpha)$$

We shall mention below a result without proof, called **Leibnitz's rule of differentiation under the integral sign.**

Let $f(x, \alpha)$ be a function possessing continuous first order partial derivative

$$\frac{\partial}{\partial x}\left[f(x, \alpha)\right] \text{ and } F(\alpha) = \int_a^b f(x, \alpha)\, dx$$

Then $\frac{d}{d\alpha}F(\alpha) = \int_a^b \frac{\partial}{\partial x} f(x, \alpha)\, dx$**, where a and b are constants independent of α.**

This result is useful to evaluate the definite integral when the limits of integration are constants.

Example 1. Evaluate $\int_0^1 \frac{x^a - 1}{\log x}\, dx$**, where a is a parameter.**

Solution : Let, $F(a) = \int_0^1 \frac{x^a - 1}{\log x}\, dx$ .... (1)

By Leibnitz's rule, we have,

$$F'(a) = \int_0^1 \frac{\partial}{\partial a}\left[\frac{x^a - 1}{\log x}\right] dx \quad \Rightarrow \quad F'(a) = \int_0^1 \frac{1}{\log x}\left[x^a \cdot \log x\right] dx$$

$$\Rightarrow \quad F'(a) = \int_0^1 x^a\, dx$$

$$\Rightarrow \quad F'(a) = \left.\frac{x^{a+1}}{a+1}\right]_0^1 x^a\, dx \quad \Rightarrow F'(a) = \frac{1}{a+1}$$

Integrating both sides w.r.t. a, we get,

$$F(a) = \int \frac{1}{a+1}\, da + c \quad \Rightarrow \quad F(a) = \log(a+1) + c \qquad \text{.... (2)}$$

where c is a constant of integration. To find this constant of integration put $a = 0$ in (1), we get,

$$F(0) = \int_0^1 \frac{x^0 - 1}{\log x}\, dx = 0$$

Now put $a = 0$ and $f(a) = 0$ in (2), we get, $0 = \log 1 + c \Rightarrow c = 0.$

Thus, $F(a) = \log(1 + a) \Rightarrow \int_0^1 \frac{x^a - 1}{\log x}\, dx =$ **log (1 + a)**

Example 2. By applying differentiation under the integral sign, show that

$$\int_0^{\pi/2} \frac{\log\left(1 + y\sin^2 x\right)}{\sin^2 x}\, dx = \pi\left(\sqrt{1+y} - 1\right).$$

Solution : Let, $F(y) = \int_0^{\pi/2} \frac{\log\left(1 + y\sin^2 x\right)}{\sin^2 x}\, dx$ .... (1)

Differentiating both sides w.r.t. y partially, we get

$$F'(y) = \int_0^{\pi/2} \frac{\partial}{\partial y}\left[\frac{\log\left(1 + y\sin^2 x\right)}{\sin^2 x}\right] dx$$

$$= \int_0^{\pi/2} \frac{1}{\sin^2 x}\left[\frac{\sin^2 x}{1 + y\sin^2 x}\right] dx$$

$$= \int_0^{\pi/2} \left[\frac{1}{1 + y \sin^2 x} \right] dx$$

$$= \int_0^{\pi/2} \left[\frac{1}{\cos^2 x + \sin^2 x + y \sin^2 x} \right] dx$$

$$= \int_0^{\pi/2} \left[\frac{1}{\cos^2 x + (1 + y) \sin^2 x} \right] dx$$

$$= \int_0^{\pi/2} \left[\frac{\sec^2 x}{1 + (1 + y) \tan^2 x} \right] dx \qquad \text{(dividing by } \cos^2 x\text{)}$$

$$= \int_0^{\infty} \frac{dt}{1 + (1 + y)}, \text{ where } t = \tan x \Rightarrow dt = \sec^2 x \, dx \text{ and } x = 0, t = \infty,$$

when $x = \frac{\pi}{2}$

$$= \frac{1}{1 + y} \int_0^{\infty} \frac{dt}{\left(\frac{1}{1 + y}\right) + t^2}$$

$$= \frac{1}{1 + y} \left(\sqrt{1 + y}\right) \tan^{-1} \left[t \sqrt{1 + y} \right]_0^{\infty} \qquad \left[\because \int \frac{dx}{a^2 + x^2} = \frac{1}{a} \tan^{-1} \frac{x}{a} \right]$$

$$= \frac{1}{\sqrt{1 + y}} \left[\tan^{-1} \infty - \tan^{-1} 0 \right] = \frac{\pi}{2\sqrt{1 + y}}$$

Thus we have, $F'(y) = \frac{\pi}{2\sqrt{1 + y}}$. Integrating both sides w.r.t. y, we have,

$$F(y) = \frac{\pi}{2} \cdot 2\sqrt{1 + y} + c \qquad \text{.... (2)}$$

Now, from (1), we have, $F(0) = 0$ when $y = 0$

Now, (2) $\Rightarrow$ $F(0) = \pi + c \Rightarrow c = -\pi$

Thus, $F(y) = \pi \cdot \sqrt{1 + y} - \pi \Rightarrow F(y) = \pi [\sqrt{1 + y} - 1]$

This is the required result.

Example 3. Evaluate $\int_0^{\infty} \frac{\tan^{-1} x}{x(1 + x^2)} dx$ $(a \geq 0)$, by applying differentiation under integral sign and hence evaluate $\int_0^{\infty} \frac{\tan^{-1} ax}{x(1 + x^2)} dx$.

Solution : Let

$$F(a) = \int_0^{\infty} \frac{\tan^{-1} ax}{x(1 + x^2)} dx \qquad \text{.... (1)}$$

Differentiating both sides partially w.r.t. a, we get,

$$F'(a) = \int_0^\infty \frac{\partial}{\partial a} \frac{\tan^{-1} ax}{x\left(1+x^2\right)} dx$$

$$= \int_0^{\pi/2} \frac{1}{x\left(1+x^2\right)} \frac{x}{1+a^2x^2} dx = \int_0^{\pi/2} \frac{1}{\left(1+x^2\right)\left(1+a^2x^2\right)} dx$$

Resolving the quantity under integral sign into partial fraction (Hint : put $x^2 = t$) we get,

$$F'(a) = \frac{1}{1-a^2} \int_0^\infty \left[\frac{1}{\left(1+x^2\right)} - \frac{a^2}{1+a^2x^2}\right] dx$$

$$= \frac{1}{1-a^2}\left[\int_0^\infty \frac{dx}{\left(1+x^2\right)} - \int_0^\infty \frac{dx}{\frac{1}{a^2}+x^2}\right]$$

$$= \frac{1}{1-a^2}\left[\tan^{-1} x - a\tan^{-1}(ax)\right]_0^\infty = \frac{1}{1-a^2}\left[\frac{\pi}{2} - a\frac{\pi}{2}\right] = \frac{1}{1-a^2}(1-a)\frac{\pi}{2}$$

Thus, $F'(a) = \dfrac{\pi}{2(1+a)}$. Integrating both sides, w.r.t. a, we get,

$$F(a) = \frac{\pi}{2}\int \frac{1}{(1+a)} da + c \;\Rightarrow\; F(a) = \frac{\pi}{2}\log(1+a) + c \qquad \text{.... (2)}$$

Now from (1), we have, $F(0) = 0$, when $a = 0$.

Now, (2) $\Rightarrow$ $F(0) = \frac{\pi}{2}\log 1 + c \Rightarrow c = 0$. Thus, $F(a) = \frac{\pi}{2}\log(1+a)$.

Hence, $$\int_0^\infty \frac{\tan^{-1} ax}{x\left(1+x^2\right)} dx = \frac{\pi}{2}\log(1+1) = \boldsymbol{\frac{\pi}{2}\log 2}$$

In particular when $a = 1$, we have,

$$\int_0^\infty \frac{\tan^{-1} ax}{x\left(1+x^2\right)} dx = \frac{\pi}{2}\log(1+1) = \boldsymbol{\frac{\pi}{2}\log 2}$$

Example 4. Evaluate $\int_0^\infty e^{-x^2}\cos\alpha x\,dx$, given that $\int_0^\infty e^{-x^2}dx = \frac{\sqrt{\pi}}{2}$.

Solution : Let, $$F(\alpha) = \int_0^\infty e^{-x^2}\cos\alpha x\,dx \qquad \text{.... (1)}$$

Differentiating partially w.r.t. α, we get,

$$F'(\alpha) = \int_0^\infty \frac{\partial}{\partial\alpha}\left[e^{-x^2}\cos\alpha x\right] dx \;\Rightarrow\; F'(\alpha) = \int_0^\infty e^{-x^2}(-\sin\alpha x)(x)\,dx$$

$$\Rightarrow F'(\alpha) = \int_0^\infty -xe^{-x^2}\sin\alpha x\,dx$$

Integrating by parts by taking $\sin \alpha x$ as first function and $\left(-xe^{-x^2}\right)$ as second function. Further $\int -xe^{-x^2}\,dx = \frac{e^{-x^2}}{2}$. We have,

$$F'(\alpha) = \left[\sin \alpha x \cdot \frac{e^{-x^2}}{2}\right]_0^\infty - \int_0^\infty \frac{e^{-x^2}}{2}\,\alpha \cos \alpha x\,dx$$

$$= -\frac{\alpha}{2}\int_0^\infty \frac{e^{-x^2}}{2}\cos \alpha x\,dx = -\frac{\alpha}{2}F(\alpha) \Rightarrow \frac{F'(\alpha)}{F(\alpha)} = \frac{-\alpha}{2}$$

Integrating w.r.t. α, both sides we get,

$$\log F(\alpha) = -\frac{\alpha^2}{4} + c \qquad \text{.... (2)}$$

Now from (1), we have, $\log F(0) = \int_0^\infty e^{-x^2} = \frac{\sqrt{\pi}}{2}$ (given)

Now, (2) $\Rightarrow$ $\log [F(0)] = c \Rightarrow \log \frac{\sqrt{\pi}}{2} = c$

Thus, $\log F(\alpha) = -\frac{\alpha^2}{2} + \log \frac{\sqrt{\pi}}{2}$

Now, $\log F(\alpha) - \log \frac{\sqrt{\pi}}{2} = -\frac{\alpha^2}{4}$

$$\Rightarrow \quad \log \frac{F(\alpha)}{\left(\frac{\sqrt{\pi}}{2}\right)} = \frac{\alpha^2}{4} \Rightarrow F(\alpha) = \frac{\sqrt{\pi}}{2}e^{-\alpha^2/4}$$

Thus, $\int_0^\infty e^{-x^2}\cos \alpha x\,dx = \frac{\sqrt{\pi}}{2}e^{-\alpha^2/4}$

Example 5. From the value of $\int_0^1 x^m\,dx$, deduce the value of $\int_0^1 x^m (\log x)^n\,dx$, when $m \geq 0$ and n is a positive integer.

Solution : We have,

$$\int_0^1 x^m\,dx = \left[\frac{x^{m+1}}{m+1}\right]_0^1 = \frac{1}{m+1}$$

Differentiating w.r.t. m, we get

$$\int_0^1 x^m \log x\,dx = -\frac{1}{(m+1)^2}$$

Again differentiating w.r.t. m, we get,

$$\int_0^1 x^m (\log x)^2 \, dx = (-1) \cdot (-2) \cdot \frac{1}{(m+1)^3}$$

Repeating the process of differentiation w.r.t. m, n times we get,

$$\int_0^1 x^m (\log x)^n \, dx = (-1) \cdot (-2) \ldots\ldots (-n) \cdot \frac{1}{(m+1)^{n+1}}$$

$$\int_0^1 x^m (\log x)^n \, dx = \frac{(-1) \cdot n!}{(m+1)^{n+1}}$$

EXERCISES

6. Differentiating $\int_0^x \frac{dx}{x^2 + a^2} = \frac{1}{a} \tan^{-1} \frac{x}{a}$, under the integral sign show that

$$\int_0^x \frac{dx}{x^2 + a^2} = \frac{1}{2a^3} \tan^{-1}\left(\frac{x}{a}\right) + \frac{x}{2a^2\left(x^2 + a\right)}.$$

7. Starting from $\int_0^\infty e^{-ax} \, dx = \frac{1}{a}$, for $a > 0$, deduce that $\int_0^\infty x^m e^{-ax} \, dx = \frac{m}{a^{m+1}}$.

8. From $\int_0^\infty \frac{dx}{x^2 + a^2} = \frac{\pi}{2a}$, $a > 0$, deduce that

$$\int_0^\infty \frac{dx}{\left(x^2 + a^2\right)^n} = \frac{\pi}{2a^{2n-1}} \cdot \frac{1 \cdot 3 \cdot 5 \ldots\ldots\ldots (2n-3)}{2 \cdot 4 \cdot 6 \ldots\ldots\ldots (2n-2)},$$ where n is a positive integer.

9. Show that, if $e^2 < 1$, $\int_0^{\pi/2} \log\left(1 - e^2 \sin^2 x\right) dx = \pi \log \frac{1 + \sqrt{1 - e^2}}{2}$.

10. Show that $\int_0^\alpha \log(1 + \alpha \cos x) \, dx = \pi \log \frac{1 + \sqrt{1 - \alpha^2}}{2}$

Answers

1. $\tan^{-1} \frac{m}{a}$ **2.** $\log (1 + \alpha)$

BSM – 3.2

Chapter 1. Riemann Integration

Chapter 2. Differential Equations

1 Riemann Integration

1.1 Introduction

In this chapter we shall study the arithmetic treatment of definite integral, which is free from geometrical concepts. This type of treatment was first introduced by G.F.B. Riemann (1826-1866). Riemann definition of integration covered only bounded functions. In the present chapter we shall see Riemann integral for real valued bounded functions defined on a closed interval.

As a premilary, we shall recapitulate the real number system, which is the foundation of the real analysis.

1.2 Real Number System

The set R of all real numbers, together with two fundamental binary operations, addition (+) and multiplication ($\cdot$), satisfy the following axioms :

1. $a, b \in R \Rightarrow a + b \in R$ and $a \cdot b \in R$ (Closure law)
2. $a, b \in R \Rightarrow a + b = b + a$ and $a \cdot b = b \cdot a$ (Commutative law)
3. $a, b, c \in R \Rightarrow a + (b + c) = (a + b) + c$ and $a \cdot (b \cdot c) = (a \cdot b) \cdot c$ (Associative law)
4. $a, b, c \in R \Rightarrow a \cdot (b + c) = a \cdot b + a \cdot c$ (Distributive law)
5. $\exists\, 0 \in R$, such that $\forall\, a \in R,\ a + 0 = 0 + a = a$ (Existence of zero)
6. $\forall\, a \in R,\ \exists -a \in R$, such that $a + (-a) = 0 = (-a) + a$ (Existence of inverse w.r.t. adddition)
7. $\exists\, 1 \in R$, such that $\forall\, a \in R,\ a \cdot 1 = 1 \cdot a = a$ (Existence of unity)
8. $\exists\, a (\neq 0) \in R,\ \exists\, \frac{1}{a} \in R$, such that, $a \cdot \frac{1}{a} = \frac{1}{a} \cdot a = 1$

(Existence of inverse w.r.t. multiplication)

The set R of all real numbers, together with + and $\cdot$, and with above axioms is said to be a **field**. The above axioms are called **field axioms**.

The following results can be deduced from the above axioms :

1. The number 0, is unique.
2. $\forall\, a \in R$, the number $-a$ is unique.
3. $\forall\, a \neq 0 \in R$, the number $\frac{1}{a}$ is unique.
4. $\forall\, a \in R,\ -(-a) = a,\ a \cdot 0 = 0,\ (a^{-1})^{-1} = a \ \ (a \neq 0)$
5. $a \neq 0,\ b \neq 0$, then $(a \cdot b)^{-1} = b^{-1} \cdot a^{-1} = a^{-1} \cdot b^{-1}$.
6. $a + b = a + c \Rightarrow b = c$; $a \neq 0,\ a \cdot b = a \cdot c \Rightarrow b = c$.

7. $(-a) \cdot b = a \cdot (-b) = -(a \cdot b)$; $(-a) \cdot (-b) = a \cdot b$.
8. $a \neq 0$, $b \neq 0$ then $a \cdot b \neq 0$; $a \cdot b = 0 \Rightarrow a = 0$ or $b = 0$.
9. The equation $x + a = b$, has unique solution, $x = b - a$, $\forall\, a, b \in \boldsymbol{R}$.
10. The equation $a \cdot x = b$, has unique solution, $x = \dfrac{b}{a}$, for $a, b \in \boldsymbol{R}$, $a \neq 0$.

In the set $\boldsymbol{R}$ of real numbers, an **ordered relation** '>' (greater than) is defined for pair of real numbers, with the following properties called ordered axioms :

1. For $\forall\, a, b \in \boldsymbol{R}$, one and only one of the following is true,
$$a > b,\ a = b \text{ or } b > a \qquad \text{(Law of trichotomy)}$$
2. For $a, b, c \in \boldsymbol{R}$, $a > b$, $b > c \Rightarrow a > c$ (Transitivity)
3. For $a, b, c \in \boldsymbol{R}$, $a > b \Rightarrow a + c > b + c$ (Monotone property for addition)
4. For $a, b, c \in \boldsymbol{R}$, $a > b$, $c > 0 \Rightarrow ac > bc$ (Montone property for multiplication)

The $(\boldsymbol{R}, +, \cdot, >)$ is called an **ordered filed or linearly ordered field.**

Note :

1. The order relation '<' (less than) is such that $a < b \Leftrightarrow b > a$, for all $a, b \in \boldsymbol{R}$.
2. A real number 'a' is said to be positive, if $a > 0$.
3. A real number 'a' is said to be negative, if $a < 0$.
4. The set all positive real numbers is denoted by $\boldsymbol{R}^+$ and the set of all negative real numbers is denoted by $\boldsymbol{R}^-$ and therefore $\boldsymbol{R} = \boldsymbol{R}^+ \cup \boldsymbol{R}^- \cup \{0\}$.
5. $a \in \boldsymbol{R}^+$ and $b \in \boldsymbol{R}^- \Rightarrow a > b$.
6. The set of all real numbers x such that $a < x < b$, is called an **open interval** and it is denoted by (a, b). That is,
$$(a, b) = \{x : a < x < b\}.$$
7. The set of all real numbers x such that $a \le x \le b$ is called a **closed interval** and it is denoted by $[a, b]$. That is,
$$[a, b] = [x : a \le x \le b].$$
8. The set of all real numbers x such that, $a \le x < b$ is called half closed (or half open) interval and it is denoted by $[a, b)$. That is,
$$[a, b) = \{x : a \le x < b\}$$
Similarly,
$$(a, b] = \{x : a < x \le b\}.$$

In an open interval (a, b) the end points are not included, whereas in the closed interval $[a, b]$ the end points are included. In an half open (or half closed) intervals one end point is included and the other is not.

Following are few properties of order relation :

1. $\forall\, a \in \boldsymbol{R}$, one and only one of $a > 0$, $a = 0$, $a < 0$ is true.
2. $a, b \in \boldsymbol{R}^+$, then $a + b \in \boldsymbol{R}^+$ and $a \cdot b = \boldsymbol{R}^+$.
3. $a, b \in \boldsymbol{R}^-$, then $a + b \in \boldsymbol{R}^-$ and $a \cdot b = \boldsymbol{R}^+$.
4. $a > 0 \Leftrightarrow -a < 0$; $a < 0 \Leftrightarrow -a > 0$.
5. $a < b \Leftrightarrow -a > -b$; $a > b \Leftrightarrow -a < -b$.
6. $a < 0 \Leftrightarrow \dfrac{1}{a} < 0$; $a > 0 \Leftrightarrow \dfrac{1}{a} > 0$.
7. $a > b \Leftrightarrow a - b > 0$; $a < b \Leftrightarrow a - b < 0$.

8. $a \cdot b > 0,\ b > 0 \Rightarrow a > 0$; $a \cdot b < 0,\ b < 0 \Rightarrow a > 0$.

9. $a > b > 0 \Rightarrow \frac{1}{a} < \frac{1}{b}$; $a < b \Rightarrow \frac{1}{a} > \frac{1}{b}$.

10. $a \in R \Rightarrow a^2 \geq 0$; $a > b > 0 \Rightarrow a^2 > b^2$; $a < b < 0 \Rightarrow a^2 > b^2$.

11. $a \leq b$ and $b \leq a \Rightarrow a = b$.

The following are subsets of R :

1. The set N of natural numbers.
2. The set Z of integers.
3. The set Q of rational numbers.
4. The set of irrational numbers.

The following axioms for the set N of natural numbers, are known as Peano's axioms :

1. $1 \in N$ 2. $n \in N \Rightarrow n + 1 \in N$ 3. $n \in N \Rightarrow n + 1 \neq 1$

4. $m, n \in N$ and $m + 1 = n + 1$, then $m = n$

5. If $A \subset N$, $1 \in A$ and $(n \in A \Rightarrow n + 1 \in A)$ then $A = N$.

The following statement is known as **Well ordering principle :**

The set of R of real numbers together with two symbols $-\infty$ and ∞, is called the **extended real number system.**

For every $a \in R$, we define $-\infty < a < \infty$

Also we assume the following conventions :

1. $a \in R \Rightarrow a + \infty = \infty$ and $a - \infty = -\infty$.
2. $a \in R$ and $a < 0$ then $a \cdot \infty = -\infty$, $a \cdot (-\infty) = \infty$.
3. $a \in R$ and $a > 0$ then $a \cdot \infty = \infty$, $a \cdot (-\infty) = -\infty$.

1.3 Absolute Value of Real Number

Definition (Modulus of a real number : The modulus or the absolute value of real number x is denoted and defined as

$$|x| = \begin{cases} x & x \geq 0 \\ -x & x < 0 \end{cases}$$

The following are the properties of modulus of a real number.

1. $x \in R$, then $|x| \geq 0$
2. $x \in R$, then $|x| = \max\{x, -x\}$
3. $x \in R$, then $|x|^2 = x^2$
4. $x \in R$, then $|-x| = |x|$
5. $x, y \in R$, then $|x \cdot y| = |x| \cdot |y|$
6. $x, y \in R$, then $\left|\frac{x}{y}\right| = \frac{|x|}{|y|}$
7. $x, y \in R$, then $|x + y| \leq |x| + |y|$
8. $x, y \in R$, then $|x - y| \geq |x| - |y|$
9. $x \in R$ and $\delta(>0) \in R$, then $|x| < \delta \Rightarrow -\delta < x < \delta$
10. $x, y \in R$ and $\delta(>0) \in R$, then $|x - a| < \delta \Rightarrow a - \delta < x < a + \delta$
11. $x \in R$ and $\delta(>0) \in R$, then $|x| > \delta \Rightarrow x > \delta$ or $x < -\delta$.

1.4 Boundedness of Subsets of Real Numbers

Definition : A non empty subset of the set R of real numbers is called an aggregate.

That is any finite subset and every infinite subset of R are aggregates.

Definition : An aggregate S is said to be bounded above, if there exists a real number u, such that $a \leq u$ for all $a \in S$. The real number u is called an upper bound of S.

Note :

1. If u is an upper bound of the aggregate S of R, then any real number greater than u is an upper bound of S.
2. A real number u is not an upper bound of S, if there exists an element $a \in S$ such that $a > u$.
3. An aggregate S is said to be unbounded above, if there exists no real number u such that $a \leq u$, for all $a \in S$.

   That is for every real number u, there exists atleast one element $x \in S$ such that $x > u$.

   In this case we say the aggregate S is **unbounded above**.

Examples :

1. The set R^+ is not bounded above.
2. The set R^- is bounded above and 0 is an upper bound.
3. Every finite set is bounded above.
4. The interval (4, 7) is bounded above by 7.

Definition : An aggregate S is said to be bounded below, if there exists a real number v such that $a \geq v$ for all $a \in S$. The real number v is called a lower bound of S.

Note :

1. If v a lower bound of the aggregate S, then any real number less than v is a lower bound of S.
2. A real number v is not a lower bound of S, if there exists an element $a \in S$ such that $a < v$.
3. An aggregate S is said to be unbounded below, if there exists no real number v such that $a \geq v$ for all $a \in S$.

   That is for every real number v, there exists at least one element $x \in S$ such that $x < v$.

   In this case we say the aggregate S is **unbounded below** :

Examples :

1. The set R^+ is bounded below and 0 is a lower bound.
2. The set R^- is unbounded below.
3. Every finite set is bounded below.
4. The interval (5, 7) is bounded below by 5.

Definition : An aggregate S is said to be bounded, if S is bounded both above and below :

That is, an aggregate S is said to be bounded, if there exists $u, v \in R$ such that $v \leq a \leq u$, for all $a \in S$.

Examples :

1. The set N of natural numbers, is bounded below but not above.
2. The set Z, Q and R are not bounded sets.
3. The interval (7, 11) is a bounded set, with an upper bound 11 and a lower bound 7. Infact $\forall\ x \in (7, 11)$, we have $7 < x < 11$.
4. Every finite aggregate is bounded.

Theorem : An aggregate S is bounded if and only if, there exists $k \in R^+$ such that $|x| \leq k$ for all $x \in S$.

Proof : Let S be bounded. Then there exists $k_1, k_2 \in \boldsymbol{R}$, such that $k_1 \leq x \leq k_2$, for all $x \in S$. Let $k > \max\{|k_1|, |k_2|\}$. Then,

$$-k < k_1 \leq x \leq k_2 < k, \ \forall\, x \in S \quad \Rightarrow \quad -k < x < k \ \ \forall\ x \in S \quad \Rightarrow \quad |x| < k, \quad \forall\, x \in S$$

Conversely, let there exist $k \in R^+$ such that $|x| < k$ for all $x \in S$.

$$\Rightarrow \qquad -k < x < k, \forall\, x \in S \quad \Rightarrow \quad S \text{ is bounded}$$

Definition : Let S be an aggregate. A real number u is said to be least upper bound (l.u.b) or supremum (sup.) of S, if

(i) u is an upper bound of S (ii) $x < u \Rightarrow x$ is not an upper bound of S.

Definition : Let S be an aggregate. A real number v is said to be greatest lower bound (g.l.b.) or infimum (inf.) of S, if

(i) v is a lower bound of S (ii) $x > v \Rightarrow x$ is not a lower bound of S.

Theorem :

(i) If an aggregate S has least upper bound (supremum) then it is unique.

(ii) If an aggregate S has greatest lower bound (infimum) then it is unique.

Proof : (i) Let u_1 and u_2 be two least upper bounds of the aggregate S.

Now, u_1 is a l.u.b. and u_2 is an upper bound $\Rightarrow u_1 \leq u_2$.

Again, u_2 is a l.u.b and u_1 is an upper bound $\Rightarrow u_2 \leq u_1$.

Now, $u_1 \leq u_2$ and $u_2 \leq u_1 \Rightarrow u_1 = u_2$. Thus, l.u.b. of S is unique.

(ii) Left as an exercise.

Theorem :

(i) If u is an upper bound of the aggregate S and $u \in S$, then u is the l.u.b. of S.

(ii) If v is a lower bound of the aggregate S and $v \in S$, then v is the g.l.b. of S.

Proof : (i) If $x \in S$ and $x < u$, then x is not an upper bound of S. Thus u is the least upper bound.

(ii) Left as an exercise.

Note : If u and v are respectively l.u.b. and g.l.b. of the aggregate S, and $u \in S$ and $v \in S$, the u is the greatest element and v is the least element of S.

Theorem :

(i) If u = l.u.b. of S and $y < u$, there exists $x \in S$ such that $y < x \leq u$.

(ii) If v = g.l.b. of S and $y > v$, then there exists $x \in S$ such that $y > x \geq l$.

Proof : (i) Let if possible, $\exists$ no $x \in A$ such that $y < x$. Then $x \leq y$ for all $x \in S$.

Now, $x \leq y$ for all $S \Rightarrow y$ is an upper bound and $y < u$.

This contradicts the fact that u is l.u.b. Hence, $\exists\, x \in S$, such that $y < x \leq u$.

(ii) Left as an exercise.

Theorem : Let S be an aggregate, which is bounded above. The real number u is the supremum (l.u.b.) of S, if and only if

(a) $x \leq u, \ \forall\, x \in S$ (b) for $\forall\, \varepsilon > 0$, $\exists$ a real number $x \in S$, such that $x > u - \varepsilon$.

Proof : Let u be the supremum of S. Then the condition (a), implies that u is an upper bound. Thus (a) is true.

Now, u is the supremum and $u - \varepsilon < u$. Thus $u - \varepsilon$ cannot be an upper bound. Thus there must exist an $x \in S$ such that $x > u - \varepsilon$. Hence the condition (b) is true.

Conversely, let (a) and (*b*) are true.

Now, (a) $\Rightarrow$ u is an upper bound of S.

Let if possible $u' < u$ be an upper bound of S. Let $\varepsilon = u - u' > 0$. Now, the condition (*b*) says that there exists $x \in S$ such that,

$$x > u - \varepsilon = u' \Rightarrow u' \text{ is not an upper bound of } S$$

This is a contradiction. Hence u is the least upper bound of S. That is u is the supremum of S.

Theorem : Let S be an aggregate, which is bounded below. The real number v is the infimum (g.l.b.) of S, if and only if,

(a) $x \geq v$, $\forall x \in S$ (b) $\forall \varepsilon > 0$, $\exists$ a real number $x \in S$ such that $x < v + \varepsilon$.

Proof : Left as an exercise.

Following are some of the properties of boundedness.

1. If an aggregate is bounded above then it has supremum.
2. If an aggregate is bounded below then it has infimum.
3. If v is a lower bound and u is an upper bound of an aggregate then $v \leq u$.
4. If S and T are bounded aggregate then
   (i) sup. $(S + T)$ = sup. S + sup. T **(ii)** inf. $(S + T)$ = inf. S + inf. T
5. Let S and T are two aggregates such that $a \leq b$, for all $a \in S$ and $b \in T$. If T has supremum then S has also supremum and sup. $S \leq$ sup. T.
6. If S and T are two aggregates of $\mathbf{R}$, then
   $$\text{sup. } (S \cup T) = \text{max. [sup. } S, \text{ sup. } T]$$
7. If S and T are two aggregates of $\mathbf{R}$, such that $S \subseteq T$, then
   $$\text{inf. } T \leq \text{inf. } S \leq \text{sup. } S \leq \text{sup. } T$$

Examples :

1. If $S = [a, b]$, then sup. $S = b$ and inf. $S = a$. Further sup. S and inf. S belongs to S.
2. If $S = (a, b]$, here sup. $S = b \in S$ and inf. $S = a \notin S$
3. If $S = [a, b)$, here sup. $S = b \notin S$ and inf. $S = a \in S$
4. If $S = (a, b)$, here sup. $S = b \notin S$ and inf. $S = a \notin S$
5. If $S = (a, \infty)$, here inf. $S = a \notin S$, sup. does not exist as S is not bounded above.
6. If $S = (\infty, b]$, here sup. $S = b \in S$, inf. does not exist as S is not bounded below.
7. Let $S = \left\{\dfrac{n+1}{n} : n \in N\right\}$.

   Clearly, $S = \left\{\left(1 + \dfrac{1}{n}\right) : n \in N\right\}$. Also $1 + \dfrac{1}{n} > 1$ for all $n \in N$ an $1 + \dfrac{1}{n} \leq 2$ for all $n \in N$.

   Thus 1 is a lower bound of S and 2 is an upper bound of S. Further 1 is the greatest lower bound - i.e., 1 is inf. of S and 2 is the least upper bound - i.e., 2 is sup. of S.
8. Let $S = \{x \in \mathbf{R} : x \leq 5\}$. Clearly 5 is an upper bound and $5 \in S$. Thus 5 is sup. S. S is not bounded below.

EXERCISE

1. Find the supremum and the infimum (if exists) of the following sets.

(a) $S = \left\{\frac{5n+2}{n} : n \in \mathbf{N}\right\}$ **(b)** $S = \left\{\frac{n-1}{n} : n \in \mathbf{N}\right\}$ **(c)** $S = \{x \in \mathbf{R} : x \leq 3\}$

(d) $S = \{x \in \mathbf{R} : x > 0\}$ **(e)** $S = \{2^n : n \in \mathbf{N}\}$ **(f)** $S = \{x \in \mathbf{Q}^+ : 0 < x^2 - 1 \leq 2)$

Answers

1. **(a)** 7, 5 **(b)** 1, 0 **(c)** 3, does not exist

(d) does not exist, 0 **(e)** does not exist, 2 **(f)** $\sqrt{3}$, 1

1.5 Partition of a Closed Interval

Consider a closed interval $[a, b]$, i.e., $a \leq x \leq b$. Let,

$$a = x_0 < x_1 < x_2 < \ldots\ldots < x_n = b$$

Then the set $P = [x_0 < x_1 < x_2 < \ldots\ldots < x_n]$ is called a **partition** of the closed interval $[a, b]$. The $(n + 1)$ points, $x_0 < x_1 < x_2 < \ldots\ldots < x_n$ are called partition points of P. The n-sub intervals.

$$[x_0, x_1], [x_1, x_2], \ldots\ldots, [x_{r-1}, x_r], \ldots\ldots [x_{n-1}, x_n]$$

are called the **segments** of the partition P or **components** of the partition P.

The length $x_r - x_{r-1}$ of the rth subinterval $I_r = [x_{r-1}, x_r]$, is denoted by δ_r or Δ_r.

The maximum of the lengths of subintervals of the partition P is called the **norm of the partition** P and is denoted by $\| P \|$ or $\Delta(P)$. Thus,

$$\| P \| = \max\{\delta_1, \delta_2, \delta_3, \ldots\ldots, \delta_n\}$$

Given a closed interval $[a, b]$, it can be partitioned in many ways. The set of all partitions of $[a, b]$ is denoted by $\phi[a, b]$.

Examples :

1. Let $[a, b] = [2, 7]$ and $P = \{2, 3, 4, 5, 6, 7\}$. Then subintervals of the partition P are

$$[2, 3], [3, 4], [4, 5], [5, 6] \text{ and } [6, 7]$$

The lengths of subintervals are $\delta_1 = 1, \delta_2 = 1, \delta_3 = 1, \delta_4 = 1, \delta_5 = 1$.

Also $\| P \| = 1$, i.e., norm of $P = 1$.

2. Let the closed interval be $[1, 2]$ and $P = \{1, 1.2, 1.3, 1.5, 1.8, 2\}$. Then the sub intervals are

$[1, 1.2], [1.2, 1.3], [1.3, 1.5], [1.5, 1.8]$ and $[1.8, 2]$. The lengths of sub intervals are

$$\delta_1 = .2,\ \delta_2 = 0.1,\ \delta_3 = 0.2,\ \delta_4 = 0.3,\ \delta_5 = 0.2$$

Now, norms of $P = \| P \| = \max\{.2, .1, .2, .3, .2\} = .3$. That is, $\| P \| =$ **0.3.**

Note :

1. If $\delta_1, \delta_2, \delta_3, \ldots\ldots, \delta_n$ are the lengths of subintervals of the partition P, then $\delta_n \leq \| P \|$ for $r = 1, 2, \ldots\ldots, n$.
2. As $\| P \| \to 0$, then the number of points of the partition P tends to infinity.
3. If $P_1, P_2 \in \phi\,[a, b]$ and $P_1 \subset P_2$ then $\| P_2 \| \leq \| P_1 \|$.
4. $\sum_{r=1}^{n} \delta_r = \delta_1 + \delta_2 + \ldots. + \delta_n = (x_1 - a) + (x_2 - x_1) + \ldots. + (b - x_{n-1}) = b - a$

1.6 Upper and Lower Riemann Sums

Let f be a bounded real valued function defined on $[a, b]$.

Let $P = \{a = x_0, x_1, x_2, \ldots, x_n = b\}$ be a partition of $[a, b]$.

By data f is bounded on $[a, b]$. Thus f is bounded on each of the sub intervals of the partition P. Let M_r and m_r be the supremum (least upper bound) and infimum (greatest lower bound) of f in the rth sub interval, $I_r = [x_{r-1}, x_r]$, $r = 1, 2, \ldots, n$.

Thus,
$$M_r = \text{l.u.b.}\, f(x), \text{ in } x_{r-1} \le x \le x_r$$
$$m_r = \text{g.l.b.}\, f(x), \text{ in } x_{r-1} \le x \le x_r$$

The sum, denoted and defined by,

$$U(P, f) = M_1 \cdot \delta_1 + M_2 \cdot \delta_2 + \ldots\ldots + M_n \cdot \delta_n = \sum_{r=1}^{n} M_r \cdot \delta_r$$

is called **upper Riemann sum or upper Darbox sum of** f, corresponding to the partition P.

The sum, denoted and defined by

$$L(P, f) = m_1 \cdot \delta_1 + m_2 \cdot \delta_2 + \ldots\ldots + m_n \cdot \delta_n = \sum_{r=1}^{n} m_r \cdot \delta_r$$

is called **Lower Riemann sum** or **Lower Darbox sum of** f corresponding to the partition P.

Example 1. If $f(x) = x^2$ is defined on $[0, 1]$ and $P = \left\{0, \frac{1}{4}, \frac{2}{4}, \frac{3}{4}, 1\right\}$ find $U(P, f)$ and $L(P, f)$.

Solution : The subintervals of P are

$$I_1 = \left[0, \frac{1}{4}\right], I_2 = \left[\frac{1}{4}, \frac{2}{4}\right], I_3 = \left[\frac{2}{4}, \frac{3}{4}\right], I_4 = \left[\frac{3}{4}, 1\right]$$

Further the length of each subinterval is $\frac{1}{4}$. That is, $\delta_1 = \delta_2 = \delta_3 = \delta_4 = \frac{1}{4}$.

Clearly, the $f(x) = x^2$ is an increasing function is $[0, 1]$. Thus, sup. and inf. of f in

$$I_1 \text{ are, } M_1 = \frac{1}{16}, m_1 = 0; \qquad I_2 \text{ are } M_2 = \frac{4}{16}, m_1 = \frac{1}{16}$$

$$I_3 \text{ are, } M_2 = \frac{9}{16}, m_3 = \frac{4}{16}; \qquad I_4 \text{ are } M_4 = 1, \quad m_1 = \frac{9}{16}$$

Now,
$$U(P, f) = M_1 \cdot \delta_1 + M_2 \cdot \delta_2 + M_3 \cdot \delta_3 + M_4 \cdot \delta_4$$
$$= \frac{1}{16} \cdot \frac{1}{4} + \frac{4}{16} \cdot \frac{1}{4} + \frac{9}{16} \cdot \frac{1}{4} + 1 \cdot \frac{1}{4}$$

$$\Rightarrow \quad U(P, f) = \frac{1}{64}(1 + 4 + 9 + 16) \Rightarrow \mathbf{U(P, f) = \frac{15}{32}}$$

Now,
$$L(P, f) = m_1 \cdot \delta_1 + m_2 \cdot \delta_2 + m_3 \cdot \delta_3 + m_4 \cdot \delta_4$$
$$= 0 \cdot \frac{1}{4} + \frac{4}{16} \cdot \frac{1}{4} + \frac{4}{16} \cdot \frac{1}{4} + \frac{9}{16} \cdot \frac{1}{4}$$

$$\Rightarrow \quad L(P, f) = \frac{1}{64}(1 + 4 + 9) \Rightarrow \mathbf{L(P, f) = \frac{7}{32}}$$

Theorem : If $f(x)$ is real valued bounded function defined on $[a, b]$ and $P \in f[a, b]$, then $U(P, -f) = -L(P, f)$ and $L(P, -f) = -U(P, f)$.

Proof : Let $P = [a = x_0, x_1, x_2, ..., x_n = b]$ be a partition of $[a, b]$.

Let M_r and m_r be the supremum and infimum of $f(x)$ on

$$I_r = [x_{r-1}, x_r] \text{ for } r = 1, 2, 3, ..., n$$

By data $f(x)$ is bounded on $[a, b]$. Thus the function $-f(x)$ is also bounded on $[a, b]$.

Thus, $-m_r$ and $-M_r$ are the supremum and infimum of $-f(x)$ on $I_r = [x_{r-1}, x_r]$. Let δ_r be the length of I_r.

Now, by definition, we have,

$$U(P, -f) = \sum_{r=1}^{n} (-m_r) \cdot \delta_r \Rightarrow U(P, -f) = -\sum_{r=1}^{n} m_r \cdot \delta_r \quad \Rightarrow \mathbf{U(P, -f) = -L(P, f)}$$

Again by definition, we have,

$$L(P, -f) = \sum_{r=1}^{n} (-M_r) \cdot \delta_r \Rightarrow L(P, -f) = -\sum_{r=1}^{n} M_r \cdot \delta_r \quad \Rightarrow \mathbf{L(P, -f) = -U(P, f)}$$

Theorem : If $f(x)$ is a real valued bounded function defined on $[a, b]$ and $P \in \phi(a, b]$, then

$$\mathbf{m(b-a) \le L(P, f) \le U(P, f) \le M(b-a)}$$

where M and m are respectively the supremum and infimum of $f(x)$ on $[a, b]$.

Proof : Let $P\{a = x_0, x_1, x_2, ..., x_n = b\}$ be a partition of $[a, b]$. Let M_r and m_r be supremum and infimum of $f(x)$ on the interval $I_r = [x_{r-1}, x_r]$, for $r = 1, 2, 3, ..., n$. Let δ_r be the length of the interval I_r.

We have for $r = 1, 2, ..., n$,

$$m \le m_r \le M_r \le M \Rightarrow m \cdot \delta_r \le m_r \cdot \delta_r \le M_r \cdot \delta_r < M \cdot \delta_r$$

$$\Rightarrow \quad \sum_{r=1}^{n} m \cdot \delta_r \le \sum_{r=1}^{n} m_r \cdot \delta_r \le \sum_{r=1}^{n} M_r \cdot \delta_r \le \sum_{r=1}^{n} M \cdot \delta_r$$

$$\Rightarrow \quad m \cdot \sum_{r=1}^{n} \delta_r \le L(P, f) \le U(P, f) \le M \sum_{r=1}^{n} \delta_r$$

$$\Rightarrow \quad \mathbf{m(b-a) \le L(P, f) \le U(P, f) \le M(b-a)} \qquad \left(\because \sum_{r=1}^{n} \delta_r = b - a \right)$$

Note : From the above theorem we have, $U(P, f) \ge L(P, f)$.

Definition : If $f(x)$ is a real valued bounded function defined on $[a, b]$ and $P \in \phi\ [a, b]$, then $U(P, f) - L(P, f)$ is called the oscillatory sum of f corresponding to the partition P, and it is denoted by $W(P, f)$ or $O(P, f)$.

Note :

1. Clearly $W(P, f) = U(P, f) - L(P, f) \ge 0$.
2. $W(P, f) = \sum_{r=1}^{n} (M_r - m_r)\, \delta_r$

Example 2. If $f(x)$ is the function defined on $[a, b]$ by

$$f(x) = \begin{cases} 1 & \text{if } x \text{ is rational} \\ -1 & \text{if } x \text{ is irrational} \end{cases}$$

then find the oscillation of $f(x)$ in $[a, b]$.

Solution : Clearly $-1 \leq f(x) \leq 1$, for all $x \in [a, b]$. Thus $f(x)$ is bounded on $|a, b|$. Also sup. $f = 1$ and inf. $f = -1$.

We have, $$U(P, f) = \sum_{r=1}^{n} M_r \cdot \delta_r = \sum_{r=1}^{n} 1 \cdot \delta_r = (b - a)$$

and $$U(P, f) = \sum_{r=1}^{n} m_r \cdot \delta_r = \sum_{r=1}^{n} (-1) \cdot \delta_r = -(b - a)$$

Now, $$W(P, f) = U(P, f) - L(P, f) = (b - a) + (b - a) = \mathbf{2(b - a)}$$

Thus the oscillation of f on $[a, b]$ is $\mathbf{2(b - a)}$.

Theorem 1 : If $f(x)$ and $g(x)$ are real valued functions defined on $[a, b]$ and $P \in \phi[a, b]$, then

(a) $U(P, f + g) \leq U(P, f) + U(P, g)$ **(b) $L(P, f + g) \geq L(P, f) + L(P, g)$**

(c) $W(P, f + g) \leq W(P, f) + W(P, g)$

Proof : Let $P = \{a = x_0, x_1, x_2, \ldots, x_n = b\}$ be a partition of $[a, b]$. Since f and g are bounded on $[a, b]$, the function $f + g$ is also bounded on $[a, b]$.

Let m_r' and M_r' be infimum and supremum on f on I_r.

Let m_r'' and M_r'' be infimum and supremum on g on I_r.

Also, let m_r and M_r be infimum and supremum of $f + g$ on I_r.

(a) Now, M_r' and M_r'' are suprema of f and g on I_r

$\Rightarrow$ $f(x) \leq M_r'$, $g(x) \leq M_r''$ for all $x \in I_r$

$\Rightarrow$ $f(x) + g(x) \leq M_r' + M_r''$, for all $x \in I_r$

$\Rightarrow$ $M_r' + M_r''$ is an upper bound of $f + g$ on I_r

Since, M_r is the least upper bound of $f + g$ on I_r, we have,

$$M_r \leq M_r' + M_r'' \text{ on } I_r, \quad r = 1, 2, 3, \ldots, n$$

Now, $$U(P, f+g) = \sum_{r=1}^{n} M_r \cdot \delta_r \leq \sum_{r=1}^{n} (M_r' + M_r'') \cdot \delta_r = \sum_{r=1}^{n} M_r' \cdot \delta_r + \sum_{r=1}^{n} M_r'' \cdot \delta_r$$

$\Rightarrow$ $\mathbf{U(P, f + g) \leq U(P, f) + U(P, g)}$

(b) Left as an exercise.

(c) We have,

$$\begin{aligned} W(P, f + g) &= U(P, f + g) - L(P, f + g) \\ &\leq \{U(P, f) + U(P, g)\} - \{L(P, f) + L(P, g)\} \\ \Rightarrow W(P, f + g) &\leq \{U(P, f) - U(P, f)\} + \{U(P, g) - L(P, g)\} \\ \Rightarrow \mathbf{W(P, f + g)} &\leq \mathbf{W(P, f) + W(P, g)} \end{aligned}$$

Theorem : If $f(x)$ is a real valued bounded function defined on $[a, b]$ and $P_1, P_2 \in f[a, b]$ such that $P_1 \subset P_2$, then

(a) $U(P_1, f) \geq U(P_2, f)$ **(b) $L(P_1, f) \leq L(P_2, f)$** **(c) $W(P_1, f) \geq W(P_2, f)$**

Proof : By data $f(x)$ is bounded on $[a, b]$. Thus there exists $k \in \mathbf{R}^+$ such that $|f(x)| \leq k$ for all $x \in [a, b]$.

Thus, sup. f on $[a, b]$ will be $\leq k$.

Let $P_1 = [a = x_0, x_1, ..., x_n = b]$ and $P_2 = [a = x_0, x_1, ..., x_{t-1}, y_1, x_t, ..., x_n = b]$ be two partitions of $[a, b]$. The partition P_2 contains one partition point y_1 more than P_1.

Let m_r and M_r be infimum and supremum of f in $[x_{r-1}, x_r]$.

Let λ_1, λ_2 be the suprema of f in $[x_{r-1}, y_1]$, $[y_1, x_r]$ respectively.

Let μ_1, μ_2 be the infima of f in $[x_{r-1}, y_1]$, $[y_1, x_r]$ respectively.

Clearly, $M_r \geq \lambda_1$, $M_r \geq \lambda_r$, $m_r \leq \mu_1$ and $m_r \leq \mu_2$.

(a) Now, $U(P_1, f) - U(P_2, f)$

$$= \sum_{r=1}^{n} M_r\left(x_r - x_{r-1}\right) - \left[\sum_{r=1}^{t} M_r\left(x_r - x_{r-1}\right) + \lambda_1\left(y_1 - x_{t-1}\right) + \lambda_2\left(x_t - y_1\right) + \sum_{r=t+1}^{n} M_r\left(x_r - x_{r-1}\right)\right]$$

$$= M_t\left(x_t - x_{t-1}\right) - \lambda_1\left(y_1 - x_{t-1}\right) - \lambda_2\left(x_t - y_1\right)$$

$$\Rightarrow \quad U(P_1, f) - U(P_2, f) \geq M_t(x_t - x_{t-1}) - M_1(y_1 - x_{t-1}) - M_t(x_t - y_t)$$

$$\rightarrow \quad U(P_1, f) - U(P_2, f) \geq M_t(x_t - x_{t-1} - y_1 + x_{t-1} - x_t - y_t) = 0$$

$$\Rightarrow \quad U(P_1, f) \geq U(P_2, f)$$

If P_2 contains m partition points more than P_1, then also we can show by induction that $U(P_1, f) \geq U(P_2, f)$.

Similarly one can prove the other two results.

Note :

1. If $P_1 \subset P_2$ and P_2 contains p partition points more than P_1, then
   (a) $U(P_1, f) - U(P_2, f) \leq 2kp \ \|P\|$ **(b)** $L(P_2, f) - L(P_1, f) \leq 2kp \ \|P\|$
   (c) $W(P_1, f) - W(P_2, f) \leq 4kp \ \|P\|$
2. From the results $U(P_1, f) \geq U(P_2, f)$ and $L(P_1, f) \leq L(P_2, f)$, we observe that by taking finer partitions upper sum does not increase and lower sum does not decrease.

Theorem : If P_1 and $P_2 \in f[a, b]$, then $L(P_1, f) \leq U(P_2, f)$.

Proof : Let $P = P_1 \cup P_2$. Then $P_1 \subset P$ and $P_2 \subset P$.

$\Rightarrow$ $L(P_1, f) \leq L(P, f)$ and $U(P, f) \leq U(P_2, f)$

But, $L(P, f) \leq U(P, f)$ for any partition P

Thus, $L(P_1, f) \leq U(P_2, f) \Rightarrow$ every lower sum $\leq$ an upper sum.

Theorem : If $f(x)$ is a real valued bounded function defined on $[a, b]$ $P \in \phi[a, b]$, the set $\{L(P, f) : P \in f[a, b]\}$ is bounded above and the set $[U(P, f) : P \in \phi[a, b]$ is bounded below :

Proof : Let $P = \{a = x_0, x_1, x_2, ..., x_n = b\}$ be a partition of $[a, b]$. Let M and n be the supremum and infimum of $f(x)$ on $[a, b]$. Also M_r and n_r be the supremum and infimum of $f(x)$ on $[x_{r-1}, x_r]$ for $r = 1, 2, ..., n$ and δ_r be the length of $[x_{r-1}, x_r]$.

Clearly, $m \leq m_r \leq M_r \leq M$, for $r = 1, 2, ..., n$

$$\Rightarrow \quad m \cdot \delta_r \le m_r \cdot \delta_r \le M_r \cdot \delta_r \le M \cdot \delta_r, \quad \text{for } r = 1, 2, ..., n$$

$$\Rightarrow \quad \sum_{r=1}^{n} m_r \cdot \delta_r \le \sum_{r=1}^{n} m_r \cdot \delta_r \le \sum_{r=1}^{n} M_r \cdot \delta_r \le \sum_{r=1}^{n} M \cdot \delta_r$$

$$\Rightarrow \quad \sum_{r=1}^{n} m_r \cdot \delta_r \le L(P, f) \le U(P, f) \le M \sum_{r=1}^{n} \delta_r$$

$$\Rightarrow \quad m(b-a) \le L(P, f) \le U(P, f) \le M(b-a)$$

This shows that for each $P \in \phi[a, b]$.

$$L(P, f) \le M(b-a) \text{ and } U(P, f) \ge m(b-a)$$

Thus, $M(b-a)$ is an upper bound of the set $\{L(P, f) : P \in \phi[a, b]\}$ and $m(b-a)$ is a lower bound of the set $\{U(P, f) : P \in \phi[a, b]\}$.

Hence the theorem.

Note : For each partition P of $[a, b]$.

$$\mathbf{m(b-a) \le L(P, f) \le U(P, f) \le M(b-a)}$$

EXERCISES

1. If $f(x) = x, 0 \le x \le 1$ and P is the partition $\left\{0, \frac{1}{4}, \frac{2}{4}, \frac{3}{4}, 1\right\}$ then find the value of $U(P, f)$.
2. If $f(x) = x, 0 \le x \le 1$ and $P = \left\{0, \frac{1}{3}, \frac{2}{3}, 1\right\}$ find $U(P, f)$ and $L(P, f)$.
3. If $f(x) = 2x - 1, 0 \le x \le 1$ and $P = \left\{0, \frac{1}{3}, \frac{2}{3}, 1\right\}$, find $U(P, f)$ and $L(P, f)$.
4. Find the lower and upper Riemann sums of $f(x) = \sin x$ on $[0, \pi]$ for the partition $P = \left\{0, \frac{\pi}{3}, \frac{2\pi}{3}, \pi\right\}$.

Answers

1. $\frac{8}{5}$ 2. $\frac{2}{3}, \frac{1}{3}$ 3. $\frac{1}{3}, -\frac{1}{3}$ 4. $\frac{\pi\sqrt{3}}{6}, \frac{\pi\left(\sqrt{3}+1\right)}{6}$

1.7 Upper and Lower Riemann Integral

Definition : Let $f(x)$ be a real valued bounded function defined on $[a, b]$ and $P = \{a = x_0, x_1, x_2, ..., x_n = b]$ be a partition of $[a, b]$. Then

(i) The supremum of the set $\{L(P, f) : P \in \phi[a, b]\}$ is called lower Riemann integral of $f(x)$ on $[a, b]$. This is denoted by $\int_{-a}^{b} f\, dx$ or $\int_{-a}^{b} f(x)\, dx$.

(ii) The infimum of the set $[U(P, f) : P \in \phi[a, b]\}$ is called upper Riemann integral of $f(x)$ on $[a, b]$. This is denoted by $\int_{a}^{-b} f\, dx$ or $\int_{a}^{-b} f\, dx$.

That is,
$$\int_{-a}^{b} f(x)\, dx = \text{sup. } [L(P, f) : P \in \phi[a, b])$$

and
$$\int_{a}^{-b} f(x)\, dx = \text{inf. } [U(P, f) : P \in \phi[a, b]\}$$

Note : We have that $M(b-a)$ is an upper bound of the set $\{L(P, f) : P \in \phi[a, b]\}$ and $m(b-a)$ is a lower bound of the set $\{U(P, f) : P \in \phi[a, b]\}$. Thus we have,

$$\underline{\int}_a^b f(x)\,dx \le M(b-a) \quad \text{and} \quad \overline{\int}_a^b f(x)\,dx \ge m(b-a)$$

Theorem : If $f(x)$ is bounded function on $[a, b]$, then $\underline{\int}_a^b f(x)\,dx \le \overline{\int}_a^b f(x)\,dx$.

Proof : Let P_1, P_2 be any two partition of $[a, b]$. Let $P = P_1 \cup P_2$.

Then, $P_1 \subset P_1 \cup P_2$ and $P_2 \subset P_1 \cup P_2$. Thus,

$$L(P_1, f) \le L(P, f) \text{ and } U(P, f) \le U(P_2, f)$$

But, $\quad L(P, f) \le U(P, f)$ for any partition P

Thus, we have, $\quad L(P_1, f) \le U(P_2, f) \qquad \text{.... (1)}$

Since (1) is true for all partition P_1 of $[a, b]$, we have that the set of all lower sums has an upper bound $U(P_2, f)$. Thus,

$$\sup.\{L(P, f) : P \in \phi[a, b]\} \le U(P_2, f)$$

$$\Rightarrow \quad \underline{\int}_a^b f(x)\,dx \le U(P_2, f), \text{ for all } P_2 \in \phi[a, b].$$

That is $\underline{\int}_a^b f(x)\,dx$ is a lower bound of the set of all upper sums.

Thus, $\quad \text{Inf}\,[U(P, f) : P \in \phi[a, b] \ge \underline{\int}_a^b f(x)\,dx$

$$\Rightarrow \quad \overline{\int}_a^b f(x)\,dx \ge \underline{\int}_a^b f(x)\,dx \quad \left[\because \text{inf. }\{U(P, f) : P \in f[a, b]\} = \overline{\int}_a^b f(x)\,dx\right]$$

Note : If m and M are infimum and supremum of the bounded function defined on $[a, b]$, then

$$m(b-a) \le \underline{\int}_a^b f(x)\,dx \le \overline{\int}_a^b f(x)\,dx \le M(b-a)$$

Also, $$L(P, f) \le \underline{\int}_a^b f(x)\,dx \le \overline{\int}_a^b f(x)\,dx \le U(P, f).$$

1.8 Riemann Integral

Definition : A bounded function $f(x)$ defined on $[a, b]$, is said to be Riemann integrable on $[a, b]$ if,

$$\underline{\int}_a^b f(x)\,dx = \overline{\int}_a^b f(x)\,dx$$

The common value in denoted by $\int_a^b f(x)\,dx$, and is called the **Riemann integral** (or simply integral of $f(x)$ on $[a, b]$).

The real numbers a and b are called lower and upper limits of the integral.

The set of all Riemann integral functions on $[a, b]$ is denoted by $R[a, b]$. If f is Riemann integral on $[a, b]$ then we write $f \in R[a, b]$.

Theorem : If $f \in R[a, b]$ and m, M are respectively the infimum and supremum of $f(x)$ on $[a, b]$ then,

$$m(b-a) \le \int_a^b f(x)\,dx \le M(b-a)$$

Proof : By data, $f(x)$ is a Riemann integrable function.

$$\Rightarrow \quad \int_{-a}^{b} f(x)\,dx = \int_{a}^{-b} f(x)\,dx = \int_{a}^{b} f(x)\,dx \quad \text{.... (1)}$$

Let, $P = \{a = x_0, x_1, x_2, ..., x_n = b\}$ be a partition of $[a, b]$.}

Let M_r an m_r be supremum and infimum of f on $[x_{r-1}, x_r]$.

We have, $\quad m \leq m_r \leq M_r \leq M_t$, for $r = 1, 2, ..., n$

$$\Rightarrow \quad m \cdot \delta_r \leq m_r \cdot \delta_r \leq M_r \cdot \delta_r \leq M \cdot \delta_r, \text{ for } r = 1, 2, ..., n$$

$$\Rightarrow \quad \sum_{r=1}^{n} m \cdot \delta_r \leq \sum_{r=1}^{n} m_r \cdot \delta_r \leq \sum_{r=1}^{n} M_r \cdot \delta_r \leq \sum_{r=1}^{n} M \cdot \delta_r$$

$$\Rightarrow \quad m(b-a) \leq L(P,f) \leq U(P,f) \leq M(b-a) \qquad \left(\because \sum_{r=1}^{n} \delta_r = b-a \right)$$

This result is true for all $P \in \phi[a, b]$.

Also, $\quad \int_{-a}^{b} f(x)\,dx = \sup. \{L(P,f) : P \in \phi[a,b]\} \Rightarrow \int_{-a}^{b} f(x)\,dx \geq L(P,f)$

and $\quad \int_{a}^{-b} f(x)\,dx = \inf. \{U(P,f) : P \in \phi[a,b]\} \Rightarrow \int_{a}^{-b} f(x)\,dx \leq U(P,f)$

Thus we have,

$$m(b-a) \leq L(P,f) \leq \int_{-a}^{b} f(x)\,dx = \int_{a}^{-b} f(x)\,dx \leq U(P,f) \leq M(b-a)$$

$$\Rightarrow \quad m(b-a) \leq \int_{a}^{b} f(x)\,dx \leq M(b-a) \qquad \text{(from (1))}$$

Theorem (Darboux's theorem) : If $f(x)$ is a real valued bounded function defined on $[a, b]$, then for $\forall\, \varepsilon > 0, \exists\, \delta > 0$ such that,

(i) $U(P,f) < \int_{a}^{-b} f(x)\,dx + \varepsilon \qquad$ **(ii)** $L(P,f) > \int_{-a}^{b} f(x)\,dx - \varepsilon$

for each $P \in \phi[a, b]$, with $\| P \| < \delta$.

Proof : By data $f(x)$ is bounded on $[a, b]$. Thus there exists $k > 0$ such that

$$|f(x)| < k, \quad \forall\, x \in [a,b] \quad \text{.... (1)}$$

By definition, we have,

$$\int_{a}^{-b} f(x)\,dx = \inf. \{U(P,f) : P \in \phi[a,b]\}$$

Thus there exists $P_1 \in \phi[a, b]$, such that

$$U(P_1, f) < \int_{a}^{-b} f(x)\,dx + \frac{\varepsilon}{2} \quad \text{.... (2)}$$

Let P_1 have $(p-1)$ partition points excluding the end points a and b.

Now, choose $\delta > 0$, such that $2k(p-1)\delta = \frac{\varepsilon}{2}$ .... (3)

Now, let P be any partition of $[a, b]$ with $\| P \| < \delta$ and $P_2 = P \cup P_1$.

Then P_2 contains at most $(p-1)$ points more than P.

$$\Rightarrow \quad U(P,f) - U(P_2,f) \leq 2k(p-1)\delta$$

$$\Rightarrow \quad U(P,f) \leq U(P_2,f) + 2k(p-1)\delta \quad \text{.... (4)}$$

Further, $P_1 \subset P_2 \Rightarrow U(P_2, f) \leq U(P_1, f)$ .... (5)

Thus, (4) $\Rightarrow$ $U(P, f) \leq U(P_1, f) + 2k(p-1)\delta$

$$\Rightarrow \quad U(P,f) < \int_a^{-b} f(x)\,dx + \frac{\varepsilon}{2} + \frac{\varepsilon}{2} \qquad \text{(from (2) and (3))}$$

$$\Rightarrow \quad U(P,f) < \int_a^{-b} f(x) + \varepsilon$$

Hence, for every $\varepsilon > 0$, there exists $\delta > 0$, such that

$$U(P,f) < \int_a^{-b} f(x) + \varepsilon,$$

for $P \in \phi[a, b]$ with $\| P \| < \delta$.

Similarly, the result **(ii)** can be proved.

Corollary : If $f(x)$ is a real valued bounded function defined on $[a, b]$ and $P \in \phi$ $[a, b]$, then

(a) $\lim\limits_{\|P\| \to 0} L(P, f) = \int_{-a}^{b} f(x)\,dx$ **(b)** $\lim\limits_{\|P\| \to 0} M(P, f) = \int_{a}^{-b} f(x)\,dx$

Also, as $\| P \| \to 0$, the number of partitions, becomes larger and larger, i.e., $n \to \infty$.

The results of the above corollary, which are in the limit from, are very useful in working out the problems on integrability.

Now we give below the criterion for integrability, in the form of a theorem.

Theorem (Necessary and Sufficient Condition) : A bounded function $f(x)$ defined on $[a, b]$ is Riemann integrable on $[a, b]$, if and only if for each $\varepsilon > 0$, there exists a partition P of $[a, b]$ such that,

$$\mathbf{0 < U(P, f) - L(P, f) - \varepsilon}$$

Proof : Let $f(x)$ be Riemann integrable on $[a, b]$. Then,

$$\int_{-a}^{b} f(x)\,dx = \int_{a}^{-b} f(x)\,dx = \int_a^b f(x)\,dx$$

By Darboux's theorem, we have for every $\varepsilon > 0$, there exists $\delta > 0$, such that,

$$U(P,f) < \int_a^{-b} f(x)\,dx + \frac{\varepsilon}{2} \text{ and } L(P,f) > \int_{-a}^{b} f(x)\,dx - \frac{\varepsilon}{2}$$

for each $P \in \phi[a, b]$ with $\| P \| < \delta$.

$$\Rightarrow \quad U(P,f) < \int_a^b f(x)\,dx + \frac{\varepsilon}{2} \text{ and } L(P,f) + \frac{\varepsilon}{2} > \int_a^b f(x)\,dx$$

$$\Rightarrow \quad U(P,f) < \int_a^b f(x)\,dx + \frac{\varepsilon}{2} \text{ and} -L(P,f) < -\int_a^b f(x)\,dx + \frac{\varepsilon}{2}$$

$$\Rightarrow \quad U(P,f) - L(P,f) < \int_a^b f(x)\,dx + \frac{\varepsilon}{2} - \int_a^b f(x)\,dx + \frac{\varepsilon}{2}$$

$$\Rightarrow \quad \mathbf{U(P,f) - L(P,f) < \varepsilon}$$

Conversely, let for each $\varepsilon > 0$, there exist $P \in \phi[a, b]$, such that

$$\mathbf{U(P,f) - L(P,f) < \varepsilon} \qquad \text{.... (1)}$$

We have, $\int_a^{-b} f(x)\,dx \leq U(P,f)$ and $\int_{-a}^{b} f(x)\,dx \geq L(P,f)$

$$\Rightarrow \quad \overline{\int_a^b} f(x)\,dx \le U(P,f) \text{ and } -\underline{\int_a^b} f(x)\,dx \le -L(P,f)$$

$$\Rightarrow \quad \overline{\int_a^b} f(x)\,dx - \underline{\int_a^b} f(x)\,dx \le U(P,f) - L(P,f)$$

$$\Rightarrow \quad \overline{\int_a^b} f(x)\,dx - \underline{\int_a^b} f(x)\,dx \le \varepsilon \qquad \text{(from (1))}$$

Since $\varepsilon > 0$ is arbitrary, letting $\varepsilon \to 0$, we get,

$$\overline{\int_a^b} f(x)\,dx - \underline{\int_a^b} f(x)\,dx = 0 \Rightarrow \overline{\int_a^b} f(x)\,dx = \underline{\int_a^b} f(x)\,dx$$

Hence $f(x)$ is Riemann integrable.

Note : $f \in \mathbf{R}[a, b]$, if an only if for each $\varepsilon > 0$, $\exists\, P \in \phi[a, b]$ such that $0 \le W(P, f) < \varepsilon$. This result follows from the fact $W(P, f) = U(P, f) - L(P, f)$.

Example 1. A constant function defined on a closed interval is Riemann integrable.

Solution : Let $f(x) = k$ for all $x \in [a, b]$, where k is a constant. Clearly $f(x)$ is bounded on $[a, b]$ and inf. $f = k$ and sup. $f = k$.

Let $P = \{a = x_0, x_1, x_2, \ldots\ldots, x_n = b\}$ be a partition of $[a, b]$.

The infimum m_r and supremum M_r of f on $[x_{r-1}, x_r]$ are equal and equal to k. That is $m_r = M_r = k$ for $r = 1, 2, \ldots\ldots, n$.

Now, $\quad L(P,f) = \sum_{r=1}^{n} m_r \cdot \delta_r = k \sum_{r=1}^{n} \delta_r \qquad (\because m_r = k)$

$\Rightarrow \quad L(P,f) = k(b-a) \qquad \left(\because \sum_{r=1}^{n} \delta_r = b - a\right)$

Similarly, $\quad U(P,f) = k(b-a)$

Now, $\quad \underline{\int_a^b} f(x)\,dx = \text{sup. } \{L(P,f) : P \in \phi[a, b]\} = k(b-a)$

and $\quad \overline{\int_a^b} f(x)\,dx = \text{inf. } \{U(P,f) : P \in \phi[a, b]\} = k(b-a)$

Thus, $\quad \underline{\int_a^b} f(x)\,dx = \overline{\int_a^b} f(x)\,dx \Rightarrow$ **$f(x)$ is Riemann integrable on $[a, b]$**

Also, $\quad \int_a^b f(x)\,dx = \mathbf{k(b-a)}$

Example 2. Show that the function $f(x) = x^2$ is integrable on $[0, a]$ and $\int_0^a x^2\,dx = \frac{a^3}{3}$.

Solution : Since $f(x) = x^2$ is an increasing function, it is bounded on $[0, a]$.

Consider, the partition

$$P_n = \left\{0, \frac{a}{n}, \frac{2a}{n}, \ldots\ldots, \frac{ra}{n}, \ldots\ldots, \frac{na}{n}\right\}$$

of $[0, a]$, formed by dividing $[0, a]$, into n equal parts.

$$I_r = r^{\text{th}} \text{ subinterval} = \left[\frac{(r-1)a}{n}, \frac{ra}{n}\right], \text{ length of } I_r = \delta_r = \frac{a}{n}$$

Now, $\quad M_r = \text{sup. of } f \text{ in } I_r = \left(\frac{ra}{n}\right)^2$ and $m_r = \text{inf. } f \text{ of in } I_r = \left[\frac{(r-1)a}{n}\right]^2$

$$\text{Now, } U(P, f) = \sum_{r=1}^{n} M_r \cdot \delta_r = \sum_{r=1}^{n} \frac{r^2 a^2}{n^2} \cdot \frac{a}{n}$$

$$= \frac{a^3}{n^3} \sum_{r=1}^{n} r^2$$

$$= \frac{a^3}{n^3} \frac{n(n+1)(2n+1)}{6} = \frac{a^3}{6}\left(1 + \frac{1}{n}\right)\left(2 + \frac{1}{n}\right)$$

$$L(P, f) = \sum_{r=1}^{n} m \cdot \delta_r = \sum_{r=1}^{n} \frac{(r-1)^2 \cdot a^2}{n^2} \cdot \frac{a}{n}$$

$$= \frac{a^3}{n^3} \sum_{r=1}^{n} (r-1)^2$$

$$= \frac{a^3}{n^3} \frac{(n-1) \cdot n \cdot (2n-1)}{6} = \frac{a^3}{6}\left(1 - \frac{1}{n}\right)\left(2 - \frac{1}{n}\right)$$

$$\text{Now, } \int_{-0}^{b} f(x)\, dx = \lim_{n \to \infty} L(P, f) = \lim_{n \to \infty} \frac{a^3}{6}\left(1 - \frac{1}{n}\right)\left(2 - \frac{1}{n}\right) = \frac{a^3}{3}$$

$$\text{Now, } \int_{0}^{-a} f(x)\, dx = \lim_{n \to \infty} U(P, f) = \lim_{n \to \infty} \frac{a^3}{6}\left(1 + \frac{1}{n}\right)\left(2 + \frac{1}{n}\right) = \frac{a^3}{3}$$

Thus, $\int_{\underline{0}}^{b} f(x)\, dx = \int_{0}^{-a} f(x)\, dx$ and therefore $f(x) = x^2$ **is integrable on [0, a]. Also,**

$$\int_0^a x^2\, dx = \frac{a^3}{3}.$$

Example 3. Show that the function $f(x) = 3x + 1$ is integrable on [1, 2] and

$$\int_1^2 (3x + 1)\, dx = \frac{11}{2}.$$

Solution : Clearly $f(x) = 3x + 1$, is bounded on [1, 2]. Consider the partition

$$P = \left\{1, 1 + \frac{1}{n}, 1 + \frac{2}{n}, \ldots\ldots, 1 + \frac{r}{n}, \ldots\ldots, 2\right\}$$

of [1, 2], formed by dividing [1, 2], into n equal parts.

The, $I_r = r^{\text{th}}$ sub interval $= \left[1 + \frac{r-1}{n}, 1 + \frac{r}{n}\right]$ and length of $I_r = \delta_r = \frac{1}{n}$

The function $f(x) = 3x + 1$ is an increasing function in [1, 2]. Thus

$$M_r = \text{sup. of } f \text{ in } I_r = 3\left(1 + \frac{r}{n}\right) + 1 = 4 + \frac{3r}{n}$$

$$\text{and} \qquad m_r = \text{inf. of } f \text{ in } I_r = 3\left(1 + \frac{r-1}{n}\right) + 1 = 4 + \frac{3(r-1)}{n}$$

$$\text{Now, } U(P, f) = \sum_{r=1}^{n} M_r \cdot \delta_r = \sum_{r=1}^{n} \left(4 + \frac{3r}{n}\right)\frac{1}{n}$$

$$= \frac{4}{n}\sum_{r=1}^{n} 1 + \frac{3}{n^2}\sum_{r=1}^{n} r$$

$$= \frac{4}{n}\cdot n + \frac{3}{n^2}\frac{n(n+1)}{2} = \mathbf{4 + \frac{3}{2}\left(1 + \frac{1}{n}\right)}$$

$$L(P, f) = \sum_{r=1}^{n} m_r \cdot \delta_r = \sum_{r=1}^{n}\left[4 + \frac{3(r-1)}{n}\right]\frac{1}{n}$$

$$= \frac{4}{n}\sum_{r=1}^{n} 1 + \frac{3}{n^2}\sum_{r=1}^{n}(r-1)$$

$$= \frac{4}{n}\cdot n + \frac{3}{n^2}\frac{(n-1)n}{2} = \mathbf{4 + \frac{3}{2}\left(1 - \frac{1}{n}\right)}$$

Now, $$\underline{\int_{-1}^{2}} f(x)\,dx = \lim_{n\to\infty} L(P, f)$$

$$= \lim_{n\to\infty}\left[4 + \frac{3}{2}\left(1 - \frac{1}{n}\right)\right] = 4 + \frac{3}{2} = \frac{11}{2}$$

$$\overline{\int_{1}^{2}} f(x)\,dx = \lim_{n\to\infty} U(P, f)$$

$$= \lim_{n\to\infty}\left[4 + \frac{3}{2}\left(1 + \frac{1}{n}\right)\right] = 4 + \frac{3}{2} = \frac{11}{2}$$

Thus, $\underline{\int_{-1}^{2}} f(x)\,dx = \overline{\int_{1}^{2}} f(x)\,dx$ and hence $\mathbf{f(x) = 3x + 1}$ **is integrable on [1, 2].**

Also, $\int_{1}^{2}(3x + 1)\,dx = \mathbf{\frac{11}{2}}$.

Example 4. Show that the function $f(x)$ defined by $f(x) = \begin{cases} 1 & x \text{ is rational} \\ 0 & x \text{ is irrational}\end{cases}$ is not integrable over [0, 1].

This function is known as Dirichlets' function.

Solution : Let $P = \{a, x_0, x_1, x_2, \ldots, x_n = b\}$ be a partition [0, 1]. Clearly we have,

$$0 \le f(x) \le 1, \text{ for all } x \in [0, 1]$$

$I_r = r^{th}$ sub interval $= [x_{r-1}, x_r]$ and length of $I_r = \delta_r$

Now, $M_r = 1$ and $m_r = 0$, for $r = 1, 2, \ldots n$

where, $M_r =$ sup. of f in I_r and $m_r =$ inf. of f in I_r

$$U(P, f) = \sum_{r=1}^{n} M_r \cdot \delta_r = \sum_{r=1}^{n} 1\cdot\delta_r = 1 \qquad \left(\because \sum_{r=1}^{n}\delta_r = 1 - 0 = 1\right)$$

$$L(P, f) = \sum_{r=1}^{n} m_r \cdot \delta_r = \sum_{r=1}^{n} 0\cdot\delta_r = 0$$

Thus, $$\underline{\int_{0}^{1}} f(x)\,dx = 0 \ne \overline{\int_{0}^{1}} f(x)\,dx = 1$$

Hence, $\mathbf{f(x)}$ **is not integrable on [0, 1].**

Example 5. Show that the function $f(x)$ defined by $f(x) = \begin{cases} 1 & x \in Q \\ -1 & x \in R - Q \end{cases}$ is not integrable over $[a, b]$. Here Q is the set of rationals and R the set of reals.

Solution : Clearly, $-1 \le f(x) \le 1$, for all $x \in [a, b]$. Thus $f(x)$ is bounded on $[a, b]$. Also sup. $f = 1$ and inf. $f = -1$.

Let, $P = \{a = x_0, x_1, x_2, ..., x_n = b\}$ be a partition of $[a, b]$.

The, $I_r = r^{th}$ sub interval $= [x_{r-1}, x_r]$ and length of $I_r = \delta_r$.

Now, $M_r =$ sup. of f in $I_r = 1$ and $m_r =$ inf. of f in $I_r = -1$.

$$\text{Thus, } U(P, f) = \sum_{r=1}^{n} M_r \cdot \delta_r = \sum_{r=1}^{n} 1 \cdot \delta_r = b - a \qquad \left(\because \sum_{r=1}^{n} \delta_r = b - a \right)$$

$$L(P, f) = \sum_{r=1}^{n} m_r \cdot \delta_r = \sum_{r=1}^{n} (-1) \cdot \delta_r = -(b - a)$$

$$\text{Thus, } \int_{-a}^{b} f(x)\,dx = -(b-a) \neq (b-a) = \int_{a}^{-b} f(x)\,dx$$

Hence **$f(x)$ is not integrable** over $[a, b]$.

Note :

1. The Riemann integral is defined subjected to two conditions : **(i)** f is bounded on the interval $[a, b]$ and **(ii)** the interval is a closed interval.
2. The given function need not be Riemann integrable on a closed interval $[a, b]$ even it is bounded on $[a, b]$ (see examples 4 and 5 above].

The following alternative condition for $f(x)$ to be integrable is useful.

Let $f(x)$ be a bounded function defined on the closed interval $[a, b]$ and

$$P[a = x_0, x_1, ..., x_n = b] \in \phi[a, b]$$

Let $\{\xi_1, \xi_2, \xi_3, ..., \xi_n\} \in [a, b]$ be such that $x_{r-1} \le \xi_r \le x_r$, for $r = 1, 2, ..., n$. Then,

$$f(x) \textbf{ is integrable on } [a, b] = \int_{a}^{b} f(x)\,dx = \lim_{||P|| \to 0} \sum_{r=1}^{n} f(\xi)\,\delta_r$$

EXERCISES

1. If $f(x)$ is defined on $[3, 4]$ by $f(x) = 8$, $\forall\, x \in [3, 4]$, is $f(x)$ is R-integrable on $[3, 4]$; why?
2. Show that $f(x) = x$ on $[0, 1]$ is R-integrable.
3. If $f(x) = x^2$ for all $x \in [0, 1]$, show that $f \in R[0, 1]$ and $\int_{0}^{1} f(x)\,dx = \frac{1}{3}$.
4. Give an example of a bounded function which is not integrable.
5. If $f(x) = \cos x$, $\forall\, x \in \left[0, \frac{\pi}{2}\right]$, show that f is integrable on $\left[0, \frac{\pi}{2}\right]$ and $\int_{0}^{1/2} \cos x\,dx = 1$.

1.9 Algebra of Integrable Functions

We shall see properties of Riemann integrable functions in the form of the theorems.

Definition : If $f(x)$ is a bounded function defined on $[a, b]$, then we define

$$\textbf{(a)} \int_{a}^{a} f(x)\,dx = 0 \textbf{ and } \int_{a}^{b} f(x)\,dx = -\int_{b}^{a} f(x)\,dx$$

Theorem 1 : If $f \in R[a, b]$, then $-f \in R[a, b]$ and $\int_a^b (-f)(x)\,dx = -\int_a^b f(x)\,dx$.

Proof : By data, $f \in R[a, b]$. Thus we have,

$$\underline{\int_a^b} f(x)\,dx = \overline{\int_a^b} f(x)\,dx = \int_a^b f(x)\,dx \qquad \text{.... (1)}$$

Let, $P = [a = x_0, x_1, x_2, \ldots\ldots, x_n = b]$ be a partition of $[a, b]$. Let M_r and m_r be supremum and infimum of $f(x)$ in $I_r = [x_{r-1}, x_r]$ and δ_r = length of I_r.

Since $f(x)$ is bounded on $[a, b]$, the function $-f(x)$ is also bounded on $[a, b]$.

Further, $\quad$ sup. $(-f)$ on I_r = $-$inf. f on $I_r = -m_r$

and, $\quad$ inf. $(-f)$ on I_r = $-$sup. f on $I_r = -M_r$

Now, $$U(P, -f) = \sum_{r=1}^{n} (-m_r) \cdot \delta_r = -\sum_{r=1}^{n} m_r \cdot \delta_r = -L(P, f)$$

and $$L(P, -f) = \sum_{r=1}^{n} (-M_r) \cdot \delta_r = -\sum_{r=1}^{n} M_r \cdot \delta_r = -U(P, f)$$

Consider,
$$\begin{aligned}\underline{\int_a^b} (-f)(x)\,dx &= \text{sup. } \{L(P, -f) : P \in \phi[a, b]\} \\ &= \text{sup. } \{-U(P, f) : P \in \phi[a, b]\} \\ &= -\text{inf. } \{U(P, f) : P \in \phi[a, b]\} \\ &= -\overline{\int_a^b} f(x)\,dx = -\int_a^b f(x)\,dx \qquad \text{(from (1))}\end{aligned}$$

Again consider,
$$\begin{aligned}\overline{\int_a^b} f(x)\,dx &= \text{inf. } \{U(P, f) : P \in \phi[a, b\} \\ &= \text{inf. } \{-L(P, f) : P \in \phi[a, b]\} \\ &= -\text{sup. } \{L(P, f) : P \in \phi[a, b]\} \\ &= -\underline{\int_a^b} f(x)\,dx = -\int_a^b f(x)\,dx \qquad \text{(from (1))}\end{aligned}$$

Thus, $$\underline{\int_a^b} (-f)(x)\,dx = \overline{\int_a^b} f(x)\,dx = -\int_a^b f(x)\,dx$$

Hence $-f$ is Riemann integrable and $\underline{\int_a^b} (-f)(x)\,dx = -\int_a^b f(x)\,dx$.

Theorem 2 : If $f, g \in R[a, b]$ and k be any scalar, then

(a) $kf \in [a + b]$ and $\int_a^b (kf)(x)\,dx = k\int_a^b f(x)\,dx$

(b) $f + g \in R[a, b]$ and $\int_a^b (f + g)(x)\,dx = \int_a^b f(x)\,dx + \int_a^b f(x)\,dx$

Proof : By data $f \in R[a, b]]$ and $g \in R[a, b]$. Thus we have,

$$\underline{\int_a^b} f(x)\,dx = \overline{\int_a^b} f(x)\,dx = \int_a^b f(x)\,dx \qquad \text{.... (1)}$$

$$\overline{\int_a^b} g(x)\,dx = \underline{\int_a^b} g(x)\,dx = \int_a^b g(x)\,dx \qquad \text{.... (2)}$$

Let $P = \{a = x_0, x_1, \ldots\ldots, x_n = b\}$ be a partition of $[a, b]$. Let M_r and m_r be supremum and infimum of $f(x)$ in $I_r = [x_{r-1}, x_r]$ and δ_r = length of I_r.

(a) Since $f(x)$ is bounded on $[a, b]$, the function $k \cdot f(x)$ is also bounded on $[a, b]$. Further,

$$\text{sup. } kf \text{ on } I_r = k \cdot \text{sup. } f \text{ on } I_r = kM_r$$

and

$$\text{imf. } kf \text{ on } I_r = k \cdot \text{inf. } f \text{ on } I_r = km_r$$

Now,

$$U(P, kf) = \sum_{r=1}^{n} (kM_r) \cdot \delta_r = k \sum_{r=1}^{n} M_r \cdot \delta_r = k \cdot U(P, f)$$

$$L(P, kf) = \sum_{r=1}^{n} (km_r) \cdot \delta_r = k \sum_{r=1}^{n} m_r \cdot \delta_r = k \cdot L(P, f)$$

Consider,

$$\begin{aligned} \underline{\int_a^b} (kf)(x)\, dx &= \text{sup. } \{L(P, kf) : P \in \phi[a, b]\} \\ &= \text{sup. } \{k \cdot L(P, f) : P \in \phi[a, b]\} \\ &= k \cdot \text{sup. } \{L(P, f) : P \in \phi[a, b]\} \\ &= k \cdot \underline{\int_a^b} f(x)\, dx = k \cdot \int_a^b f(x)\, dx \end{aligned} \qquad \text{(by (1))}$$

Again consider,

$$\begin{aligned} \overline{\int_a^b} (kf)(x)\, dx &= \text{inf. } \{U(P, kf) : P \in \phi[a, b]\} \\ &= \text{inf. } \{k \cdot U(P, f) : P \in \phi[a, b]\} \\ &= k \cdot \text{inf. } \{U(P, f) : P \in \phi[a, b]\} \\ &= k \cdot \overline{\int_a^b} f(x)\, dx = k \cdot \int_a^b f(x)\, dx \end{aligned} \qquad \text{(by (1))}$$

Thus,

$$\underline{\int_a^b} (kf)(x)\, dx = \overline{\int_a^b} (kf)(x)\, dx = k \int_a^b f(x)\, dx$$

Hence, $\boldsymbol{kf \in R[a, b]}$ **and** $\boldsymbol{\int_a^b (kf)(x)\, dx = k \int_a^b f(x)\, dx}$.

(b) Since $f(x)$ and $g(x)$ are bounded on $[a, b]$ the function $(f + g)(x)$ is also bounded on $[a, b]$. Let $\varepsilon > 0$, then there exists $\delta_1 > 0$ and $\delta_2 > 0$ such that

$$U(P_1, f) - L(P_1, f) < \frac{\varepsilon}{2}, \text{ with } \| P_1 \| < \delta_1 \qquad \text{.... (2)}$$

and

$$U(P_2, g) - U(P_2, g) < \frac{\varepsilon}{2}, \text{ with } \| P_2 \| < \delta_2 \qquad \text{.... (3)}$$

Let, $P = (P_1 \cup P_2)$. Then $\| P \| < \| P_1 \| < \delta_1$ and $\| P \| < \| P_2 \| < \delta_2$.

Thus (2) and (3) are true for the partition P.

Now we have,

$$\begin{aligned} & W(P, f+g) = U(P, f+g) - L(P, f+g) \\ \Rightarrow\ & W(P, f+g) \le \{U(P, f) + U(P, g) - L(P, f) - L(P, g)\} \\ \Rightarrow\ & W(P, f+g) \le \{U(P, f) - L(P, f)\} + \{U(P, g) - L(P, g)\} \\ \Rightarrow\ & W(P, f+g) \le \frac{\varepsilon}{2} + \frac{\varepsilon}{2} \qquad \text{[from (1) and (2)]} \\ \Rightarrow\ & W(P, f+g) \le \varepsilon \end{aligned}$$

Thus for each $\varepsilon > 0$, there exists $\delta = \max \{\delta_1, \delta_2\}$, such that,

$$0 \le W(P, f+g) < \varepsilon, \quad \text{with } \| P \| < \delta$$

Thus the function $(f+g)(x)$ is Riemann integrable. That is $f + g \in R[a, b]$.

Now, $$f \in R[a,b] \Rightarrow \int_a^b f(x)\,dx = \lim_{\|P\|\to 0} \sum_{r=1}^{n} f(\alpha_r)\cdot\delta_r$$

and, $$g \in R[a,b] \Rightarrow \int_a^b g(x)\,dx = \sum_{r=1}^{n} g(\alpha_r)\cdot\delta_r$$

Consider, $$\int_a^b (f+g)(x)\,dx = \lim_{\|P\|\to 0} \sum_{r=1}^{n} [f(\alpha_r)+g(\alpha_r)]\cdot\delta_r$$

$$= \lim_{\|P\|\to 0} \sum_{r=1}^{n} f(\alpha_r)\cdot\delta_r + \lim_{\|P\|\to 0} \sum_{r=1}^{n} g(\alpha_r)\cdot\delta_r$$

$$\Rightarrow \quad \int_a^b (f+g)(x)\,dx = \int_a^b f(x)\,dx - \int_a^b g(x)\,dx.$$

Corollary : 1. If $f, g \in R[a, b]$ then $f - g \in R[a, b]$ and

$$\int_a^b (f-g)(x)\,dx = \int_a^b f(x)\,dx - \int_a^b g(x)\,dx.$$

2. If $f, g \in R[a, b]$ and k, l are non zero scalars, then $kf + lg \in R[a, b]$ and,

$$\int_a^b (kf+lg)(x)\,dx = k\int_a^b f(x)\,dx + l\int_a^b g(x)\,dx.$$

Proof : By data, $f \in R[a,b] \Rightarrow k\cdot f \in R[a,b]$ and $g \in R[a,b] \Rightarrow l\cdot g \in R[a,b]$

Again, $k\cdot f \in R[a,b]$ and $l\cdot g \in R[a,b] \Rightarrow k\cdot f + l\cdot g \in R[a,b]$

Also, $$\int_a^b (kf+lg)(x)\,dx = \int_a^b (kf)(x)\,dx + \int_a^b (lg)(x)\,dx$$

$$= k\int_a^b f(x)\,dx + l\int_a^b g(x)\,dx.$$

Theorem : If $f \in R[a, b]$ then $|f| \in R[a, b]$.

Proof : By $f \in R[a, b]$. Thus for every $\varepsilon > 0$, there exists a partition $P = \{a = x_0, x_1, ..., x_n = b\}$ such that

$$0 \le U(P,f) - L(P,f) < \varepsilon \qquad .... (1)$$

Again $f \in R[a, b] \Rightarrow f$ is bounded on $[a, b]$.

Thus there exists $k > 0$ such that

$$|f(x)| < k, \ \forall x \in [a,b] \Rightarrow |f| \text{ bounded on } [a,b]$$

Let m_r, M_r be infimum and supremum on $I_r = [x_{r-1}, x_r]$.

Also, let m_r', M_r' be infimum and supremum of $|f(x)|$ on I_r.

Now, for $a, b \in I_r$, we have,

$$||f|(\alpha) - |f|(\beta)| = ||f|(\alpha) - |f|(\beta)| \le |f(\alpha) - f(\beta)|$$

$$\Rightarrow \quad M_r' - m_r' \le M_r - m_r, \quad \text{for } n = 1, 2, ..., n$$

Now, $$U(P, |f|) - L(P, |f|) = \sum_{r=1}^{n} (M_r' - m_r')\cdot\delta_r \le \sum_{r=1}^{n} (M_r - m_r)\cdot\delta_r$$

$\Rightarrow \quad U(P, |f|) - L(P, |f|) \leq U(P,f) - L(P,f) < \varepsilon$ [from (1)]

Hence, $|f| \leq R[a, b]$.

Note : The converse of the above theorem, is not true. That is if $|f|$ is Riemann integrable on $[a, b]$, then f need not be Riemann integrable on $[a, b]$.

For example, consider the function $f(x)$ defined on $[a, b]$, by

$$f(x) = \begin{cases} 0 & x \in \mathbf{Q} \\ 1 & x \in \mathbf{R} - \mathbf{Q} \end{cases}$$

Now, $|f(x)| = 1, \forall x \in R$ and $|f|$ is a constant function. Thus it is Riemann integrable. But $f(x)$ is not Riemann integrable on $[a, b]$, (see example 5).

Theorem : If $f, g \in R[a, b]$ then $f \cdot g \in R[a, b]$.

Proof : By data, $f, g \in R[a, b] \Rightarrow f + g \in R[a, b]$ and $f - g \in R[a, b]$

Thus, $\quad (f+g)^2 \in R[a, b]$ and $(f-g)^2 \in R[a, b]$

$\Rightarrow \quad (f+g)^2 - (f-g)^2 \in R[a, b]$

$\Rightarrow \quad \frac{1}{4}[(f+g)^2 - (f-g)^2] \in R[a, b] \Rightarrow f \cdot g \in \mathbf{R}[a, b]$

Note : Even, if f and g are not Riemann integrable on $[a, b]$, the function $f \cdot g$ may be integrable on $[a, b]$.

For example consider the functions f and g defined on $[a, b]$ by

$$f(x) = \begin{cases} 0 & x \in \mathbf{Q} \\ 1 & x \in \mathbf{R} - \mathbf{Q} \end{cases} \text{ and } g(x) = \begin{cases} 1 & x \in \mathbf{Q} \\ 0 & x \in \mathbf{R} - \mathbf{Q} \end{cases}$$

Now, $\quad (fg)(x) = f(x) \cdot g(x) = 0, \ \forall x \in R$

Thus fg is a constant function and therefore integrable. But f and g are not integrable on $[a, b]$.

Theorem : If $f, g \in R[a, b]$ and there exists $\lambda > 0$ such that $|g(x)| \geq \lambda$, for all $x \in [a, b]$ then $\frac{f}{g} \in R[a, b]$.

Proof : By data, $f, g \in R[a, b]$. Thus f and g are bounded on $[a, b]$. Therefore, there exists $k > 0$, such that

$$|f(x)| < k, \quad |g(x)| < k, \quad \text{for all } x \in [a, b]$$

$$\Rightarrow \quad \forall x \in [a, b], \quad \left|\frac{f(x)}{g(x)}\right| \leq \frac{k}{\lambda} \quad \text{(by data) .... (1)}$$

$$\Rightarrow \quad \forall x \in [a, b], \quad \left|\frac{f}{g}(x)\right| \leq \frac{k}{\lambda} \Rightarrow \frac{f}{g} \text{ is bounded on } [a, b]$$

Let, $\quad P = \{a = x_0, x_1, x_2, ..., x_n = b\}$ be a partition of $[a, b]$.

Let, $\quad M_r', m_r'$ be supremum and infimum of f on $I_r = [x_{r-1}, x_r]$

$\quad M_r'', m_r''$ be supremum and infimum of g on $I_r = [x_{r-1}, x_r]$

Also, $\quad$ let M and m be supremum and infimum of $\frac{f}{g}$ on I_r.

Now for all $x, y \in [a, b]$ we get,

$$\left|\frac{f(y)}{g(y)} - \frac{f(x)}{g(x)}\right| = \left|\frac{g(x)\cdot f(y) - f(x)\cdot g(y)}{g(y)\cdot g(x)}\right|$$

$$= \frac{g(x)\cdot f(y) - g(x)\cdot f(x) - f(x)\cdot g(y) + g(x)\cdot f(x)}{g(y)\cdot g(x)}$$

$$= \frac{g(x)\left[f(y) - f(x)\right] - f(x)\left[g(y) - g(x)\right]}{g(y)\cdot g(x)}$$

$$\leq \frac{|g(x)|}{|g(x)||g(y)|}|f(y) - f(x)| + \frac{|-f(x)|}{|g(x)||g(y)|}|g(y) - g(x)|$$

$$\leq \frac{k}{\lambda^2}|f(y) - f(x)| + \frac{k}{\lambda^2}|g(y) - g(x)|$$

Thus,
$$M_r - m_r \leq \frac{k}{\lambda^2}(M_r' - m_r') + \frac{k}{\lambda^2}(M_r'' - m_r'')]$$

By data, $f, g \in R[a, b]$. Thus for $\varepsilon > 0$, there exists $P \in \phi[a, b]$, such that,

$$U(P, f) - L(P, f) < \frac{\lambda^2\varepsilon}{2k} \text{ and } U(P, g) - L(P, g) < \frac{\lambda^2\varepsilon}{2k}$$

Now, we have,

$$\sum_{r=1}^{n}(M_r - m_r)\cdot\delta_r \leq \frac{k}{\lambda^2}\sum_{r=1}^{n}\left(M_r'' - m_r'\right)\cdot\delta_r + \frac{k}{\lambda^2}\sum_{r=1}^{n}\left(M_r'' - m_r''\right)\cdot\delta_r$$

$$\Rightarrow \sum_{r=1}^{n}(M_r - m_r)\cdot\delta_r \leq \frac{k}{\lambda^2}\left[U(P, f) - L(P, f)\right] + \frac{k}{\lambda^2}\left[U(P, g) - L(P, g)\right]$$

$$\Rightarrow \sum_{r=1}^{n}(M_r - m_r)\cdot\delta_r \leq \frac{k}{\lambda^2}\left(\frac{\lambda^2\varepsilon}{2k}\right) + \frac{k}{\lambda^2}\left(\frac{\lambda^2\varepsilon}{2k}\right) = \varepsilon$$

Thus for each $\varepsilon > 0$, there exists $P \in [a, b]$ such that

$$U\left(P, \frac{f}{g}\right) - L\left(P, \frac{f}{g}\right) < \varepsilon$$

Hence, $\frac{f}{g}$ **is Riemann integrable on $[a, b]$.**

Theorem : If $f \in [a, b]$ and $a < c < b$, then $f \in R[a, c]$ and $f \in R[c, b]$. Also

$$\int_a^b f(x)\,dx = \int_a^c f(x)\,dx + \int_c^b f(x)\,dx$$

Proof : By data $f \in R[a, b]$. Thus f is bounded on $[a, b]$. Hence f is bounded on $[a, c]$ and $[c, b]$.

Let P be a partition of $[a, b]$ and $Q = = P \cup \{c\}$, be another partition. Clearly $P \subset Q$. Thus $U(Q, f) \leq U(P, f)$ and $L(P, f) \leq L(Q, f)$.

Let $Q_1 = Q \cap [a, c]$ and $Q_2 = Q \cap [c, b]$. Then Q_1 is a partition of $[a, b]$ and Q_2 is a partition of $[c, b]$. Further $Q_1 \cup Q_2 = Q$ and $Q_1 \cap Q_2 = \{c\}$.

Now, $\quad L(P,f) \le L(Q,f)$

$\Rightarrow \quad L(P,f) \le L(Q_1,f) + L(Q_2,f) \qquad (\because Q = Q_1 \cup Q_2)$

$$\Rightarrow \quad L(P,f) \le \underline{\int_a^c} f(x)\,dx + \underline{\int_c^b} f(x)\,dx$$

$$\Rightarrow \quad L(P,f) \in \overline{\int_a^c} f(x)\,dx + \overline{\int_c^b} f(x)\,dx$$

$\Rightarrow \quad L(P,f) \le U(Q_1,f) + U(Q_2,f)$

$\Rightarrow \quad L(P,f) \le U(Q,f) \Rightarrow L(P,f) \le U(P,f)$

$$\text{Hence,} \quad L(P,f) \le \underline{\int_a^c} f(x)\,dx + \underline{\int_c^b} f(x)\,dx \le \overline{\int_a^c} f(x)\,dx + \overline{\int_c^b} f(x)\,dx \le U(P,f)$$

This is true for all $P \in \phi[a, b]$.

$$\Rightarrow \quad \underline{\int_a^b} f(x)\,dx \le \underline{\int_a^c} f(x)\,dx + \underline{\int_c^b} f(x)\,dx$$

$$\Rightarrow \quad \underline{\int_a^b} f(x)\,dx \le \overline{\int_a^c} f(x)\,dx + \overline{\int_c^b} f(x)\,dx \le \overline{\int_a^b} f(x)\,dx$$

$$\text{But, } f \in R[a,b]. \text{ Thus, } \quad \underline{\int_a^b} f(x)\,dx = \overline{\int_a^b} f(x)\,dx = \int_a^b f(x)\,dx$$

$$\Rightarrow \underline{\int_a^c} f(x)\,dx + \underline{\int_c^b} f(x)\,dx = \overline{\int_a^c} f(x)\,dx + \overline{\int_c^b} f(x)\,dx = \int_a^b f(x)\,dx \qquad \text{.... (1)}$$

$$\Rightarrow \quad \left[\overline{\int_a^c} f(x)\,dx - \underline{\int_a^c} f(x)\,dx\right] + \left[\overline{\int_c^b} f(x)\,dx - \underline{\int_c^b} f(x)\,dx\right] = 0$$

Since each term is non negative, we have,

$$\overline{\int_a^c} f(x)\,dx - \underline{\int_a^c} f(x)\,dx = 0 \text{ and } \overline{\int_c^b} f(x)\,dx - \underline{\int_c^b} f(x)\,dx = 0$$

$$\Rightarrow \quad \overline{\int_a^c} f(x)\,dx = \underline{\int_a^c} f(x)\,dx \text{ and } \overline{\int_c^b} f(x)\,dx = \underline{\int_c^b} f(x)\,dx$$

$\Rightarrow \quad$ **$f \in R[a, c]$ and $f \in R[c, b]$**

Also from (1) we have, $\int_a^b f(x)\,dx = \int_a^c f(x)\,dx + \int_c^b f(x)\,dx$.

Theorem : If $f \in R[a, c]$, $f \in R[c, b]$ and $a < c < b$ then $f \in [a, b]$.

Proof : By data $f \in R[a, c]$ and $f \in R[c, b]$. Thus f is a bounded on $[a, c]$ and $[c, b]$. Hence f is bounded on $[a, b]$.

Now, for $\varepsilon > 0$, there exists $P_1 \in \phi[a, c]$ and $P_2 \in \phi[c, b]$, such that

$$U(P_1,f) - L(P_1,f) < \frac{\varepsilon}{2} \quad \text{and} \quad U(P_2,f) - L(P_2,f) < \frac{\varepsilon}{2}$$

Let, $P = (P_1 \cup P_2)$. Then $P \in \phi[a, b]$.

Consider, $\quad U(P,f) - L(P,f) = [U(P_1,f) - U(P_2,f)] - [L(P_1,f) - L(P_2,f)]$

$\qquad\qquad = [U(P_1,f) - L(P_2,f)] + [U(P_2,f) - L(P_2,f)]$

$\Rightarrow \quad U(P,f) - L(P,f) < \frac{\varepsilon}{2} + \frac{\varepsilon}{2} = \varepsilon \Rightarrow$ **$f \in R[a, b]$**

Corollary : If $f \in R[a, b]$ and $[a_1, b_1] \subset [a, b]$ then $f \in R[a_1, b_1]$.

Note : The above corollary implies that if f is integrable on $[a, b]$, then it is integrable in any sub interval of $[a, b]$.

1.10 Integrability of Continuous and Mototonic Functions

Theorem : If $f(x)$ is a continuous function defined on $[a, b]$, then $f(x)$ is R-integrable.

Proof : By data $f(x)$ is continuous on $[a, b]$. Thus $f(x)$ is bounded on $[a, b]$ and it is uniformly continuous on $[a, b]$.

Thus, given $\varepsilon > 0$, there exists a $\delta > 0$, such that for $x, y \in [a, b]$, we have,

$$|f(x) - f(y)| < \frac{\varepsilon}{a-b}, \quad \text{such that } |x - y| < \delta$$

Consider a partition $P = \{a = x_0, x_1, ..., x_n = b\}$ of $[a, b]$.

Let r, such that $1 \le r \le n$, be arbitrary and $x', y' \in [x_{r-1}, x_r]$. Then,

$$|x' - y'| \le \frac{b-a}{n}$$

We shall choose $n \in N$, such that $\frac{b-a}{n} < \delta$. Then, $|x' - y'| < \delta$

$$\Rightarrow \qquad |f(x') - f(y')| < \frac{\varepsilon}{a-b}, \quad \forall\, x', y' \in [x_{r-1}, x_r]$$

Let m_r and M_r be infimum and supremum of $f(x)$ in I_r.

Since f is continuous on I_r, there exists $\alpha_r, \beta_r \in I_r$, such that $m_r = f(\alpha_r)$ and $M_r = f(\beta_r)$.

$$\therefore \qquad M_r - m_r = |f(\alpha_r) - f(\beta_r)| < \frac{\varepsilon}{a-b}, \quad \text{for } r = 1, 2, ... , n$$

Now,

$$\begin{aligned} W(P, f) &= U(P, f) - L(P, f) \\ &= \sum_{r=1}^{n} (M_r - m_r)\, \delta_r \\ &< \sum_{r=1}^{n} \left(\frac{\varepsilon}{a-b}\right) \delta_r \qquad \left(\because M_r - m_r < \frac{\varepsilon}{a-b}\right) \end{aligned}$$

$$\Rightarrow \qquad U(P, f) - L(P, f) < \frac{\varepsilon}{a-b} \sum_{r=1}^{n} \delta_r$$

$$\Rightarrow \qquad U(P, f) - L(P, f) < \frac{\varepsilon}{a-b} \cdot (a-b) \qquad \left(\because \sum_{r=1}^{n} \delta_r = a - b\right)$$

$$\Rightarrow \qquad U(P, f) - L(P, f) < \varepsilon$$

Thus, for all $\varepsilon > 0$, there exists $P \in \phi[a, b]$, such that, $U(P, f) - L(P, f) < \varepsilon$.

Hence $f(x)$, is R-integrable on $[a, b]$.

Note : The converse of the above theorem is not true. That is an integrable function need not be continuous. Thus the continuity of a function is only a sufficient condition, for integrability.

We know that $f(x) = x^n$, $n \in N$, is continuous on $[a, b] \subset R$, thus $f(x) = x^n$ is integrable on $[a, b]$. Again $f(x) = e^x$ is continuous on $[a, b] \subset R$, and therefore it is integrable on $[a, b]$.

Theorem : If $f(x)$ is a monotonic function on $[a, b]$, then f is R-integrable on $[a, b]$.

Proof : Let $f(x)$ be a monotonically increasing function on $[a, b]$.

Then $f(a) \leq f(x) \leq f(b)$, $\forall\, x \in [a, b]$. Also $f(x)$ is bounded on $[a, b]$ and inf. $f(x) = f(a)$ and sup. $f(x) = f(b)$.

Let $\varepsilon > 0$ and $P = \{a = x_0, x_1, x_2, ..., x_n = b\}$ be a partition of $[a, b]$ such that,

$$\delta_r < \frac{\varepsilon}{f(b) - f(a) + 1}, \quad r = 1, 2, ..., n \qquad (\delta_r = x_{r-1} - x_r)$$

Let m_r, M_r be the infimum and supremum of $f(x)$ on $I_r = [x_{r-1} - x_r]$. Then, $m_r = f(x_{r-1})$, $M_r = f(x_r)$.

$$\text{Now, } U(P, f) - L(P, f) = \sum_{r=1}^{n} (M_r - m_r)\,\delta_r = \sum_{r=1}^{n} \left[f(x_r) - f(x_{r-1})\right] \delta_r$$

$$\Rightarrow \quad U(P, f) - L(P, f) < \sum_{r=1}^{n} \left[f(x_r) - f(x_{r-1})\right] \frac{\varepsilon}{f(b) - f(a) + 1}$$

$$\Rightarrow \quad U(P, f) - L(P, f) < \frac{\varepsilon}{f(b) - f(a) + 1} \sum_{r=1}^{n} \left[f(x_r) - f(x_{r-1})\right]$$

$$\Rightarrow \quad U(P, f) - L(P, f) < \frac{\varepsilon}{f(b) - f(a) + 1} [f(x_1) - f(x_0) + f(x_2) - f(x_1) + ... + f(x_n) - f(x_{n-1})]$$

$$\Rightarrow \quad U(P, f) - L(P, f) < \frac{\varepsilon}{f(b) - f(a) + 1} [f(x_n) - f(x_0)]]$$

$$\Rightarrow \quad U(P, f) - L(P, f) < \frac{\varepsilon}{f(b) - f(a) + 1} [f(b) - f(a)] < \varepsilon$$

Thus for every $\varepsilon > 0$, there exists a partition $P \in \phi[a, b]$, such that

$$U(P, f) - L(P, f) < \varepsilon$$

Hence **$f(x)$ is integrable on $[a, b]$.**

Similarly, one can prove that f is integrable when $f(x)$ is a monotonic decreasing function on $[a, b]$.

Examples :

1. The function $f(x) = x^2$ is increasing on $[a, b]$, where $b > a > 0$ and therefore x^2 is integrable on $[a, b]$.

2. The function $f(x) = \dfrac{1}{x}$ is a decreasing function on $[2, 3]$ and therefore $\dfrac{1}{x}$ is integrable on $[2, 3]$.

Theorem : A bounded function $f(x)$ defined on $[a, b]$, having a finite number of points of discontinuity in $[a, b]$ is integrable on $[a, b]$.

Proof : By data $f(x)$ is bounded on $[a, b]$. Thus $f(x)$ has supremum and infimum on $[a, b]$. Let $m = \text{inf.}\, f$ and $M = \text{sup.}\, f$ on $[a, b]$.

Let $c_1, c_2, c_3, ..., c_p$ be the finite number of points of discontinuity of $f(x)$ on $[a, b]$ taken in order.

Let $\varepsilon > 0$, also let,

$$v_1 = [a_1, b_1],\ v_2 = [a_2, b_2], \ldots\ldots, v_p = [a_p, b_p]$$

be p non-over lapping intervals containing these p points respectively.

Further let the sum of their lengths is less than $\dfrac{\varepsilon}{2(M-m)}$.

The oscillation of $f(x)$ in each interval v_r is $\leq M - m$, for $r = 1, 2, \ldots, p$.

$\Rightarrow$ the sum of the oscillations of $f(x)$ on these p intervals

$$\leq \frac{\varepsilon}{2(M-m)}(M-m) = \frac{\varepsilon}{2} \quad \ldots (1)$$

Now, let $u_1 = [a, a_1],\ u_2 = [b_1, a_2], \ldots,\ u_{p+1} = [b_p, b]$

These are $p + 1$ sub intervals in $[a, b]$, that are formed by deleting the above p subintervals.

Clearly f is continuous in each u_r, $r = 1, 2, 3, \ldots, p + 1$.

$\Rightarrow$ f is integrable on u_r, for $r = 1, 2, 3, \ldots, p + 1$

$\Rightarrow$ $\exists$ a partition $P_r \in \phi[u_r]$ such that

$$W(P_r, f) < \frac{\varepsilon}{2(p+1)},\quad r = 1, 2, \ldots, p+1$$

$\Rightarrow$ the sum of all oscillations of f on these $(p + 1)$ intervals

$$< \frac{\varepsilon}{2(p+1)}(p+1) = \frac{\varepsilon}{2} \quad \ldots (2)$$

From (1) and (2), the sum of oscillation of f is $<\varepsilon$ for the partition

$$P = \{a, \ldots, a_1, \ldots, b_1, \ldots, a_p, \ldots, b_p, \ldots, b]$$

of $[a, b]$.

Thus for every $\varepsilon > 0$, there exists $P \in \phi[a, b]$, such that $W(P, f) < \varepsilon$.

Hence, **f is R-integrable, on $[a, b]$.**

Corollary : If the set of points of discontinuity of a bounded function $f(x)$ on $[a, b]$, can be enclosed in a finite number of intervals whose total length does not exceed given $\varepsilon > 0$, then f is integrable on $[a, b]$.

Proof : Let the set of points of discontinuity be enclosed in p non-overlapping intervals.

$$v_1 = [a_1, b_1],\ v_2 = [a_2, b_2], \ldots\ldots, v_p = [a_p, b_p]$$

such that the sum of their lengths is less than ε.

Clearly the oscillation in each $v_r < M - m$.

$\Rightarrow$ the sum of the oscillations in all the p subintervals $<(M - m)\varepsilon$.

The rest of the proof is same as in the theorem.

Example 1. If $f(x) \in R[a, b]$, g is bounded on $[a, b]$ and $f(x) = g(x)$ except for finite number of points in $[a, b]$, then show that $f(x) - g(x) \in R[a, b]$.

Solution : Define, $\varphi : [a, b] \to R$,

by, $\varphi(x) : f(x) - g(x),\ \forall\, x \in [a, b]$

By data $f(x) = g(x)$, except for finite number of points in $[a, b]$.

Thus $\varphi(x) = 0$, except for a finite number of points in $[a, b]$.

$\Rightarrow$ $\varphi(x)$ has a finite number of points of discontinuity in $[a, b]$.

Now, $f \in R[a, b] \Rightarrow f$ is bounded on $[a, b]$

$\Rightarrow f$ is bounded on $[a, b]$ and g is bounded on $[a, b] \Rightarrow f - g$ is bounded on $[a, b]$.

Thus $\varphi(x)$ is bounded on $[a, b]$ and further has a finite number of points of discontinuity in $[a, b]$.

Hence, **$\varphi(x)$ is R-integrable on $[a, b]$.**

Example 2. If $f(x) = \dfrac{1}{2^n}$, where $\dfrac{1}{2^{n+1}} < x \leq \dfrac{1}{2^n}$ for $n = 0, 1, 2, ...$ and $f(0) = 0$, then show that f is R-integrable on $[0, 1]$.

Solution : Now,

$$n = 0 \Rightarrow f(x) = 1, \quad \frac{1}{2} < x \leq 1$$

$$n = 1 \Rightarrow f(x) = \frac{1}{2}, \quad \frac{1}{2^2} < x \leq \frac{1}{2}$$

$$n = 2 \Rightarrow f(x) = \frac{1}{2^2}, \quad \frac{1}{2^3} < x \leq \frac{1}{2^2}$$

....

....

$$n = n \Rightarrow f(x) = \frac{1}{2^n}, \quad \frac{1}{2^{n+1}} < x \leq \frac{1}{2^n}$$

....

Also, $f(x) = 0$, when $x = 0$.

Thus, $\forall\, x \in [0, 1]$, we have $f(x) \in [0, 1] \Rightarrow f$ is bounded on $[0, 1]$.

Also $f(x)$ is monotonically increasing on $[0, 1]$. Hence f is R-integrable on $[0, 1]$.

1.11 Fundamental Theorem of Integral Calculus

Theorem : If $f(x) \in R[a, b]$ and $\phi(x)$ is *a* primitive of $f(x)$ then $\int_a^b f(x)\, dx = \varphi(b) - \varphi(a)$.

Proof : By data $\varphi(x)$ is a primitive of $f(x)$.

$$\Rightarrow \quad \varphi'(x) = f(x), \quad x \in [a, b] \Rightarrow \varphi'(x) \in R[a, b] \quad .... (1)$$

Again by data, $f(x) \in R[a, b]$. Thus for

$$P = \{a = x_0, x_1, x_2, ..., x_n = b\}$$

of $[a, b]$ and $x_{r-1} \leq \lambda_r \leq x_r$, $r = 1, 2, 3, ..., r$, we have,

$$\lim_{\|P\| \to 0} \sum_{r=1}^{n} f(\lambda_r)\, \delta_r = \int_a^b f(x)\, dx \quad .... (2)$$

Now, $\varphi(x)$ is differentiable on $[a, b]$, therefore it is continuous and differentiable on $[x_{r-1}, x_r]$, $r = 1, 2, 3, ..., n$.

Thus by Lagrange's mean-value theorem

$$\varphi(x_r) - \varphi(x_{r-1}) = (x_r - x_{r-1})\, \varphi'(\lambda_1), \text{ for } \lambda_r \in (x_{r-1}, x_r),\ r = 1, 2, ... n$$

$$\Rightarrow \quad \sum_{r=1}^{n} \left[\varphi(x_r) - \varphi(x_{r-1})\right] = \sum_{r=1}^{n} \varphi'(\lambda_r) \cdot \delta_r = \sum_{r=1}^{n} f(\lambda_r) \cdot \delta_r \quad \text{[by (1)]}$$

$$\Rightarrow \quad [\varphi(x_1) - \varphi(x_0) + \varphi(x_2) - \varphi(x_1) + ... + \varphi(x_n) - \varphi(x_{n-1})] = \sum_{r=1}^{n} f(\lambda_r) \cdot \delta_r$$

$$\Rightarrow \quad \varphi(x_n) - \varphi(x_0) = \sum_{r=1}^{n} f(\lambda_r) \cdot \delta_r$$

$$\Rightarrow \quad \lim_{\|P\| \to 0} [\varphi(x_n) - \varphi(x_0)] = \lim_{\|P\| \to 0} \sum_{r=1}^{n} f(\lambda_r) \cdot \delta_r$$

$$\Rightarrow \quad \varphi(b) - \varphi(a) = \int_a^b f(x)\, dx \qquad \text{(by (2))}$$

which is the required result.

Note :

1. If $\varphi'(x)$ is a continuous function on $[a, b]$ then, $\int_a^b \varphi'(x)\, dx = \varphi(b) - \varphi(a)$.
2. $\varphi(b) - \varphi(a)$ is denoted by $\left[\varphi(x)\right]_a^b$.

Example 1. Show that $\int_1^2 x^3\, dx = \frac{15}{4}$.

Solution : Let $f(x) = x^3$. Clearly $f(x) = x^3$ is continuous on $[1, 2]$. Thus $\int_1^2 x^3\, dx$ exists.

Now consider the function $\varphi(x) = \frac{x^4}{4}$ defined on $[1, 2]$.

Clearly $\varphi(x)$ is differentiable on $[1, 2]$ and $\varphi'(x) = x^3 = f(x)$ on $[1, 2]$.

Thus $\varphi(x)$ is a primitive of $f(x)$ on $[1, 2]$.

Hence by fundamental theorem of integral calculus, we have,

$$\int_1^2 x^3\, dx = \int_1^2 f(x)\, dx = \varphi(2) - \varphi(1) = \frac{2^4}{4} - \frac{(1)^4}{4} = \frac{16}{4} - \frac{1}{4} = \frac{15}{4}$$

Example 2. Show that $\int_a^b \cos x\, dx = \sin b - \sin a$.

Solution : Let $f(x) = \cos x$, which is continuous on $[a, b]$. Thus $\int_a^b \cos x\, dx$ exists.

Consider the function $\varphi(x) = \sin x$ defined on $[a, b]$.

Clearly, $\varphi(x)$ is differentiable on $[a, b]$ and $\varphi'(x) = \cos x = f(x)$.

Thus $\varphi(x)$ is a primitive of $f(x)$ on $[a, b]$.

Hence by fundamental theorem of integral calculus, we have,

$$\int_a^b \cos x\, dx = \varphi(b) - \varphi(a) = \sin b - \sin a$$

Example 3. Evaluate $\int_0^{\pi/4} \left(\sec^4 x - \tan^4 x\right) dx$.

Solution : Let,

$$f(x) = \sec^4 x - \tan^4 x = (\sec^2 x - \tan^2 x)(\sec^2 x + \tan^2 x)$$

$$\Rightarrow \quad f(x) = \sec^2 x + \tan^2 x \qquad (\because \sec^2 x - \tan^2 x = 1)$$

$$\Rightarrow \quad f(x) = 2\sec^2 x - 1 \qquad (\because \tan^2 x = \sec^2 x - 1)$$

Clearly $f(x)$ is continuous on $\left[0, \frac{\pi}{4}\right]$. Thus, $\int_0^{\pi/4} f(x)\, dx$ exists.

Now, consider, $\varphi(x) = 2\tan x - x$ on $\left[0, \frac{\pi}{4}\right]$.

Again, $\varphi(x)$ is differentiable on $\left[0, \frac{\pi}{4}\right]$ and $\varphi'(x) = 2\sec^2 x - 1 = f(x)$

Thus, $\varphi(x)$ is a primitive of $f(x)$ on $\left[0, \frac{\pi}{4}\right]$.

Hence by fundamental theorem, we have,

$$\int_0^{\pi/4} \left(\sec^2 x - \tan^2 x\right) dx = \varphi\left(\frac{\pi}{4}\right) - \varphi(0) = \left(2 - \frac{\pi}{4}\right) - 0 - 0 = 2 - \frac{\pi}{4}$$

Example 4. If $f(x) = \frac{1}{2^n}, \frac{1}{2^{n+1}} < x \le \frac{1}{2^n}$, ($n = 0, 1, 2, \ldots$) and $f(0)$, show that $\int_0^1 f(x)\,dx = \frac{2}{3}$.

Solution : We have shown already $f(x)$ is integrable on [0, 1]. Consider,

$$\int_{(1/2)^n}^1 f(x)\,dx = \int_{(1/2)^n}^{(1/2)^{n-1}} f(x)\,dx + \int_{(1/2)^{n-1}}^{(1/2)^{n-2}} f(x)\,dx + \ldots + \int_{1/2}^1 f(x)\,dx$$

$$= \int_{(1/2)^n}^{(1/2)^{n-1}} \frac{1}{2^{n-1}}\,dx + \int_{(1/2)^{n-1}}^{(1/2)^{n-2}} \frac{1}{2^{n-2}}\,dx + \ldots + \int_{1/2}^1 1\,dx$$

$$= \frac{1}{2^{n-1}}\left[\frac{1}{2^{n-1}} - \frac{1}{2^n}\right] + \frac{1}{2^{n-2}}\left[\frac{1}{2^{n-2}} - \frac{1}{2^{n-1}}\right] + \ldots + \left[1 + \frac{1}{2}\right]$$

$$= \frac{1}{2^{n-1}}\left(\frac{1}{2^n}\right) + \frac{1}{2^{n-2}}\left(\frac{1}{2^{n-1}}\right) + \ldots + \frac{1}{2}$$

$$= \frac{1}{2}\left[\frac{1}{4^{n-1}} + \frac{1}{4^{n-2}} + \ldots + 1\right]$$

$$= \frac{1}{2}\left[\frac{1 - (1/4)^n}{1 - (1/4)}\right] = \frac{2}{3}\left(1 - \frac{1}{4^n}\right) \qquad \text{(sum of } n \text{ terms of } G.P.)$$

Now,

$$\int_0^1 f(x)\,dx = \lim_{n\to\infty} \int_{(1/2)^n}^1 f(x)\,dx$$

$$= \lim_{n\to\infty} \frac{2}{3}\left(1 - \frac{1}{4^n}\right) = \frac{2}{3} \qquad \left(\because \lim_{n\to\infty} \frac{1}{4^n} = 0\right)$$

Example 5. If $f(x)$ is continuous on $[a, b]$, $f(x) \ge 0$, $\forall\, x \in [a, b]$ and $\int_a^b f(x)\,dx = 0$, then show that $f(x) = 0$.

Solution : Let if possible $f(x) \ne 0$. Then there exists $c \in [a, b]$ such that $f(c) \ne 0$ and $f(c) > 0$.

Let $\varepsilon = \frac{1}{2} f(c)$.

Since $f(x)$ is continuous at c, $\exists\, \delta > 0$, such that

$$|f(x) - f(c)| < \varepsilon, \quad \text{for } |x - c| < \delta$$

$$\Rightarrow \quad f(c) - \varepsilon < f(x) < f(x) + \varepsilon$$

$$\Rightarrow \quad f(x) > f(c) - \frac{1}{2} f(c) \qquad \left(\because \varepsilon = \frac{1}{2} f(c)\right)$$

$$\Rightarrow \quad f(x) > \frac{1}{2} f(c) > 0$$

Choose δ such that, $a < c - \delta < c < c + \delta < b$.

Now, $$\int_a^b f(x)\,dx = \int_a^{c-\delta} f(x)\,dx + \int_{c-\delta}^{c+\delta} f(x)\,dx + \int_{c+\delta}^{b} f(x)\,dx$$

$$\Rightarrow \int_a^b f(x)\,dx \geq \int_{c-\delta}^{c+\delta} f(x)\,dx$$

$$\Rightarrow \int_a^b f(x)\,dx > \int_{c-\delta}^{c+\delta} \frac{1}{2} f(c)\,dx$$

$$\Rightarrow \int_a^b f(x)\,dx > \frac{1}{2} f(c)\,[(c+\delta)-(c-\delta)] \Rightarrow \int_a^b f(x)\,dx > \delta \cdot f(c) > 0$$

This contradicts the fact that $\int_a^b f(x)\,dx = 0$.

Even if $c = a$ or $c = b$, we also get a contradiction. Thus $f(x) = 0$.

EXERCISES

1. Show that $\int_a^b e^x\,dx = e^b - e^a$

2. Show that $\int_a^b x^2\,dx = \frac{a^3}{3}$

3. Show that $\int_0^{1/2} \frac{1}{\sqrt{1-x^2}}\,dx = \frac{\pi}{6}$

4. Show that $\int_0^1 \frac{1}{1-x^2}\,dx = \frac{\pi}{4}$

1.12 Integration by Change of Variables (Integration by Substitution)

Theorem : If *(i)* $g(x)$ is a differentiable and integrable function on $[\alpha, \beta]$. *(ii)* $g(\alpha) = a$, $g(\beta) = b$, where $a < b$, *(iii)* the range of $g(x)$ is $[a, b]$ and (iv) $f(x)$ is a continuous function on $[a, b]$, then,

$$\int_a^b f(x)\,dx = \int_\alpha^\beta f[g(t)] \cdot g'(t)\,dt.$$

Proof : First we shall establish that $\int_a^b f(x)\,dx$ and $\int_\alpha^\beta f[g(t)] \cdot g'(t)\,dt$ exists.

By data $f(x)$ is continuous on $[a, b]$. Thus $f(x) \in R[a, b]$.
Again, by data $g(x)$ is differentiable on $[\alpha, \beta]$. Thus $g(x)$ is continuous on $[\alpha, \beta]$.
Now, the range of $g(x)$ is $[a, b]$ and $f(x)$ is defined on $[a, b]$.
$\Rightarrow$ $fog(x)$ is a function defined on $[\alpha, \beta]$.
Again, $g(x)$ is continuous on $[\alpha, \beta]$ and f is continuous on $[a, b]$.
$\Rightarrow$ $fog(x)$ is continuous on $[\alpha, \beta] \Rightarrow fog(x) \in R[\alpha, \beta]$.

Now, $fog \in R[\alpha, \beta]$ and $g \in R[\alpha, \beta] \Rightarrow \int_\alpha^\beta (fog) \cdot g'\,dx$ **exists.**

Now, $f(x)$ is continuous on $[a, b]$. Thus its primitive $F(x)$ exits.
That is, $F'(x) = f(x), \ \forall x \in [a, b]$
Again, for $t \in [\alpha, \beta]$, let $G(t) = F(g(t))$. Then by chain rule,

$$G'(t) = F'(g(t)) \cdot g'(t) \Rightarrow G'(t) = f(g(t)) \cdot g'(t)$$

Thus by fundamental theorem, we have,

$$\int_{\alpha}^{\beta} f[g(t)] \cdot g'(t)\, dt = \int_{\alpha}^{\beta} G'(t)\, dt$$

$$\Rightarrow \int_{\alpha}^{\beta} f[g(t)] \cdot g'(t)\, dt = G(\beta) - G(\alpha)$$

$$\Rightarrow \int_{\alpha}^{\beta} f[g(t)] \cdot g'(t)\, dt = F[g(\beta)] - F(g(\alpha)] \Rightarrow \int_{\alpha}^{\beta} f[g(t)] \cdot g'(t)\, dt = F(b) - F(a)$$

Hence, $\displaystyle\int_{\alpha}^{\beta} f[g(t)] \cdot g'(t)\, dt = \int_{a}^{b} f(x)\, dx$

Note : The above theorem can also be stated as below. If

(i) $g(x)$ is differentiable on $[\alpha, \beta]$, $g(\alpha) = a$, $g(\beta) = b$, $g'(x) \in R[\alpha, \beta]$ and $g(t) \neq 0$, $\forall\, t \in [\alpha, \beta]$

(ii) $f(x) \in R[a, b]$ and **(iii)** $fog(x) \in R[\alpha, \beta]$ then $\displaystyle\int_{a}^{b} f(x)\, dx = \int_{\alpha}^{\beta} f[g(t)] \cdot g'(t)\, dt$.

Corollary : If $f(x)$ is differentiable on $[a, b]$, $f'(x) \in R[a, b]$ and $f(x) \neq 0$, $\forall\, x \in [a, b]$, then $\displaystyle\int_{a}^{b} \frac{f'(x)}{f(x)}\, dx = \log \left|\frac{f(b)}{f(a)}\right|$.

Proof : By data $f(x)$ is differentiable $\Rightarrow$ $f(x)$ is continuous on $[a, b]$.

Again, by data $f(x) \neq 0$ on $[a, b]$ and $f(x)$ is continuous.

$\Rightarrow$ $\dfrac{1}{f(x)}$ is also continuous on $[a, b]$ $\Rightarrow$ $\dfrac{1}{f(x)} \in R[a, b]$

Now, $f'(x) \in R[a, b]$ and $\dfrac{1}{f(x)} \in R[a, b]$ $\Rightarrow$ $\dfrac{f'(x)}{f(x)} \in R[a, b]$

Let, $\varphi(x) = \log |f(x)|$, on $[a, b]$ $\Rightarrow$ $\varphi'(x) = \dfrac{f'(x)}{f(x)}$, $\forall\, x \in [a, b]$

Thus $\varphi(x)$ is a primitive of $\dfrac{f'(x)}{f(x)}$ on $[a, b]$. Hence by the fundamental theorem.

$$\int_{a}^{b} \frac{f'(x)}{f(x)}\, dx = \varphi(b) - \varphi(a) = \log |f(b)| - \log |f(a)| = \log \left|\frac{f(b)}{f(a)}\right|.$$

Working rule to evaluate $\displaystyle\int_{a}^{b} f(x)\, dx$, by the method of substitution :

1. Put $x = g(t)$, where $g(t)$ is integrable on $[\alpha, \beta]$.
2. Replace dx by $g'(t)\, dt$.
3. Write α, β in place of a and b, where $g'(\alpha) = a$, $g(\beta) = b$.

Example 1. Using the substitution $x = \pi - t$ show that,

$$\int_{0}^{\pi} x \cdot \varphi(\sin x)\, dx = \frac{\pi}{2} \int_{0}^{\pi} \varphi(\sin x)\, dx.$$

Solution : Clearly, $g(t) = \pi - t$ is defined on $[0, \pi]$, $g(t) \in R[0, \pi]$. Also, $g'(t) = -1$, $\forall\, t \in [0, \pi]$

Again, $x = \pi - t \Rightarrow t = \pi$, when $x = 0$; $t = 0$ when $x = \pi$.

Thus by change of variable theorem, we have,

$$I = \int_0^{\pi} x \cdot \varphi(\sin x)\, dx$$

$$= \int_{\pi}^{0} (\pi - t)\, \varphi[\sin(\pi - t)](-1)\, dt \qquad (\because g'(t) = -1)$$

$$= \int_{\pi}^{0} (\pi - t)\, \varphi(\sin t)(-1)\, dt \qquad (\because \sin(\pi - t) = \sin t)$$

$$\Rightarrow \quad I = \int_0^{\pi} (\pi - t) \cdot \varphi(\sin x)\, dx \qquad \left(\because \int_a^b f(x)\, dx = -\int_b^a f(x)\, dx\right)$$

$$\Rightarrow \quad I = \pi \int_0^{\pi} \varphi(\sin t)\, dt - \int_0^{\pi} t \cdot \varphi(\sin t)\, dt$$

$$\Rightarrow \quad 2I = \pi \int_0^{\pi} \varphi(\sin t)\, dt \Rightarrow \boldsymbol{I = \frac{\pi}{2} \int_0^{\pi} \varphi(\sin t)\, dt}$$

Example 2. Can we evaluate $\int_{-1}^{1} \frac{dx}{1 + x^2}$, by substituting $x = \frac{1}{t}$?

Solution : Let $f(x) = \frac{1}{1 + x^2}$, $\forall\, x \in [-1, 1]$ and $g(x) = \tan^{-1}x$, $\forall\, x \in [-1, 1]$

Now, $$g'(x) = \frac{1}{1 + x^2} = f(x), \ \forall\, x \in [-1, 1]$$

Thus by fundamental theorem, we have,

$$\int_{-1}^{1} \frac{dx}{1 + x^2} = \tan^{-1} x \Big]_{-1}^{1} = \tan^{-1}1 - \tan^{-1}(-1) = \frac{\pi}{4} - \left(\frac{\pi}{4}\right) = \frac{\pi}{2}$$

Now, put $x = \frac{1}{t} = g(t)$, then $g'(t) = -\frac{1}{t^2}$.

Also, $x = -1 \Rightarrow t = -1$ and $x = 1 \Rightarrow t = 1$

Thus, $$\int_{-1}^{1} \frac{1}{1 + x^2}\, dx = \int_{-1}^{1} \frac{1}{1 + (1/t)^2} \left(-\frac{1}{t^2}\right) dt = -\int_{-1}^{1} \frac{1}{1 + t^2}\, dt = -\left(\frac{\pi}{2}\right)$$

The substitution $x = \frac{1}{t}$, gives the value other than the correct value. Thus we cannot evaluate the integral by putting $x = \frac{1}{t}$.

The reason being that $g(t) = \frac{1}{t}$ does not possess continuous derivative on $[-1, 1]$. In fact $g(t)$ is not defined at $t = 0$.

Example 3. If $f \in R[0, a]$, show that $\int_0^a f(x)\, dx = \int_0^a f(a - x)\, dx = 0$.

Solution : Let, $g(t) = a - t$, be defined on $[0, a]$ and $g(t) \in R[0, a]$. Now, $g'(t) = -1$, $\forall\, t \in [0, a]$.

Again, $x = a - t \Rightarrow dx = -dt$; $x = 0 \Rightarrow t = a$ and $x = a \Rightarrow t = 0$

By change of variable theorem, we have,

$$\int_0^a f(x)\, dx = \int_0^a f(a - t) \cdot g'(t)\, dt = \int_0^a f(a - t) \cdot (-1)\, dt \qquad (\because g'(t) = -1)$$

$$= \int_0^a f(a - t)\, dt = \int_0^a f(a - x)\, dx$$

1.13 Integration by Parts

Theorem : If $f(x)$ and $g(x)$ are differentiable functions on $[a, b]$ and $f'(x)$ and $g'(x)$ are continuous on $[a, b]$ then,

$$\int_a^b f(x) \cdot g'(x)\, dx = f(b)\, g(b) - f(a)\, g(a) - \int_a^b f'(x) \cdot g(x)\, dx$$

Proof : By data $f(x)$ and $g(x)$ are differentiable on $[a, b]$.

$\Rightarrow$ $f(x)$ and $g(x)$ are continuous on $[a, b]$ $\Rightarrow$ $f(x), g(x)$ and $f(x) \cdot g(x) \in R[a, b]$

Now, $f'(x)$ and $g'(x)$ are continuous on $[a, b] \Rightarrow f'(x)$ and $g'(x) \in R[a, b]$

$$f(x) \in R[a, b],\ g'(x) \in R[a, b] \Rightarrow f(x) \cdot g'(x) \in R[a, b]$$

$$f'(x) \in R[a, b],\ g(x) \in R[a, b] \Rightarrow f'(x) \cdot g(x) \in R[a, b]$$

Thus, $\int_a^b f(x) \cdot g'(x)\, dx$ and $\int_a^b f'(x) \cdot g(x)\, dx$ are defined.

Now, we have, $[f(x) \cdot g(x)]' = f(x) \cdot g'(x) + f'(x) \cdot g(x)$ on $[a, b]$.

$\Rightarrow$ $f(x) \cdot g(x)$ is primitive of $f(x) \cdot g'(x) + f'(x) \cdot g(x)$ on $[a, b]$.

Thus by fundamental theorem, we have,

$$\int_a^b \left[f(x) \cdot g'(x) + f'(x) \cdot g(x)\right] dx = f(b) \cdot g(b) - f(a) \cdot g(a)$$

$$\Rightarrow \int_a^b f(x) \cdot g'(x)\, dx + \int_a^b f'(x) \cdot g(x)\, dx = f(b) \cdot g(b) - f(a) \cdot g(a)$$

$$\Rightarrow \int_a^b f(x) \cdot g'(x)\, dx = f(b) \cdot g(b) - f(a) \cdot g(a) - \int_a^b f'(x) \cdot g(x)\, dx$$

Hence the theorem.

Corollary : If $f(x)$ is differential function on $[a, b]$ and $f'(x)$ is continuous on $[a, b]$ then,

$$\int_a^b \left[f(x) + f'(x)\right] e^x\, dx = f(x) \cdot e^x \Big]_a^b .$$

Proof : Let, $g(x) = e^x$. Then $g'(x) = e^x$ and it is continuous on $[a, b]$. Thus,

$$\begin{aligned}\int_a^b f(x) \cdot e^x\, dx &= \int_a^b f(x) \cdot g'(x)\, dx \\ &= f(b) \cdot g(b) - f(a) \cdot g(a) - \int_b^a f'(x)\, g(x)\, dx \\ &= f(b) \cdot g(b) - f(a) \cdot g(a) - \int_b^a f'(x) \cdot e^x\, dx\end{aligned}$$

$$\Rightarrow \int_a^b f(x) \cdot e^x\, dx + \int_a^b f'(x) \cdot e^x\, dx = f(b) \cdot g(b) - f(a) \cdot g(a)$$

$$\Rightarrow \int_a^b \left[f(x) + f'(x)\right] e^x\, dx = f(x) \cdot g(x)\Big]_a^b = f(x) \cdot e^x\Big]_a^b$$

Example 1. Show that $\int_0^{\pi/2} x \cdot \cos x\, dx = \frac{\pi}{2} - 1$.

Solution : Let $f(x) = x$, $g(x) = \sin x$. Clearly both are differentiable on $\left[0, \frac{\pi}{2}\right]$ and $f'(x) = 1$ and $g'(x) = \cos x$. Also $f'(x)$ and $g'(x)$ are continuous on $\left[0, \frac{\pi}{2}\right]$.

$$\text{Now,} \quad \int_0^{\pi/2} x \cdot \cos x \, dx = \int_0^{\pi/2} f(x) \cdot g'(x) \, dx$$

$$= f\left(\frac{\pi}{2}\right) \cdot g\left(\frac{\pi}{2}\right) - f(0) \cdot g(0) - \int_0^{\pi/2} f'(x) \cdot g(x) \, dx$$

$$= \left[\frac{\pi}{2} \cdot 1 - 0\right] - \int_0^{\pi/2} 1 \cdot \sin x \, dx$$

$$= \frac{\pi}{2} - [-\cos x]_0^{\pi/2} = \frac{\pi}{2} + (0 - 1) = \frac{\pi}{2} - 1$$

Example 2. Evaluate $\int_0^{\pi/2} e^x (\sin x + \cos x) \, dx$.

Solution : Let $f(x) = \sin x$. Then $f'(x) = \cos x$, $\forall \, x \in \left[0, \frac{\pi}{2}\right]$.

Clearly, $f'(x)$ is continuous on $\left[0, \frac{\pi}{2}\right]$.

$$\text{Thus,} \quad \int_0^{\pi/2} e^x [\sin x + \cos x] \, dx = \left[e^x \cdot \sin x\right]_0^{\pi/2} = e^{\pi/2} - 0 = e^{\pi/2}$$

Example 3. Show that $\int_0^1 \frac{1 + x}{(2 + x)^2} e^x \, dx = \frac{e}{x} - \frac{1}{2}$.

Solution : Consider, $\quad \frac{1 + x}{(2 + x)^2} = \frac{1}{2 + x} + \frac{-1}{(2 + x)^2}$

Let, $f(x) = \frac{1}{2 + x}$ $\forall \, x \in [0, 1]$. Then $f'(x) = \frac{-1}{(2 + x)^2}$ for all $x \in [0, 1]$.

Further, $f'(x)$ is continuous on $[0, 1]$. Thus,

$$\int_0^1 \frac{1 + x}{(2 + x)^2} e^x \, dx = \int_0^1 [f(x) + f'(x)] \, e^x = \left[e^x \cdot f(x)\right]_0^1 = \left[\frac{e^x}{2 + x}\right]_0^1 = \frac{e}{3} - \frac{1}{2}$$

1.14 Mean Value Theorems

In this section we shall consider two theorems, which are in a sense similar to the mean value theorem for derivatives.

First mean value theorem : If $f(x)$ is continuous function defined on $[a, b]$, then there exists $c \in (a, b)$ such that

$$\int_a^b f(x) \, dx = f(c) \cdot (b - a)$$

Proof : By data $f(x)$ is continuous on $[a, b]$. Thus $f(x)$ is bounded on $[a, b]$ and $f(x) \in R[a, b]$.

Let m and M be the infimum and supremum of $f(x)$ on $[a, b]$. Thus we have,

$$m(b - a) \leq \int_a^b f(x) \, dx \leq M(b - a)$$

$\Rightarrow$ there exists a real number $\lambda \in [m, M]$ such that

$$\int_a^b f(x) \, dx = \lambda(b - a)$$

Again, as $f(x)$ is continuous on $[a, b]$ and $\lambda \in [m, M]$, there exists $c \in (a, b)$, such that $f(c) = \lambda$. Hence,

$$\int_a^b f(x)\,dx = (b-a)\cdot f(c).$$

Second mean value theorem : Let $f(x)$ be a monotonic function defined on $[a, b]$. Then there exists a point $c \in [a, b]$ such that

$$\int_a^b f(x)\,dx = (c-a)\,f(a) + (b-c)\,f(b)$$

Proof : By data $f(x)$ is monotonic on $[a, b]$. Thus $f(x) \in R[a, b]$. Now define the function $h(x)$ by

$$h(x) = (x-a)\,f(a) + (b-x)\,f(b), \quad \forall\, x \in [a, b]$$

Clearly, $h(x)$ is a continuous function on $[a, b]$.

Thus, by intermediate value theorem, $h(x)$ assumes every value between $h(a)$ and $h(b)$.

That is $h(x)$ assumes every value between $(b-a)\,f(b)$ and $(b-a)\,f(a)$.

Again, as $f(x)$ is monotonic $\int_a^b f(x)\,dx$, lies between $(b-a)\,f(b)$ and $(b-a)\,f(a)$.

Thus there exists a point $c \in [a, b]$, such that

$$\int_a^b f(x)\,dx = h(c)$$

$$\Rightarrow \quad \int_a^b f(x)\,dx = (c-a)\,f(a) + (b-c)\,f(b)$$

Hence the proof.

Theorem : If $f, g \in R[a, b]$ and $g(x) \geq 0$, $x \in [a, b]$, then there exists a real number λ, lying between the infimum and supremum of f on $[a, b]$, such that

$$\int_a^b f(x)\cdot g(x)\,dx = \lambda \int_a^b g(x)\,dx$$

Proof : By data $f(x) \in [a, b]$. Thus $f(x)$ is bounded on $[a, b]$.

$$\Rightarrow \quad m \leq f(x) \leq M, \ \forall\, x \in [a, b]$$

where m and M are infimum and supremum of $f(x)$ on $[a, b]$.

Since $g(x) > 0$, $\forall\, x \in [a, b]$, we have,

$$m\cdot g(x) \leq f(x)\cdot g(x) \leq M\cdot g(x), \ \forall\, x \in [a, b]$$

$$\Rightarrow \quad m\cdot \int_a^b g(x)\,dx \leq \int_a^b f(x)\cdot g(x)\,dx \leq M \int_a^b g(x)\,dx$$

Thus there eists a number $\lambda \in [m, M]$, such that

$$\int_a^b f(x)\cdot g(x)\,dx = \lambda \int_a^b g(x)\,dx$$

Hence the theorem.

Corollary : If f is continuous on $[a, b]$, $g \in R[a, b]$ and $g(x) \geq 0$ for all $x \in [a, b]$, then there exists a number $c \in (a, b)$ such that

$$\int_a^b f(x)\cdot g(x)\,dx = f(c)\int_a^b g(x)\,dx.$$

Proof : By data $f(x)$ is continuous on $[a, b]$. Thus $f(x) \in R[a, b]$. From, the theorem, it follows, that there exists $\lambda \in [m, M]$ such that

$$\int_a^b f(x)\cdot g(x)\,dx = \lambda \int_a^b g(x)\,dx$$

As $f(x)$ is continuous on $[a, b]$ and $\lambda \in [m, M]$, there exists $c \in [a, b]$, such that $f(c) = \lambda$. Hence,

$$\int_a^b f(x) \cdot g(x)\, dx = f(c) \int_a^b g(x)\, dx$$

Note : If $g(x) = 1, \forall\, x \in [a, b]$, then $g(x) \in R[a, b]$ and $g(x) > 0$.

Thus we have, $$\int_a^b f(x) \cdot 1\, dx = \lambda \int_a^b 1\, dx = \lambda(b - a).$$

Where $\lambda \in [m, M]$, where m and M are infimum and supremum of $f(x)$ on $[a, b]$.

Example 1. Show there exists $\lambda \in \left(0, \frac{\pi}{2}\right)$ such that $\int_0^{\pi/2} x \cdot \sin x\, dx = \lambda$.

Solution : Let $f(x) = x$ and $g(x) = \sin x$. Clearly $f(x)$ is continuous on $\left(0, \frac{\pi}{2}\right)$ and g is integrable on $\left(0, \frac{\pi}{2}\right)$. Also $g(x) = \sin x \geq 0$ on $\left[0, \frac{\pi}{2}\right]$.

Thus we have,

$$\int_0^{\pi/2} x \cdot \sin x\, dx = \lambda \int_0^{\pi/2} \sin x\, dx, \text{ where } \lambda \in \left(0, \frac{\pi}{2}\right)$$

$$\Rightarrow \quad \int_0^{\pi/2} x \cdot \sin x\, dx = \lambda \cdot 1 = \lambda \qquad \left(\because \int_0^{\pi/2} \sin x\, dx = 1\right)$$

Example 2. By applying mean value theorem to the integral $\int_0^{\pi/4} \sec x\, dx$, show that $\frac{\pi}{4} \leq \int_0^{\pi/4} \sec x\, dx \leq \frac{\pi}{2\sqrt{2}}$.

Solution : Let $f(x) = \sec x$, which is continuous on $\left[0, \frac{\pi}{4}\right]$ and therefore integrable on $\left[0, \frac{\pi}{4}\right]$.

By first mean value theorem, we have $a \in \left[0, \frac{\pi}{4}\right]$, such that

$$\int_0^{\pi/4} \sec x\, dx = \left(\frac{\pi}{4} - 0\right) \cdot \sec(a) = \frac{\pi}{4} \sec(a)$$

Now, $0 < a < \frac{\pi}{4}$

$$\begin{aligned}
&\Rightarrow \quad \cos 0 > \cos(a) > \cos\frac{\pi}{4} \\
&\Rightarrow \quad 1 < \sec a < \sqrt{2} \\
&\Rightarrow \quad \frac{\pi}{4} < \frac{\pi}{4} \cdot \sec a < \frac{\pi}{2\sqrt{2}} \\
&\Rightarrow \quad \frac{\pi}{4} < \int_0^{\pi/4} \sec x\, dx < \frac{\pi}{2\sqrt{2}}
\end{aligned}$$

Also, when, $a = 0$, $\frac{\pi}{4} \cdot \sec a = \frac{\pi}{4}$ and when $a = \frac{\pi}{4}$, $\frac{\pi}{4} \cdot \sec a = \frac{\pi}{2\sqrt{2}}$.

Thus, we have, $$\frac{\pi}{4} \leq \int_0^{\pi/4} \sec x\, dx \leq \frac{\pi}{2\sqrt{2}}.$$

Example 2. Show that $\frac{1}{4} < \int_0^{\pi/4} \frac{dx}{\sqrt{1-x^2}} < \frac{1}{\sqrt{15}}$.

Solution : Let $f(x) = \frac{1}{\sqrt{1-x^2}}$, for $x \in \left[0, \frac{1}{4}\right]$.

Clearly, $1 - x^2 > 0$ for $x \in \left[0, \frac{1}{4}\right]$. Thus, $f(x) = \frac{1}{\sqrt{1-x^2}}$ is a continuous function on $\left[0, \frac{1}{4}\right]$. Hence, $f(x) \in R\left[0, \frac{1}{4}\right]$.

Thus by first mean value theorem, there exists $a \in \left(0, \frac{1}{4}\right)$, such that

$$\int_0^{\pi/4} \frac{dx}{\sqrt{1-x^2}} = \left(\frac{1}{4} - 0\right) \cdot f(a) = \frac{1}{4} \frac{1}{\sqrt{1-a^2}} \quad \text{.... (1)}$$

Now, $0 < a < \frac{1}{4} \Rightarrow 0 < a^2 < \frac{1}{16} \Rightarrow 0 > -a^2 > \left(-\frac{1}{16}\right)$

$$\Rightarrow 1 > 1 - a^2 > 1 - \frac{1}{16}$$

$$\Rightarrow 1 > \sqrt{1-a^2} > \frac{\sqrt{15}}{4}$$

$$\Rightarrow 1 < \frac{1}{\sqrt{1-a^2}} < \frac{4}{\sqrt{15}}$$

$$\Rightarrow \frac{1}{4} < \frac{1}{4\sqrt{1-a^2}} < \frac{1}{\sqrt{15}}$$

$$\Rightarrow \frac{1}{4} < \int_0^{\pi/4} \frac{dx}{\sqrt{1-x^2}} < \frac{1}{\sqrt{15}} \quad \text{[from (1)]}$$

Example 3. Show that $x > \log(1+x) > \frac{x}{1+x}, x \geq 0$.

Solution : Let, $f(x) = \frac{1}{1+x}$ and $g(x) = 1$, for $x \in [0, a]$.

Clearly, f and g are bounded and integrable on $[0, a]$ and $g(x) > 0, x \in [0, a]$.

Now, $0 < x < a \Rightarrow 1 < 1 + x < 1 + a$

$$\Rightarrow 1 > \frac{1}{1+x} > \frac{1}{1+a}$$

$$\Rightarrow \frac{1}{1+a} < f(x) < 1 \text{ for } x \in [0, a]$$

i.e., $f(x)$ is bounded between $\frac{1}{1+a}$ and 1.

Now, there exists a real number λ, between the bounds of f such that,

$$\int_0^a f(x)\cdot g(x)\,dx = \lambda\int_0^a g(x)\,dx$$

$$\Rightarrow \qquad \int_0^a \frac{1}{1+x}\,dx = \lambda\int_0^a 1\,dx$$

$$\Rightarrow \qquad \log(1+a) = \lambda a, \quad \text{Also, } \frac{1}{1+a} < \lambda < 1$$

Now, $x \geq 0$, $\dfrac{1}{1+a} < \lambda < 1 \Rightarrow \dfrac{1}{1+x} < \dfrac{\log(1+x)}{x} < 1 \Rightarrow \boldsymbol{\dfrac{x}{1+x} < \log(1+x) < x}$

EXERCISES

1. By applying mean value theorem of integral calculus, show that $e^{1/4} < \dfrac{4}{3} < e^{1/3}$, by considering $\displaystyle\int_3^4 \frac{1}{x}\,dx$.

2. Show that $\dfrac{\pi^3}{24} \leq \displaystyle\int_0^\pi \frac{x^2}{5+3\cos x}\,dx \leq \frac{\pi^3}{6}$.

3. Show that $\dfrac{1}{\pi} \leq \displaystyle\int_0^1 \frac{\sin \pi x}{1+x^2}\,dx \leq \frac{2}{\pi}$.

4. Show that $\dfrac{1}{3\sqrt{2}} < \displaystyle\int_0^1 \frac{x^2}{\sqrt{1+x}}\,dx < \frac{1}{3}$.

5. Show that $\dfrac{2\pi}{13} < \displaystyle\int_0^{2\pi} \frac{dx}{10+3\cos x} < \frac{2\pi}{7}$.

2

Differential Equations

2.1 Introduction

Students are already familiar with the concept of differential equations and method of solving a first order first degree differential equation by a particular method known as by the method of separating the variables. However, we shall recaptulate these ideas and further see other methods of solving the first order first degree equations. The process of solving the differential equations of first order higher degree and determining the orthogonal trajectories as an application of differential equations are also discussed.

2.2 Definitions and Examples.

1. An equation in which, the independent, dependent variables and the differential coefficients of dependent variables w.r.t. independent variables are involved is called a differential equation.

Following are a few examples of differential equations.

(i) $(3y - 4x)\dfrac{dy}{dx} + (7y^2 + 6y) = 0$ **(ii)** $\dfrac{d^3y}{dx^3} + \dfrac{d^2y}{dx^2} + \left(\dfrac{dy}{dx}\right)^4 = e^{4x}$

(iii) $\left[1+\left(\dfrac{dy}{dx}\right)^2\right]^{3/2} = \dfrac{d^2y}{dx^2}$ **(iv)** $y = x\left(\dfrac{dy}{dx}\right) + \dfrac{1}{\left(\dfrac{dy}{dx}\right)}$

(v) $\dfrac{d^2y}{dx^2} = -a^2x$ **(vi)** $x\dfrac{\partial u}{\partial x} + y\dfrac{\partial u}{\partial y} = 2u$ **(vii)** $\dfrac{\partial^2 u}{\partial x^2} - a^2\dfrac{\partial^2 u}{\partial y^2} = 0$

2. A differential equation is said to be ordinary differential equation, if the derivatives involved in it are w.r.t. a single independent variable.

In the above examples equations (1) to (5) are ardinary differential equations.

3. A differential equation is said to be partial differential equation, if the derivative involved in it are w.r.t more than one independent variable.

In the above examples, the equations (6) and (7) are partial differential equations.

In the present book we consider only ordinary differential equations.

4. The **order** of the differential equation is the highest order derivative present in the equation.

5. The **degree** of the differential equation is the positive integral power of the highest order derivative involved in it.

In the above examples,

(1) is of order 1 and degree 1.

(2) is of order 3 and degree 1.

(3) can be written as $\left[1+\left(\frac{dy}{dx}\right)^2\right]^3 = \left(\frac{d^2y}{dx^2}\right)^2$ and therefore is of order 2 and degree 2.

(4) can be written as $y\left(\frac{dy}{dx}\right) = x\left(\frac{dy}{dx}\right)^2 + 1$ and therefore is of order 1 and degree 2.

(5) is of order 2 and degree 1. (6) is of order 1 and degree 1. (7) is of order 2 and degree 1.

6. Solution of a differential equation is a relationship between the dependent and independent variables, which satisfies the given differential equation.

Solution of a differential equation is also known as primitive or integral of the equation.

For example the relation,

$$y = a \cos x + b \sin x \quad \ldots(1)$$

is a solution of the differential equation

$$\frac{d^2y}{dx^2} + y = 0.$$

For, differentiating (1) twice we get

$$\frac{dy}{dx} = -a \sin x + b \cos x, \quad \frac{d^2y}{dx^2} = -a \cos x - b \sin x$$

$$\Rightarrow \frac{d^2y}{dx^2} = -(a \cos x + b \sin x) \Rightarrow \frac{d^2y}{dx^2} = -y \Rightarrow \frac{d^2y}{dx^2} + y = 0.$$

This is the given differential equation.

In the above example we have not only verified that $y = a \cos x + b \sin x$ is a solution of the given differential equation but also we have seen a differential equation can be formed from a given relationship between the variables x and y. This process is called the **formation of differential equation.**

We shall make a general statement (without proof)

Given a relation between the independent and dependent variables involving few constant, say n, a differential equation can be formed by differentiating the relation n times and eliminating the n constants involved in these. During the process we get a differential equation of order n.

7. A solution of a differential equation with as many constants as the order of the equation is called the **general solution** of the equation.

For example,

$$y = a \cos x + b \sin x$$

is the general solution of the equation

$$\frac{d^2y}{dx^2} + y = 0$$

8. A solution obtained from the general solution by giving particular values to the constants involved in it, is called a **particular solution** of the equation.

If we set $a = 2$ and $b = 4$ in the above solution, then

$$y = 2 \cos x + 4 \sin x$$

is a particular solution of the equation

$$\frac{d^2y}{dx^2} + y = 0.$$

In the present book we shall see few methods of solving

(i) Equations of first order first degree.

(ii) Equations of first order but not first degree.

2.3 Differential Equations of First Order and First Degree

A first order and first degree equation, in general is of the form

$$f\left(x, y, \frac{dy}{dx}\right) = 0$$

in which $\frac{dy}{dx}$ appeares with first degree.

The general solution of this equation is a relation of the form,

$$F(x, y, c) = 0$$

where c is an arbitrary constant, which satisfies the equation.

We shall see in subsequent sections, the methods of solving the following types of differential equations of first order and first degree.

1. Equations in which variables are separable.
2. Homogeneous equations.
3. Equations reducible to Homogeneous form.
4. Exact equations.
5. Equations reducible to exact form.
6. Linear equations.
7. Equations reducible to linear form.

2.4 Equations in which Variables are Separable

A differential equation is said to be of type **"variable separable"**, if it can be expressed in such a way, so that the coefficient of dx is a function of x alone and the coefficient of dy is a function of y alone.

The general form of such differential equation can be written as

$$f(x)\,dx = g(y)\,dy$$

Integrating both sides and adding an arbitrary constant c, we get the general solution as

$$\int f(x)\,dx = \int g(y)\,dy + c$$

Working rule of solving by the method of separation of variables.

1. Write the given differential equation in the form

$$f(x)\,dx = f(y)\,dy$$

That is make the coefficient of dx as an expression of x alone and that of dy as an expression of y alone.

2. Integrate both sides and add an arbitrary constant to any one side and get the general solution.

Example 1. Solve : $(x^2 + 1)\frac{dy}{dx} = 1$.

Solution : The given equation can be written as, $dy = \frac{dx}{(x^2+1)}$

On integration, we get,

$$\int dy = \int \frac{dx}{(x^2+1)} + c \quad \Rightarrow \quad y = \tan^{-1} x + c$$

This is the solution of the equation.

Example 2. Solve : $\sec^2 x \tan y\, dx + \sec^2 y \tan x\, dy = 0$

Solution : The given equation is, $\sec^2 x \tan y\, dx + \sec^2 y \tan x\, dy = 0$

Dividing throughout by tan x . tan y, we get

$$\frac{\sec^2 x}{\tan x}dx + \frac{\sec^2 y}{\tan y}dy = 0$$

$$\Rightarrow \quad \int\frac{\sec^2 x}{\tan x}dx + \int\frac{\sec^2 y}{\tan y}dy = \log c$$

$$\Rightarrow \quad \log(\tan x) + \log(\tan y) = \log c$$

$$\Rightarrow \quad \log(\tan x \,.\, \tan y) = \log c \Rightarrow \mathbf{\tan x \,.\, \tan y = c}$$

This is the required solution.

EXERCISES

Solve the following differential equations by separating the variable.

1. $x\sqrt{1+y^2}\,dx + y\sqrt{1+x^2}\,dy = 0$

2. $(1 - \cos 2x)\,dy + (1 + \cos 2y)\,dx = 0$

3. $\dfrac{dy}{dx} = \cos^3 x \sin^4 x$

4. $\dfrac{dy}{dx} + \sqrt{\dfrac{1-y^2}{1-x^2}} = 0$, given $y = 1$ when $x = 0$

5. $(e^x + e^{-x})\dfrac{dy}{dx} = (e^x - e^{-x})$

6. $\dfrac{dy}{dx} = k(y-a)(y-b)$

7. $\dfrac{dy}{dx} = \dfrac{x(2\log x + 1)}{\sin y + y\cos y}$

8. $e^x \tan y\,dx + (1 - e^x)\sec^2 y\,dy = 0$

9. $\dfrac{dy}{dx} = (1+x)(1+y^2)$, given $y = 1$ when $x = 0$

10. $\dfrac{dy}{dx} = \dfrac{y^2(x+1)}{x^2(y-1)}$

11. $3e^x \tan y\,dx + (1 - e^x)\sec^2 y\,dy = 0$

12. $x\sqrt{y}\,dx + (1+y)\sqrt{1+x}\,dy = 0$

13. $y - x\dfrac{dy}{dx} = a\left(y^2 + \dfrac{dy}{dx}\right)$

14. $a\left(x\dfrac{dy}{dx} + 2y\right) = 2xy\dfrac{dy}{dx}$

15. $(x^2 y - x^2)\,dx + (xy^2 - y^2)\,dy = 0$

16. $\sqrt{a+x}\,dy + xy\,dx = 0$

17. $(\tan^2 x + 2\tan x + 5)\dfrac{dy}{dx} = 2(1 + \tan x)\sec^2 x$

18. $\dfrac{dy}{dx} = e^{2x-3y} + 4x^2 e^{-3y}$

19. $\cos y(\log \sec x + \tan x)\,dx = \cos x \log(\sec y + \tan y)\,dy$

20. $(1-y)x\dfrac{dy}{dx} + (1+x)y = 0$, given $y = 1$ when $x = 1$

Answers

1. $\sqrt{1+x^2} + \sqrt{1+y^2} = c$

2. $\tan y = \cot x + c$

3. $y = \dfrac{\sin^5 x}{5} - \dfrac{\sin^{-1} x}{7}$

4. $\sin^{-1} x + \sin^{-1} y = \dfrac{\pi}{2}$

5. $y = \log(e^x + e^{-x}) + c$

6. $\log\left(\dfrac{y-a}{y-b}\right) = (a-b)[kx + c]$

7. $x^2 \log x = y\sin y + c$

8. $\tan y = c\,(1 - e^x)$

9. $\tan^{-1} y = x + \dfrac{x^2}{2} + \dfrac{\pi}{4}$

10. $\log\dfrac{x}{y} = \dfrac{1}{x} + \dfrac{1}{y} + c$

11. $\tan y = c(1 - e^x)^3$

12. $(x-2)\sqrt{1+x} + (y+3)\sqrt{y} = c$

13. $(1 - ay)(a + x) = cy$ **14.** $\sqrt{1+x^2} + \sqrt{1+y^2} - \log\left[\frac{(\sqrt{1+x^2}+1)(\sqrt{1+y^2}+1)}{xy}\right] = c$

15. $\frac{x^2}{2} + \frac{y^2}{2} + x + y + \log(x+1)(y-1) = c$ **16.** $\log y = -\frac{2}{3}(x - 2a)\sqrt{a+x} + c$

17. $y = \log\left[\tan^2 x + 2\tan x + 5\right]$ **18.** $3e^{2x} - 2e^{3y} + 8x^3 = c$

19. $[\log(\sec x + \tan x)]^2 - [\log(\sec y + \tan y)]^2 = c$ **20.** $\log xy = y - x$

2.5 Equations Reducible to Variable Separable Form

Some of the differential equations can be transformed, by suitable substitution, to one in which variables are separable.

One such equation is of the form,

$$\frac{dy}{dx} = \varphi(ax + by + c)$$

This equation can be transformed to one in which variables can be separated, by putting $ax + by + c = v$.

If the differential equation is of the form

$$\frac{dy}{dx} = \frac{ax + by + c}{a'x + b'y + c'}$$

where $\frac{a}{a'} = \frac{b}{b'}$, then by putting $ax + by = v$, the equation can be solved by separating the variables.

Example 1. Solve : $\frac{dy}{dx} = (3x + 2y + 4)^2$.

Solution : The given equation is, $\frac{dy}{dx} = (3x + 2y + 4)^2$

Put, $3x + 2y + 4 = v \Rightarrow 3 + 2\frac{dy}{dx} = \frac{dv}{dx} \Rightarrow \frac{dy}{dx} = \frac{1}{2}\left(\frac{dv}{dx} - 3\right)$

The given equation becomes

$$\frac{1}{2}\left(\frac{dv}{dx} - 3\right) = v^2 \Rightarrow \frac{dv}{dx} - 3 = 2v^2 \Rightarrow \frac{dv}{dx} = 2v^2 + 3$$

On separating the variables, we get

$$\frac{dv}{2v^2 + 3} = dx$$

$$\Rightarrow \int \frac{dv}{2v^2 + 3} = \int dx + c$$

$$\Rightarrow \frac{1}{2}\int \frac{dv}{(\sqrt{3/2})^2 + v^2} = \int dx + c \text{ i.e. } \frac{1}{2}\cdot\frac{\sqrt{2}}{\sqrt{3}}\tan^{-1}\left(\frac{\sqrt{2}v}{\sqrt{3}}\right) = x + c$$

or $$\frac{1}{\sqrt{6}}\tan^{-1}\left[\frac{\sqrt{2}}{\sqrt{3}}(3x + 2y + 4)\right] = x + c$$

This is the required solution.

Example 2. Solve : $\dfrac{dy}{dx} = \dfrac{6x - 2y - 7}{3x - y + 4}$.

Solution : The given equation can be written as, $\dfrac{dy}{dx} = \dfrac{2(3x-y)-7}{(3x-y)+4}$

Put, $3x - y = v \Rightarrow 3 - \dfrac{dy}{dx} = \dfrac{dv}{dx} \Rightarrow \dfrac{dy}{dx} = 3 - \dfrac{dv}{dx}$

Thus the equation becomes

$$3 - \frac{dv}{dx} = \frac{2v-7}{v+4} \Rightarrow \frac{dv}{dx} = 3 - \frac{2v-7}{v+4} \Rightarrow \frac{dv}{dx} = \frac{3v+12-2v+7}{v+4}$$

$$\Rightarrow \frac{dv}{dx} = \frac{v+19}{v+4}$$

Separating the variables, we get, $dx = \dfrac{v+4}{v+19}\,dv$

On integration, we get,

$$\int dx = \int \frac{v+4}{v+19}\,dv + c \Rightarrow x = \int \frac{v+19-15}{v+19}\,dv + c$$

$$\Rightarrow x = \int dv - 15\int \frac{1}{v+19}\,dv + c$$

$$\Rightarrow x = v - 15 \log (v + 19) + c$$

or $\quad x = 3x - y - 15 \log (3x - y + 19) + c$

$\Rightarrow \quad$ **$2x - y - 15 \log (3x - y + 19) + c = 0$**

This is the required solution.

EXERCISES

Solve the following differential equations.

1. $\dfrac{dy}{dx} = (4x + y + 1)^2$

2. $(x + y + 1)\dfrac{dy}{dx} = 1$

3. $\dfrac{dy}{dx} = \dfrac{x+y-1}{x+y+1}$

4. $\dfrac{dy}{dx} = \dfrac{4x+6y+5}{3y+2x+4}$

5. $(2x + 2y + 3)\, dy - (x + y + 1)\, dx = 0$

6. $(x - y - 2)\, dx - (2x - 2y - 3)\, dy = 0$

7. $(x + 2y - 1)\, dx = (x + 2y + 1)\, dy$

8. $(3y + 2x + 4)\, dx - (4x + 6y + 3)\, dy = 0$

9. $(4x - 6y - 1)\, dx - (2x - 3y + 2)\, dy = 0$

10. $(x + y + 1)\, dx - (2x + 2y + 1)\, dy = 0$

11. $\dfrac{dy}{dx} = \dfrac{a^2}{(x-y)^2}$

12. $\dfrac{dy}{dx} = \cos (x + y)$

13. $2 (x + y)\, dy = (1 - 3x - 3y)\, dx$

14. $\dfrac{dy}{dx} = 1 + e^{x-y}$

15. $\dfrac{dy}{dx} = (x + y)^2$

16. $\dfrac{dy}{dx} = \sin (x + y)$

17. $\dfrac{dy}{dx} = e^y \left[\dfrac{dy}{dx} - 1\right] = e^x$

18. $\dfrac{y}{x}\dfrac{dy}{dx} + \dfrac{x^2+y^2-1}{2(x^2+y^2)+1}$

Answers

1. $4x + y + 1 = 2\tan(2x + 2c)$
2. $x + y + 2 = c\, e^y$
3. $x - y + c = \log(x + y)$
4. $y - 2x + \frac{3}{8}\log(16x + 24y + 23) = c$
5. $6y - 3x + 6\log(3x + 3y + 4) = c$
6. $x - 2y - \log(x - y - 1) = c$
7. $x = y + \frac{2}{3}\log\left(x + 2y - \frac{1}{3}\right) + c$
8. $x = 2y - \frac{5}{7}\log\left(2x + 3y + \frac{18}{7}\right) + c$
9. $x = \frac{1}{2}\left[y - \frac{5}{4}\left(2x - 3y - \frac{7}{4}\right)\right] + c$
10. $6y - 3x = \log(3x + 3y + 2) + c$
11. $y = a\tan^{-1}\left(\frac{x - y}{a}\right) + c$
12. $x + c = \tan\left(\frac{x + y}{2}\right)$
13. $3x + y + 2\log(1 - x - y) = c$
14. $x - e^{-(x - y)} = c$
15. $\tan^{-1}(x + y) = x + c$
16. $\tan(x + y) - \sec(x + y) = x + c$
17. $e^{-(x - y)} = x + c$
18. $x^2 + 2y^2 = 3\log(x^2 + y^2 + 2) + c$

2.6 Homogeneous Equations

A function $f(x, y)$ is said to be homogeneous of degree n in x and y if,

$$f(x, y) = x^n g\left(\frac{y}{x}\right)$$

A differential equation of the form,

$$\frac{dy}{dx} = \frac{f(x, y)}{g(x, y)}$$

where $f(x, y)$ and $g(x, y)$ are homogeneous functions of the same degree in x and y is called a **Homogeneous differential equations of first order.**

Consider a Homogeneous differential equation

$$\frac{dy}{dx} = \frac{f(x, y)}{g(x, y)} \qquad \text{...(1)}$$

Let $f(x, y)$ and $g(x, y)$ be homogeneous functions of degree n. Then we can write the equation (1) as

$$\frac{dy}{dx} = \frac{x^n \varphi(y/x)}{x^n \psi(y/x)} \Rightarrow \frac{dy}{dx} = \frac{\varphi(y/x)}{\psi(y/x)} \qquad \text{...(2)}$$

To solve this equation, hence (1), we put

$$y = vx \Rightarrow \frac{dy}{dx} = v + x\frac{dv}{dx}$$

Then (2) becomes

$$v + x\frac{dv}{dx} = \frac{\varphi(v)}{\psi(v)}$$

which can be solved by separating the variables.

Example 1. Solve : $(x - y)\,dy - (2x - y)\,dx = 0$

Solution : The given equation can be written as

$$\frac{dy}{dx} = \frac{2x - y}{x - y}$$

This is a homogeneous equation – for $2x - y$ and $x - y$ are homogeneous equations of degree 1.

Put, $y = vx \Rightarrow \dfrac{dy}{dx} = v + x\dfrac{dv}{dx}$.

Thus the equation becomes

$$v + x\frac{dv}{dx} = \frac{2x - vx}{x - vx}$$

$$\Rightarrow \quad v + x\frac{dv}{dx} = \frac{2-v}{1-v}$$

$$\Rightarrow \quad x\frac{dv}{dx} = \frac{2-v}{1-v} - v \Rightarrow x\frac{dv}{dx} = \frac{2-v-v+v^2}{1-v}$$

$$\Rightarrow x\frac{dv}{dx} = \frac{v^2-2v+2}{1-v}$$

By separating the variables, we get,

$$\frac{1-v}{v^2-2v+2}dv = \frac{dx}{x}$$

$$\Rightarrow \quad \int \frac{1-v}{v^2-2v+2}dv = \int \frac{dx}{x} + \log c$$

$$\Rightarrow \quad -\frac{1}{2}\int \frac{2v-2}{v^2-2v+2}dv = \int \frac{dx}{x} + \log c$$

$$\Rightarrow \quad -\frac{1}{2}\log(v^2-2v+2) = \log x + \log c$$

$$\Rightarrow \quad 2\log x + \log(v^2 - 2v + 2) + 2\log c = 0$$

$$\Rightarrow \quad \log[x^2(v^2 - 2v + 2)]\,c^2 = 0$$

$$\Rightarrow \quad [x^2(v^2 - 2v + 2)]\,c^2 = 1$$

or $$x^2\left[\left(\frac{y}{x}\right)^2 - 2\left(\frac{y}{x}\right) + 2\right]c^2 = 1 \qquad \left[\text{since } v = \frac{y}{x}\right]$$

or $$\mathbf{2x^2 - 2xy + y^2 = c'} \qquad \text{where } \frac{1}{c^2} = c'$$

This is the general solution.

EXERCISES

Solve the following equations.

1. $(x^2 + y^2)\,dy = xy\,dx$
2. $\dfrac{dy}{dx} = \dfrac{x+2y}{2x+y}$
3. $\dfrac{dy}{dx} = \dfrac{x^2+y^2}{x(x+y)}$
4. $(x^2 + y^2)\,dx - 2xy\,dy = 0$
5. $(x - \sqrt{xy})\,dy = y\,dx$
6. $\dfrac{dy}{dx} - \sin\dfrac{y}{x} = \dfrac{y}{x}$
7. $x\dfrac{dy}{dx} + \dfrac{y^2}{x} = y$
8. $2xy + (y^2 - x^2)\dfrac{dy}{dx} = 0$
9. $\dfrac{dy}{dx} = \dfrac{y}{x} + \tan\dfrac{y}{x}$
10. $x(y - x)\,dy = y(x + y)\,dx$
11. $\dfrac{dy}{dx} = \dfrac{y}{x}\left(\log\dfrac{y}{x} + 1\right)$
12. $x\dfrac{dy}{dx} = yx\,\text{cosec}^2\,\dfrac{y}{x}$

13. $(x^3 + y^3)\,dx = (x^2 y + x y^2)\,dy$

14. $(x^2 + 2y^2)\,dx - xy\,dy = 0$

15. $\left(x\sin\frac{y}{x} - y\cos\frac{y}{x}\right)dx + x\cos\frac{y}{x}dy = 0$

16. $2xy^2\,dy - (x^3 + 2y^3)\,dx = 0$

17. $x^2 + (x^2 - xy)\,\frac{dy}{dx} = 0$

18. $x\,\frac{dy}{dx} = y + x\cos^2\left(\frac{y}{x}\right)$

19. $x(x - y)\,dy + y^2\,dx = 0$

Answers

1. $e^{(x^2/2y^2)} = cy$

2. $\log x = \frac{1}{2}\log\left(\frac{x+y}{x}\right) - \frac{3}{2}\log\left(\frac{x-y}{x}\right) + c$

3. $(x - y)^2 = c\,x\,e^{-y/x}$

4. $(x^2 - y^2) = k\,x$

5. $2\sqrt{\frac{x}{y}} + \log y + c = 0$

6. $\tan\left(\frac{y}{2x}\right) = c\,x$

7. $\log x - \frac{x}{y} = c$

8. $x^2 + y^2 = c\,y$

9. $y = x\sin^{-1}(c\,x)$

10. $\log(x\,y) = \frac{y}{x} + c$

11. $\log\left(\frac{y}{x}\right) = c\,x$

12. $x = c\,e^{\cos(y/x)}$

13. $\frac{y}{x} + \log(y - x) = c$

14. $x^2 + y^2 = c\,x^4$

15. $x\sin\frac{y}{x} = c$

16. $2y^3 = 3x^3(\log x + c)$

17. $y = x(\log y + c)$

18. $\log x = \tan\frac{y}{x} + c$

19. $\log x = \frac{y}{x} - \log\frac{y}{x} + c$

2.7 Equations Reducible to Homogeneous Form

The equation of the form,

$$\frac{dy}{dx} = \frac{ax + by + c}{Ax + By + C}$$

where a, b, c, A, B and C are constants such that $aB \neq bA$ $\left(\text{i.e. } \frac{a}{A} \neq \frac{b}{B}\right)$, can be reduced to homogeneous form by putting,

$$x = X + h \text{ and } y = Y + k$$

where h, k being constants to be determined.

$$x = X + h,\quad y = Y + k \Rightarrow dx = dX,\quad dy = dY \Rightarrow \frac{dy}{dx} = \frac{dY}{dX}$$

The equation takes the form,

$$\frac{dY}{dX} = \frac{(aX + bY) + (ah + bk + c)}{(AX + BY) + (Ah + Bk + C)}$$

If we choose h and k in such a way that,

$$ah + bk + c = 0$$
$$Ah + Bk + C = 0$$

then the equation reduces to,

$$\frac{dY}{dX} = \frac{aX + bY}{AX + BY}$$

which is a homogeneous equation, which can be solved by putting $Y = vX$.

Note : If $\frac{a}{A} = \frac{b}{B}$, then the above method fails to solve the equation.

$$\text{If, } \frac{a}{A} = \frac{b}{B} = \lambda\text{ (say)} \Rightarrow a = A\lambda,\quad b = B\lambda$$

The given equation becomes,

$$\frac{dy}{dx}=\frac{\lambda(Ax+By)+c}{Ax+By+C}$$

Solving of such differential equation is delt under the section equations reducible to variable separable form - by putting $Ax + By = v$.

Example 1. Solve : $(2x + 3y - 5)\dfrac{dy}{dx} + (3x + 2y - 5) = 0$.

Solution : The given equation is

$$\frac{dy}{dx}=-\left(\frac{3x+2y-5}{2x+3y-5}\right)$$

Put, $x = X + h,\ y = Y + k$, then $\dfrac{dy}{dx}=\dfrac{dY}{dX}$

The equation becomes,

$$\frac{dY}{dX}=-\left[\frac{3X+2Y+(3h+2k-5)}{2X+3Y+(2h+3k-5)}\right]$$

We shall choose h, k such that

$$3h + 2k - 5 = 0$$
$$2h + 3k - 5 = 0$$

$$\Rightarrow \quad \frac{h}{-10+15}=\frac{k}{-10+15}=\frac{1}{9-4} \Rightarrow h=1 \text{ and } k=1 \Rightarrow X=x-1,\ Y=y-1$$

The equation becomes, for these values of h and k

$$\frac{dY}{dX}=-\left[\frac{3X+2Y}{2X+3Y}\right]$$

This is a homogeneous equation. Put, $Y = vX \Rightarrow \dfrac{dY}{dX}=v+X\dfrac{dv}{dX}$

Then we have,

$$v+X\frac{dv}{dX}=-\left[\frac{3X+2vX}{2X+3vX}\right]$$

$$\Rightarrow \quad v+X\frac{dv}{dX}=-\left[\frac{3+2v}{2+3v}\right] \Rightarrow X\frac{dv}{dX}=-\frac{3+2v}{2+3v}-v \Rightarrow X\frac{dv}{dX}=\frac{-3v^2-4v-3}{2+3v}$$

On separating the variables, we get

$$\frac{dX}{X}+\frac{3v+2}{3v^2+4v+3}dv=0 \Rightarrow \int\frac{dX}{X}+\int\frac{3v+2}{3v^2+4v+3}dv=\log c_1$$

$$\Rightarrow \int\frac{dX}{X}+\frac{1}{2}\int\frac{6v+4}{3v^2+4v+3}dv=\log c_1$$

$$\Rightarrow \log X+\frac{1}{2}\log(3v^2+4v+3)=\log c_1$$

$$\Rightarrow \log X^2+\log(3v^2+4v+3)=\log c \qquad (c=c_1^2)$$

$$\Rightarrow X^2(3v^2+4v+3)=c$$

Now putting $X = x - 1$ and $v=\dfrac{Y}{X}=\dfrac{y-1}{x-1}$, we get,

$$(x-1)^2\left[3\left(\frac{y-1}{x-1}\right)^2+4\left(\frac{y-1}{x-1}\right)+3\right]=c$$

$\Rightarrow \quad 3(y-1)^2 + 4(y-1)(x-1) + 3(x-1)^2 = c$

or $\quad \mathbf{3x^2 + 4xy + 3y^2 - 10x - 10y + 10 = c}$

This is the required solution.

Second Method

The given equation is,

$$(2x + 3y - 5)\,dy + (3x + 2y - 5)\,dx = 0.$$

Observe that the coefficient of $x\,dy$ and $y\,dx$ are same, that is 2. Thus the equation can be rearranged as

$$(3x - 5)\,dx + (3y - 5)\,dy + 2(x\,dy + y\,dx) = 0$$

i.e. $\quad (3x - 5)\,dx + (3y - 5)\,dy + 2\,d(xy) = 0$

$\Rightarrow \quad \int (3x-5)dx + \int (3y-5)\,dy + 2\int d(xy) = k$

$\Rightarrow \quad \frac{3x^2}{2} - 5x + \frac{3y^2}{2} - 5y + 2xy = k \Rightarrow \mathbf{3x^2 + 4xy + 3y^2 - 10x - 10y = k}$

Observe that the solutions obtained by first and second methods differ only by constant. In fact c in the first solution and k in the second solution are related by $k = c - 10$.

EXERCISES

Solve the following differential equations.

1. $(x + y - 10)\,dx + (x - y - 2)\,dy = 0$
2. $(y + x - 2)\,dx - (y - x - 4)\,dy = 0$
3. $(x - 2y + 3)\,dx - (2x - 4y + 5)\,dy = 0$
4. $(x - 8y + 7)\,dy - (x - y)\,dx = 0$
5. $(x - 2y + 5)\,dx - (2x + y - 1)\,dy = 0$
6. $\frac{dy}{dx} = \frac{x+2y-3}{2x+y-3}$
7. $\frac{dy}{dx} = \frac{x+3y-4}{3x-y-2}$
8. $(2x + 3y - 6)\,dy = (6x - 2y - 7)\,dx$
9. $\frac{dy}{dx} = \frac{2x+y+6}{y-x-3}$, when $x = 0, y = 0$
10. $\frac{dy}{dx} + \frac{2x+3y}{y+2} = 0$
11. $\frac{dy}{dx} + \frac{7x+3y+5}{3x+11y+17} = 0$
12. $(2x + 9y - 20)\,dx - (6x + 2y - 10)\,dy = 0$
13. $(3y - 7x + 7)\,dx + (7y - 3x + 3)\,dy = 0$
14. $(2x + y - 3)\,dy = (x + 2y - 3)\,dx$
15. $(x + 2y - 4)\,dx - (2x + y - 5)\,dy = 0$
16. $\frac{dy}{dx} = \frac{x+2y-3}{2x+y-3}$

Answers

1. $x^2 + 2xy - y^2 - 20x - 4y = c$
2. $x^2 + 2xy - y^2 - 4x + 8y = c$
3. $x^2 - 4xy + 4y^2 + 6x - 10y = c$
4. $x^2 - 2xy + 8y^2 - 14y = c$
5. $x^2 - y^2 - 4xy + 10x + 2y = c$
6. $c(x - y)^3\ c.(x - y)^3 = x + y - 2$
7. $\log[(x-1)^2 + (y-1)^2] = 6\tan^{-1}\left(\frac{y-1}{x-1}\right) + c$
8. $3x^2 - 4xy - 3y^2 - 14x + 12y = c$
9. $y^2 - 2xy - 2x^2 - 6y - 12x = c$
10. $\log X - \log(v + 1) - 2\log(v + 2) = c$, where $X = x - 3, v = \frac{y+2}{x-3}$

11. $\log x + \frac{1}{2}\log(11v^2 + 6v + 7) = c$, where $X = x + \frac{1}{7}, v = \frac{y + \frac{26}{17}}{x + \frac{1}{17}}$

12. $x + 2y - 5 = c(2x - y)^2$

13. $(y + x - 1)^5 (y - x + 1)^2 = c$

14. $(y - x)^3 = c(x + y - 2)$

15. $\log X = \log\left(\frac{1+v}{1-v}\right) + \frac{1}{2}\log(1 - v^2) + c$, where $X = x - 2, v = \frac{y-1}{x-2}$

16. $\log X = \frac{1}{2}\log(1 + v) - \frac{3}{2}\log(1 - v) + c$, where $X = x - 1, v = \frac{y-1}{x-1}$

2.8 Linear Differential Equations

A differential equation of the form,

$$\frac{dy}{dx} + Py = Q \qquad \ldots(1)$$

where p and Q are functions of x, is called **a first order linear differential equation.**

A Linear differential equation of first order is also defined as one in which the dependent variable and its derivatives occur only in the first degree and are not multiplied together.

To solve the equation (1), we multiply both sides of the equation by a factor, $e^{\int P\,dx}$

This gives us,

$$\left[\frac{dy}{dx} + Py\right] e^{\int P\,dx} = Q \,.\, e^{\int P\,dx} \qquad \ldots(2)$$

Now consider,

$$\frac{d}{dx}\left[y \,.\, e^{\int P\,dx}\right] = \frac{dy}{dx} \,.\, e^{\int P\,dx} + y \,.\, Pe^{\int P\,dx}$$

$$= \left[\frac{dy}{dx} + Py\right] e^{\int P\,dx}$$

which is the L.H.S. of (2).

That is the L.H.S of (2) is the differential coefficient of a single function $y \,.\, e^{\int P\,dx}$

Thus (2) can be written as

$$\frac{d}{dx}\left[y \,.\, e^{\int P\,dx}\right] = Q \,.\, e^{\int P\,dx}$$

Integrating both sides w.r.t.x, we get

$$y \,.\, e^{\int P\,dx} = \int Q \,.\, e^{\int P\,dx}\, dx + c$$

This is the general solution of the given equation (1).

The factor $e^{\int P\,dx}$ is called an **integrating factor** (I.F) of the linear equation (1).

Working Rule.

1. Write the given equation in the form, $\frac{dy}{dx} + Py = Q$
2. Find the integrating factor (I.F) by evaluating, I.F $= e^{\int P\,dx}$
3. Write the solution of the equation as,

$$y\,(\text{I.F}) = \int Q \,.\, (\text{I.F})\, dx + c$$

Example 1. Solve : $x\frac{dy}{dx} - 2y = 2x$.

Solution : The given equation can be written as, $\frac{dy}{dx} - \frac{2}{x} \cdot y = 2$

This is of the form, $\frac{dy}{dx} + Py = Q$. Here, $P = -\frac{2}{x}$ and $Q = 2$

Now, $\int P\,dx = -2\int \frac{1}{x}dx = -2\log x = \log\frac{1}{x^2}$

$\therefore$ I.F $= e^{\int P\,dx} \Rightarrow$ I.F $= e^{\log(1/x^2)} = \frac{1}{x^2}$

The general solution is,

$$y \cdot \frac{1}{x^2} = \int 2 \cdot \frac{1}{x^2}dx + c \Rightarrow \frac{y}{x^2} = -\frac{2}{x} + c \Rightarrow \mathbf{y = c\,x^2 - 2x}$$

Example 2. Solve : $\sin x \cos x \frac{dy}{dx} = y + \sin x$.

Solution : The given equation is

$$\frac{dy}{dx} - \frac{1}{\sin x \cos x}y = \frac{1}{\cos x}$$

This is a linear differential equation, with

$$P = -\frac{1}{\sin x \cos x} \quad \text{and} \quad Q = \sec x$$

$$\therefore \quad \int P\,dx = -\int \frac{1}{\sin x \cos x}dx = -\int \frac{\sec^2 x}{\tan x}dx = -\log(\tan x)$$

Now I.F. $= e^{\int P\,dx} = e^{-\log \tan x} = e^{\log \cot x} = \cot x$

Thus the general solution is,

$$y \cdot \cot x = \int \cot x \cdot \sec x\,dx + c$$

$$\Rightarrow \quad y \cdot \cot x = \int \operatorname{cosec} x\,dx + c$$

$$\Rightarrow \quad \mathbf{y \cdot \cot x = \log(\operatorname{cosec} x - \cot x) + c}$$

Example 3. Solve : $(1 + x^2)\frac{dy}{dx} + y = e^{\tan^{-1} x}$.

Solution : The given equation can be written as,

$$\frac{dy}{dx} + \frac{1}{1+x^2}y = \frac{e^{\tan^{-1} x}}{1+x^2}$$

Here, $P = \frac{1}{1+x^2}$ and $Q = \frac{e^{\tan^{-1} x}}{1+x^2}$

Now, $\int P\,dx = \frac{1}{1+x^2}dx = \tan^{-1} x$. Thus, I.F. $= e^{\tan^{-1} x}$

The general solution is,

$$y \cdot e^{\tan^{-1} x} = \int e^{\tan^{-1} x} \cdot \frac{e^{\tan^{-1} x}}{1+x^2}dx + c$$

Consider $I = \int (e^{2\tan^{-1} x}) \cdot \frac{dx}{1+x^2}$. Put, $\tan^{-1} x = t \Rightarrow \frac{1}{1+x^2}dx = dt$

$$\therefore \qquad I = \int e^{2t}\,dt = \frac{1}{2}e^{2t} = \frac{1}{2}e^{2\tan^{-1}x}$$

Thus the solution is, $\boldsymbol{y \,.\, e^{\tan^{-1}x} = \frac{1}{2}e^{2\tan^{-1}x} + c}$

Example 4. Solve : $(1 + x^2)\,dy + (y - \tan^{-1} x)\,dx = 0$.

Solution : The given equation can be written as

$$\frac{dy}{dx} + \frac{y - \tan^{-1} x}{1+x^2} = 0$$

$$\Rightarrow \qquad \frac{dy}{dx} + y\frac{1}{1+x^2} = \frac{\tan^{-1} x}{1+x^2}$$

This is a linear differential equation, with, $P = \frac{1}{1+x^2}$, $Q = \frac{\tan^{-1} x}{1+x^2}$

Now $\quad \int P\,dx = \int \frac{1}{1+x^2}dx = \tan^{-1} x$. Thus, I.F $= e^{\int P\,dx} = e^{\tan^{-1}x}$

Thus the general solution is,

$$y \,.\, e^{\tan^{-1}x} = \int e^{\tan^{-1}x} . \frac{\tan^{-1} x}{1+x^2}dx$$

Let, $\quad I = \int \tan^{-1} x \,.\, e^{\tan^{-1}x} . \frac{dx}{1+x^2}$, put $\tan^{-1} x = t \Rightarrow \frac{1}{1+x^2}dx = dt$

$$\therefore \qquad I = \int t\,.\,e^t\,.\,dt = t\,.\,e^t - \int e^t\,dt = t\,.\,e^t - e^t = (\tan^{-1} x - 1)\,e^{\tan^{-1}x}$$

Thus the solution is,

$$\boldsymbol{y \,.\, e^{\tan^{-1}x} = (\tan^{-1} x - 1)\,.\,e^{\tan^{-1}x} + c}$$

Example 5. Solve : $\frac{dy}{dx} + y \cot x = 4x \operatorname{cosec} x$, given that $y = 0$ when $x = \frac{\pi}{2}$.

Solution : The given equation is, $\frac{dy}{dx} + y\cot x = 4x \operatorname{cosec} x$

Here, $\quad P = \cot x$ and $Q = 4x \operatorname{cosec} x$

Now, $\quad \int P\,dx = \int \cot x\,dx = \log \sin x$. Thus, I.F. $= e^{\int P\,dx} = e^{\log \sin x} = \sin x$

The general solution is,

$$y \sin x = \int \sin x \,.\, 4x \operatorname{cosec} x\,dx + c$$

$$\Rightarrow \qquad y \sin x = \int 4x\,dx + c \Rightarrow y \sin x = 2x^2 + c$$

By data, $\quad y = 0$, when $x = \frac{\pi}{2}$. Thus, $0 = 2\frac{\pi^2}{4} + c \Rightarrow c = -\frac{\pi^2}{2}$

Thus the particular solution is,

$$\boldsymbol{y \sin x = 2x^2 - \frac{\pi^2}{2}.}$$

Example 6. Solve : $dx + 2xy\,dy = ye^{-y^2}\,dy$.

Solution : The equation can be written as

$$dx + (2xy - y\,e^{-y^2})\,dy = 0$$

If we consider y as dependent variable, the equation can not be written in the form (1), for the reason here y does not appear in first degree.

However, the equation can be written as

$$\frac{dx}{dy} = y\,e^{-y^2} - 2xy \Rightarrow \frac{dx}{dy} + 2y\,.\,x = ye^{-y^2}$$

This is a linear equation, with x as dependent variable, with, $P = 2y$ and $Q = y\ e^{-y^2}$.

Now, I.F. $= e^{\int 2y\,dy} = e^{y^2}$

Thus the general solution is,

$$x\,.\,e^{y^2} = \int y\,e^{-y^2}\,.\,e^{y^2}\,.\,dy + c$$

$$\Rightarrow \quad x\,.\,e^{y^2} = \int y\,dy + c \Rightarrow \boldsymbol{x\,.\,e^{y^2} = \frac{y^2}{2} + c}$$

which is the required solution.

Example 7. Solve : $(x + \tan y)\,dy = \sin 2y\,dx$

Solution : The given equation can be written as

$$\frac{dx}{dy} = \frac{x + \tan y}{\sin 2y}$$

or

$$\frac{dx}{dy} - \frac{1}{\sin 2y}x = \frac{\tan y}{\sin 2y}$$

This is a linear equation, treating x as dependent variable, with,

$$P = -\frac{1}{\sin 2y} \text{ and } Q = \frac{\tan y}{\sin 2y}$$

Now,

$$\int p\,dy = \int -\frac{1}{\sin 2y}dy = -\frac{1}{2}\int \frac{\sec^2 y}{\tan y}dy = -\frac{1}{2}\log(\tan y)$$

$\therefore$

$$\text{I.F.} = e^{\int P\,dy} = e^{-\frac{1}{2}\log \tan y} = \frac{1}{\sqrt{\tan y}} = \sqrt{\cot y}\,.$$

Thus the solution is,

$$x\sqrt{\cot y} = \int \frac{\tan y}{\sin 2y}\,.\sqrt{\cot y}\,dy + c$$

$$= \int \frac{\sec^2 y}{2\sqrt{\tan y}}\,dy + c$$

$$= \sqrt{\tan y} + c \qquad \left[\text{since } \int \frac{1}{2\sqrt{x}}dx = \sqrt{x}\right]$$

That is, $\boldsymbol{x\sqrt{\cot y} = \sqrt{\tan y} + c}$ is the required solution.

Note : An equation of the form

$$f'(y)\frac{dy}{dx} + P\,.\,f(y) = Q$$

where p and Q are functions of x alone, can be brought to linear form by putting $f(y) = v$.

Example 8. Solve : $x\dfrac{dy}{dx} + y \log y = xy\,e^x$.

Solution : The given equation can be written as

$$\frac{1}{y}\frac{dy}{dx} + \frac{1}{x}\log y = e^x \qquad \text{(i.e. by dividing by } xy\text{)}$$

This equation is of the form,

$$f'(y)\frac{dy}{dx}+P\,.\,f(y)=Q$$

with $f(y) = \log y$.

Put, $\log y = v \Rightarrow \frac{1}{y}\frac{dy}{dx}=\frac{dv}{dx}$. Thus the given equation becomes

$$\frac{dv}{dx}+\frac{1}{x}v=e^x$$

This is a linear differential equation with, $P = \frac{1}{x}$ and $Q = e^x$.

Now, I.F. $= e^{\int 1/x\,dx} = e^{\log x} = x$

Thus the general solution is,

$$v\,x = \int x\,.\,e^x\,dx + c$$

$\Rightarrow$ $v\,x = x\,.\,e^x - \int e^x\,dx + c \Rightarrow v\,x = x\,.\,e^x - e^x + c$

or $\boldsymbol{x \log y = (x-1)\,e^x + c}$ [Since $v = \log y$]

EXERCISES

Solve the following equations.

1. $\frac{dy}{dx} + 3x^2\,y = x^5\,e^{x^3}$
2. $\sin x\frac{dy}{dx} + \cos x\,y = x\sin x$
3. $\frac{dy}{dx} + y\sec x = \tan x$
4. $\frac{dy}{dx} + y\tan x = \sin 2x$
5. $\frac{dy}{dx} - \frac{2}{x}\,y = x + x^2$
6. $\frac{dy}{dx} + y\cot x = 4x\,\text{cosec}\,x$
7. $\frac{dy}{dx} = y\cos x + \cos x\sin^2 x$
8. $(1+x^2)\,\frac{dy}{dx} + xy = \sinh^{-1} x$
9. $(1-x^2)\,\frac{dy}{dx} + 2\,xy = x\sqrt{1-x^2}$, given $y = 0$ when $x = 0$.
10. $\frac{dy}{dx} + 3y = e^{-x}\cos x$
11. $(1+x^3)\,\frac{dy}{dx} + 6\,x^2\,y = e^x$
12. $y\,dx + (3x + 2 - yx)\,dy = 0$
13. $(1+x^2)\,\frac{dy}{dx} + 2\,xy - 4x^2 = 0$
14. $(1+y^2)\,dx = (\tan^{-1} y - x)\,dy$
15. $x\log x\,\frac{dy}{dx} + y = 2\log x$
16. $(1-x^2)\,\frac{dy}{dx} - xy = 1$
17. $(2x - 10\,y^3)\,\frac{dy}{dx} + y = 0$
18. $(1+y^2) + (x - e^{\tan^{-1} y})\,\frac{dy}{dx} = 0$
19. $(1+y^2)\,dx = (\tan^{-1} y - x)\,dy$
20. $(2x - 10\,y^3)\,\frac{dy}{dx} + y = 0$
21. $y\log y\,dx + (x - \log y)\,dy = 0$
22. $(x+y+1)\,\frac{dy}{dx} = 1$
23. $\frac{dy}{dx} = \frac{x + y\cos x}{1+\sin x}$
24. $(1-x^2)\,\frac{dy}{dx} - xy = 2x\,(1-x^2)^{1/2}$
25. $(x^2-a^2)\,\frac{dy}{dx} + xy = (x+a)\sqrt{x^2-a^2}$

26. $\dfrac{dy}{dx}-\dfrac{\tan y}{1+x}=(1+x)\,e^x \sec y$

27. $\dfrac{dy}{dx}+\dfrac{y}{x}\log y=\dfrac{y}{x^2}(\log x)^2$

28. $\dfrac{dy}{dx}+\dfrac{1}{x}\tan y=\dfrac{1}{x^2}\tan y \sin y$

29. $\dfrac{dy}{dx}+\dfrac{1}{x}=\dfrac{e^y}{x^2}$

30. $\dfrac{dy}{dx}=e^{x-y}\,(e^x-e^{-y})$

Answers

1. $12\,y=e^{x^3}(2\,x^3-1)\,c\,e^{-x^3}$
2. $(y-1)\sin x+x\cos x=c$
3. $(y-1)(\sec x+\tan x)+x=c$
4. $y\sec x=c-2\cos x$
5. $y=x^2(\log x+x+c)$
6. $y\sin x=2\,x^2+c$
7. $y+\sin^2 x+2\sin x+2=c\,e^{\sin x}$
8. $y\sqrt{1+x^2}=\dfrac{1}{2}\sin h^{-1}x+c$
9. $\dfrac{y}{\sqrt{1-x^2}}=\dfrac{1}{\sqrt{1-x^2}}-1$
10. $y=c\,e^{-3x}+e^{-x}\left(\dfrac{\sin x+2\cos x}{5}\right)$
11. $y\,(x^3+1)^2=(x^3-3x^2+6x-6)\,e^x+c$
12. $xy^3=2y^2+4y+4+ce^y$
13. $y(1+x^2)=\dfrac{4x^3}{3}+c$
14. $x=\tan^{-1}y-1+ce^{-\tan^{-1}y}$
15. $y\log x=(\log x)^2+c$
16. $y\sqrt{1-x^2}=\sin^{-1}x+c$
17. $xy^2=2\,y^5+c$
18. $x\,e^{\tan^{-1}y}=\dfrac{1}{2}e^{2\tan^{-1}y}+c$
19. $x=\tan^{-1}y-1+c\,e^{\tan^{-1}y}$
20. $xy^2=2y^5+c$
21. $x=\dfrac{1}{2}\log y+\dfrac{c}{\log y}$
22. $x=c\,e^y-(y+2)$
23. $y(1+\sin x)=c-\dfrac{x^2}{2}$
24. $y\sqrt{1-x^2}=x^2+c$
25. $y\sqrt{x^2-a^2}=\dfrac{x^2}{2}+ax+c$
26. $\sin y=(1+x)(c+e^x)$
27. $\dfrac{1}{x(\log y)}=\dfrac{1}{2x^2}+c$
28. $2x=\sin y\,(1+2\,c\,x^2)$
29. $\dfrac{e^{-y}}{x}=\dfrac{1}{2x^2}+c$
30. $c^y=e^x-1+c\,e^{-e^x}$

2.9 Equations Reducible to Linear Form - Bernoulli Equation

The equation of the form,

$$\frac{dy}{dx}+Py=Q\,y^n \qquad \ldots(1)$$

where p and Q are functions of x, is called ***Bernoulli equation.***

This equation can be reduced to linear differential equation.

To this end we shall first divide the equation by y^n. We get,

$$y^{-n}\frac{dy}{dx}+Py^{1-n}=Q \qquad \ldots(2)$$

Now put, $y^{1-n}=v \Rightarrow (1-n)\,y^{-n}\dfrac{dy}{dx}=\dfrac{dv}{dx} \Rightarrow y^{-n}\dfrac{dy}{dx}=\dfrac{1}{(1-n)}\dfrac{dv}{dx}$

Thus the equation (2) becomes

$$\frac{1}{(1-n)}\frac{dv}{dx}+Pv=Q$$

or
$$\frac{dv}{dx}+p\,(1-n)\,v=(1-n)\,Q$$

This is a linear equation and can be solved by finding the integrating factor.

Example 1. Solve : $x\dfrac{dy}{dx}+(1-x)\,y=x^2y^2.$

Solution : The given equation can be written as

$$y^{-2}\frac{dy}{dx}+\left(\frac{1-x}{x}\right)\frac{1}{y}=x$$

Put, $\dfrac{1}{y}=v \;\Rightarrow\; -\dfrac{1}{y^2}\dfrac{dy}{dx}=\dfrac{dv}{dx} \;\Rightarrow\; y^{-2}\dfrac{dy}{dx}=-\dfrac{dv}{dx}$

Thus the given equation becomes,

$$\frac{dv}{dx}-\left(\frac{1-x}{x}\right)v=-x$$

This is a linear differential equation, with, $P=\left(\dfrac{x-1}{x}\right)$ and $Q=-x$

Now, $\displaystyle\int P\,dx=\int\left(1-\frac{1}{x}\right)dx=x-\log x$

$\therefore$ I.F. $=e^{\int P\,dx}=e^{x-\log x}=e^x.e^{-\log x}=\dfrac{1}{x}e^x$

The solution is,

$$v\frac{1}{x}e^x=-\int x.\frac{1}{x}e^x dx+c$$

$\Rightarrow$ $v\dfrac{e^x}{x}=-e^x+c \;\Rightarrow\; \dfrac{e^x}{xy}+e^x=c$ $\left[\text{Since } v=\dfrac{1}{y}\right]$

or $\quad$ $\boldsymbol{e^x\,(1+xy)=c\,xy}$

Example 2. Solve : $\dfrac{dy}{dx}+\dfrac{y}{x-1}=x\,.\,y^{1/3}.$

Solution : The given equation can be written as, $\dfrac{1}{y^{1/3}}\dfrac{dy}{dx}+\dfrac{1}{x-1}y^{2/3}=x$

Putting, $y^{2/3}=v \;\Rightarrow\; \dfrac{2}{3}y^{-1/3}\dfrac{dy}{dx}=\dfrac{dv}{dx} \;\Rightarrow\; \dfrac{1}{y^{1/3}}\dfrac{dy}{dx}=\dfrac{3}{2}\dfrac{dv}{dx}$

Thus the equation becomes,

$$\frac{3}{2}\frac{dv}{dx}+\frac{1}{x-1}v=x$$

or
$$\frac{dv}{dx}+\frac{2}{3(x-1)}v=\frac{2}{3}x$$

This is a linear equation, with, $P=\dfrac{2}{3\,(x-1)}$ and $Q=\dfrac{2}{3}x$

Now, $\displaystyle\int P\,dx=\frac{2}{3}\int\frac{1}{x-1}\,dx=\frac{2}{3}\log\,(x-1)$

$\therefore$ I.F. $=e^{\int P\,dx}=e^{\frac{2}{3}\log\,(x-1)}=(x-1)^{2/3}$

Thus the solution is,

$$v(x-1)^{2/3} = \int \frac{2}{3}(x-1)^{2/3}.\, x\, dx$$

Integrating by parts

$$= \frac{2}{3}\left\{x.\frac{3}{5}(x-1)^{5/3} - \int \frac{3}{5}(x-1)^{5/3} dx\right\} + c$$

$$= \frac{2}{3}\left\{\frac{3}{5}x\,(x-1)^{5/3} - \frac{9}{40}(x-1)^{8/3} dx\right\} + c$$

$$= \frac{2}{5}x\,(x-1)^{5/3} - \frac{3}{40}(x-1)^{8/3} + c$$

Thus, $$\mathbf{y^{2/3}(x-1)^{2/3} = \frac{2}{5}x\,(x-1)^{5/3} - \frac{3}{20}(x-1)^{8/3} + c} \qquad \left[\text{Since } v = y^{2/3}\right]$$

This is the required equation.

Example 3. Solve : $(x\,y^2 - x\,e^{1/x^2})\,dx - x^2\,y\,dy = 0$.

Solution : The given equation can be written as,

$$\frac{dy}{dx} = \frac{xy^2 - xe^{1/x^2}}{x^2 y}$$

or $$\frac{dy}{dx} - \frac{1}{x}y = -\frac{1}{yx}e^{1/x^2} \Rightarrow y\frac{dy}{dx} - \frac{1}{x}y^2 = -\frac{e^{-1/x^2}}{x}$$

Put, $$y^2 = v \Rightarrow 2y\frac{dy}{dx} = \frac{dv}{dx} \Rightarrow y\frac{dy}{dx} = \frac{1}{2}\frac{dv}{dx}$$

Thus the equation becomes,

$$\frac{1}{2}\frac{dv}{dx} - \frac{1}{x}v = -\frac{e^{-1/x^2}}{x} \Rightarrow \frac{dv}{dx} - \frac{2}{x}v = -2\frac{e^{-1/x^2}}{x}.$$

This is a linear differential equation, with, $P = -\frac{2}{x}$ and $Q = -2\frac{e^{-1/x^2}}{x}$.

Now, $\int P\,dx = -2\int \frac{1}{x}dx = -2\log x = \log\frac{1}{x^2}$. Thus, I.F. $= e^{\int P\,dx} = e^{\log(1/x^2)} = \frac{1}{x^2}$

Now, the solution is

$$v.\frac{1}{x^2} = \int -2\frac{e^{-1/x^2}}{x}.\frac{1}{x^2}dx + c$$

$$\Rightarrow \quad v.\frac{1}{x^2} = \int e^{1/x^2}\left(-\frac{2}{x^2}\right)dx + c.\ \text{Put } \frac{1}{x^2} = t \Rightarrow -\frac{2}{x^3}dx = dt$$

$$\therefore \quad v.\frac{1}{x^2} = \int e^t\,dt + c \Rightarrow v.\frac{1}{x^2} = e^{1/x^2} + c \Rightarrow \mathbf{\frac{y^2}{x^2} = e^{1/x^2} + c}$$

EXERCISES

Solve the following equations.

1. $x^3\frac{dy}{dx} - x^2\,y = -\,y^4\cos x$

2. $2\frac{dy}{dx} - y\sec x = y^3\tan x$

3. $x\frac{dy}{dx} + y = x^3\,y^6$

4. $\frac{dy}{dx} + y\tan x = y^3\sec x$

5. $x\dfrac{dy}{dx}+y=y^3x^3\cos x$

6. $\dfrac{dy}{dx}-\dfrac{1}{x}y=3xe^{x^3}y^2$

7. $\dfrac{dy}{dx}-2y\tan x=y^2\tan^2 x$

8. $\dfrac{dy}{dx}+\dfrac{2y}{3}=\dfrac{x}{\sqrt{y}}$

9. $\dfrac{dy}{dx}+y\cot x=y^2\sin^2 x\cos^2 x$

10. $y(2xy+e^x)\,dy-e^x\,dx=0$

11. $y\,dx-x\,dy+3x^2y^2e^{x^3}dx=0$

12. $\dfrac{dy}{dx}(x^2y^3+xy)=1$

13. $y-\cos x\dfrac{dy}{dx}=y^2(1-\sin x)\cos x$, given $y=2$, when $x=0$

14. $\dfrac{dy}{dx}+2y\tan x=y^2$

15. $\dfrac{dy}{dx}+xy=xy^3$

16. $3y^2\dfrac{dy}{dx}+2xy^3=4xe^{-x^2}$

17. $\dfrac{dy}{dx}+y\tan x=y^3\sec x$

18. $y(2xy+e^x)\,dx-xe^x\,dy=0$

19. $2\dfrac{dy}{dx}-\dfrac{y}{x}=\dfrac{y^2}{x^2}$

20. $x^3\dfrac{dy}{dx}+y^4\cos x=x^2y$

21. $y(2xy+e^x)\,dx-e^x\,dy=0$

22. $\dfrac{dy}{dx}+xy=y^2e^{x^2/2}\log x$

23. $\cos x\,dy=y(\sin x-y)\,dx$

24. $xy^2\,dy-(x^3+y^3)\,dx=0$, given $y=0$ when $x=1$.

25. $\sin x\dfrac{dy}{dx}-y\cos x+y^2=0$

Answers

1. $x^3=y^3(3\sin x+c)$

2. $\dfrac{1}{y^2}(\sec x+\tan x)=x-\sec x-\tan x+c$

3. $\left(\dfrac{5}{2}+cx^2\right)x^3y^5=1$

4. $\cos^2 x=y^2(c+2\sin x)$

5. $xy(x\sin x+\cos x+c)+1=0$

6. $y^{-1}x=e^{-x^3}+c$

7. $y^{-1}\sec^2 x=-\dfrac{\sec^3 x}{3}+c$

8. $y^{3/2}=\dfrac{3}{2}(x-1)+ce^{-x}$

9. $y^{-1}=\dfrac{1}{3}\sin x\cos^3 x+c\sin x$

10. $y^{-1}e^x=c-x^2$

11. $xy^{-1}=-e^{x^3}+c$

12. $-\dfrac{1}{x}=y^3-1+ce^{-y^2/2}$

13. $(\tan x+\sec x)=y\sin x+\dfrac{y}{2}$

14. $\dfrac{\cos^2 x}{2}+\dfrac{1}{2}x+\dfrac{1}{4}\sin 2x=c$

15. $y^{-2}=ce^{x^2}-(1+x^2)$

16. $y^3=(2x^2+c)e^{-x^2}$

17. $\cos^2 x=y^2(2\sin x+c)$

18. $\dfrac{x}{y}=2(x+1)e^{-x}-c$

19. $x=y+cy\sqrt{x}$

20. $x^3=(c+3\sin x)y^3$

21. $y(c+x^2)+e^x=0$

22. $y(x-cx\log x+c)=e^{-x^2/2}$

23. $\sec x=y(c+\tan x)$

24. $y^3=3x^3\,.\log x$

25. $y(x+c)=\sin x$

2.10 Differential Equations of the First Order Higher Degree Equations

In this section we shall consider the method of solving the ordinary differential equations in which the power of $\frac{dy}{dx}$ is greater than unity. We shall denote $\frac{dy}{dx}$ by p. The equations of the first order and higher degree can be expressed as $f(x, y, p) = 0$.

Following are the types of equations which are of the first order but not first degree are considered.

1. Equations solvable for p
2. Equations solvable for x
3. Equations solvable for y
4. Clairaut's form

1. Equations solvable for p

Let the given equation be $f(x, y, p) = 0$ is of degree n.

Supposing this differential equation can be expressed as the product of the n factors of first degree equation; that is as

$$[p - f_1(x, y)][p - f_2(x, y)] \ldots\ldots\ldots [p - f_n(x, y)] = 0$$

then we have, $\quad p - f_1(x, y) = 0, \quad p - f_2(x, y) = 0, \quad \ldots\ldots \quad p - f_n(x, y) = 0$

Here each equation is of first degree and first order.

Let the solutions of these n equations be

$$\varphi_1(x, y, c_1) = 0, \quad \varphi_2(x, y, c_2) = 0, \ldots\ldots\ldots\ldots \varphi_n(x, y, c_n) = 0,$$

Then the solution of the given equation is

$$\varphi_1(x, y, c) \cdot \varphi_2(x, y, c) \ldots\ldots \varphi_n(x, y, c) = 0$$

where, $c = c_1 = c_2 = \ldots\ldots = c_n$.

Note : We know that the first order equation must have only one arbitrary constant. Thus we have taken $c = c_1 = c_2 = \ldots\ldots = c_n$ in the general solution of the equation.

Example 1. Solve : $p^2 - 5p - 6 = 0$.

Solution : The given equation is, $p^2 - 5p - 6 = 0 \Rightarrow (p - 6)(p + 1) = 0$

$$\Rightarrow \quad p - 6 = 0 \quad \text{or} \quad p + 1 = 0$$

$$\Rightarrow \quad \frac{dy}{dx} = 6 \quad \text{or} \quad \frac{dy}{dx} = -1$$

$$\Rightarrow \quad y = 6x + c_1 \quad \text{or} \quad y = -x - c_2$$

$$\Rightarrow \quad y - 6x + c_1 = 0 \quad \text{or} \quad y + x + c_2 = 0$$

The complete solution is, $(y - 6x + c)(y + x + c) = 0$

Example 2. Solve : $x\left(\frac{dy}{dx}\right)^2 + (y - x)\frac{dy}{dx} - y = 0$.

Solution : The given differential equation is

$$x p^2 + (y - x) p - y = 0$$

$$\Rightarrow \quad x p^2 + y p - x p - y = 0$$

$$\Rightarrow \quad p(x p + y) - 1(x p + y) = 0$$

$$\Rightarrow \quad (x p + y)(p - 1) = 0 \Rightarrow x p + y = 0 \quad \text{or} \quad p - 1 = 0$$

Now consider the equation, $x p + y = 0$

i.e. $\quad x\frac{dy}{dx} + y = 0 \Rightarrow \frac{dy}{y} + \frac{dx}{x} = 0$

$$\Rightarrow \int \frac{dy}{y} + \int \frac{dx}{x} = \log c \Rightarrow \log y + \log x = \log c \Rightarrow xy = c$$

Now consider the equation, $p - 1 = 0$

$$\Rightarrow \quad \frac{dy}{dx} - 1 = 0 \quad \Rightarrow \quad dy - dx = 0 \Rightarrow \int dy - \int dx = c \Rightarrow y - x = c$$

Thus the general solution is $\mathbf{(xy - c)\,(y - x - c) = 0}$

Example 3. Solve : $p^3 - p\,(x^2 + xy + y^2) + xy\,(x + y) = 0$.

Solution : The given equation is,

$$p^3 - p\,(x^2 + xy + y^2) + xy\,(x + y) = 0$$

By inspection, $p = x$ and $p = y$ will satisfy the equation.
Thus $(p - x)$ and $(p - y)$ are the factors of L.H.S.
Therefore we have, $(p - x)\,(p - y)\,(p + x + y) = 0$
This yields three equations

$$p - x = 0, \qquad p - y = 0, \qquad p + x + y = 0$$

Consider $\quad p - x = 0 \Rightarrow \dfrac{dy}{dx} - x = 0 \Rightarrow dy - x\,dx = 0$

$$\Rightarrow \int dy - \int x\,dx + c = 0 \Rightarrow y - \frac{x^2}{2} + c = 0$$

Consider, $p - y = 0$

$$\Rightarrow \quad - y = 0 \Rightarrow \frac{dy}{y} - x\,dx = 0 \Rightarrow \int \frac{dy}{y} - \int x\,dx + c = 0$$

$$\Rightarrow \log y - x + c = 0$$

Again consider the equation, $p + x + y = 0$

i.e. $\quad \dfrac{dy}{dx} + x + y = 0 \Rightarrow \dfrac{dy}{dx} + y = -x$

This is a linear differential equation, $\Rightarrow$ I.F. $= e^{\int dx} = e^x$

Thus the solution is,

$$y\,e^x = \int -x\,e^x\,dx + c$$

$$\Rightarrow \quad y\,e^x = -[x\,e^x - e^x] + c$$

$$\Rightarrow \quad y\,e^x = -(x - 1)\,e^x + c$$

$$\Rightarrow \quad (y + x - 1)\,e^x - c = 0$$

Hence the complete solution is

$$\left(y - \frac{x^2}{2} + c\right)(\log y - x + c)\left[(y + x - 1)\,e^x - c\right] = 0.$$

Example 4. Solve : $p^2 + 2\,py\cot x = y^2$.

Solution : The given equation is,

$$p^2 + 2\,p\,y\cot x = y^2$$

$$\Rightarrow \quad (p + y\cot x)^2 - y^2\cot^2 x = y^2$$

$$\Rightarrow \quad (p + y\cot x)^2 = y^2\,(1 + \cot^2 x)$$

$$\Rightarrow \quad (p + y\cot x)^2 = y^2 \operatorname{cosec}^2 x$$

$$\Rightarrow \quad p + y\cot x = \pm\, y \operatorname{cosec} x$$

$$\text{or} \quad p = -y\cot x \pm y \operatorname{cosec} x$$

Thus we have two equations,

$$p = y\,(-\cot x + \operatorname{cosec} x) \quad \text{and} \quad p = y\,(-\cot x - \operatorname{cosec} x)$$

Consider $\quad p = y\,(-\cot x + \operatorname{cosec} x) \Rightarrow \dfrac{dy}{dx} = y\,(-\cot x + \operatorname{cosec} x)$

Separating the variables, we get,

$$\frac{dy}{y} = (-\cot x + \operatorname{cosec} x)\,dx$$

$$\Rightarrow \quad \int \frac{dy}{y} = \int (-\cot x + \operatorname{cosec} x)\,dx + \log c$$

$$\Rightarrow \quad \log y = \int \left(\frac{1}{\sin x} - \frac{\cos x}{\sin x} \right) dx + \log c$$

$$\Rightarrow \quad \log y = \int \left(\frac{1 - \cos x}{\sin x} \right) dx + \log c$$

$$\Rightarrow \quad \log y = \int \frac{2\sin^2 \frac{x}{2}}{2\sin\frac{x}{2}\cos\frac{x}{2}}\,dx + \log c$$

$$\Rightarrow \quad \log y = \int \tan\frac{x}{2}\,dx + \log c \Rightarrow \log y = 2\log\sec\frac{x}{2} + \log c$$

$$\Rightarrow \quad y = c\sec^2\frac{x}{2} \Rightarrow y - c\sec^2\frac{x}{2} = 0$$

Consider, $p = y\,(-\cot x - \operatorname{cosec} x) \Rightarrow \dfrac{dy}{dx} = y\,(\cot x + \operatorname{cosec} x)$

Separating the variables, we get,

$$\frac{dy}{y} = -(\cot x + \operatorname{cosec} x)\,dx$$

$$\Rightarrow \quad \int \frac{dy}{y} = -\int (\cot x + \operatorname{cosec} x)\,dx + \log c$$

$$\Rightarrow \quad \log y = -\int \left(\frac{\cos x}{\sin x} + \frac{1}{\sin x} \right) dx + \log c$$

$$\Rightarrow \quad \log y = -\int \left(\frac{1 + \cos x}{\sin x} \right) dx + \log c$$

$$\Rightarrow \quad \log y = -\int \frac{2\cos^2\frac{x}{2}}{2\sin\frac{x}{2}\cos\frac{x}{2}}\,dx + \log c \Rightarrow \log y = -\int \cot\frac{x}{2}\,dx + \log c$$

$$\Rightarrow \log y = -2\log\sin\frac{x}{2} + \log c$$

$$\Rightarrow y\sin^2\frac{x}{2} = c$$

$$\Rightarrow y\sin^2\frac{x}{2} - c = 0$$

Thus the complete solution is, $\left(y - c\sec^2\frac{x}{2} \right)\left(y\sin^2\frac{x}{2} - c \right) = 0$.

EXERCISES

1. $p^2 + 2px - 3x^2 = 0$
2. $p^2 - 2px + x^2 - y^2 = 0$
3. $x^2 p^2 + xyp + 2y^2 = 0$
4. $x^2 p^2 + 3\,xyp + 2y^2 = 0$
5. $x^2 p^2 - 2\,xyp + 2y^2 - x^2 = 0$
6. $p^2 + px + py + xy = 0$
7. $xyp^2 + p\,(3x^2 - 2y^2) - 6xy = 0$
8. $p^2 - 3p\cosh x + 1 = 0$

Answers

1. $\left(y - \frac{x^2}{2} - c\right)\left(y + \frac{3}{2}x^2 - c\right) = 0$
2. $(x + y + 1 - ce^x)\,(x - y - 1 - ce^{-x}) = 0$
3. $(y - cx^2)\,(x^3 y - c) = 0$
4. $(x\,y - c)\,(y\,x^2 - c) = 0$
5. $\sin^{-1}\left(\frac{y}{x}\right) = \pm\log(cx)$
6. $\left(y + \frac{x^2}{x} - c\right)(x + \log y - c) = 0$
7. $(y - cx^2)\,(y^2 + 3x^2 - c) = 0$
8. $(y - e^x - c)\,(y + e^{-x} - c) = 0$

2. Equations Solvable for x

Supposing that the equation

$$f(x, y, p) = 0$$

can be solved for x, in terms of y and p say, $x = f(y, p)$ ...(1)

Differentiating both sides w.r.t.y we get

$$\frac{dx}{dy} = \varphi\left(y, p, \frac{dp}{dy}\right) \quad \text{or} \quad \frac{1}{p} = \varphi\left(y, p, \frac{dp}{dy}\right)$$

This is a first order differential equation in the variables y and p. Solving this equation we get a relation between y and p with arbitrary constant c, say

$$F(p, y, c) = 0 \qquad \text{...(2)}$$

Eliminating p between (1) and (2) we get the complete solution of the given equation.

Example 1. Solve : $y - 2px + yp^2 = 0$

Solution : The given equation is, $y - 2px + yp^2 = 0$ ...(1)

Solving for x, we get, $2x = \frac{y}{p} + yp$

Differentiating both sides w.r.t.y we get

$$2\frac{dx}{dy} = y\left(-\frac{1}{p^2}\right)\frac{dp}{dy} + \frac{1}{p} + y\frac{dp}{dy} + p$$

$$\Rightarrow \quad \frac{2}{p} = -\frac{y}{p^2}\frac{dp}{dy} + \frac{1}{p} + y\frac{dp}{dy} + p$$

$$\Rightarrow \quad \frac{1}{p} - p = y\left(1 - \frac{1}{p^2}\right)\frac{dp}{dy} \Rightarrow p\left(\frac{1}{p^2} - 1\right) + y\left(\frac{1}{p^2} - 1\right)\frac{dp}{dy} = 0$$

$$\Rightarrow \left(\frac{1}{p^2} - 1\right).\left(p + y\frac{dp}{dy}\right) = 0$$

Ignoring the first factor (which do not involve $\frac{dp}{dy}$) we have,

$$p + y\,\frac{dp}{dy} = 0$$

This is a differential equation in the variables y and p.
Separating the variables, we get,

$$\frac{dp}{p}+\frac{dy}{y}=0 \Rightarrow \int\frac{dp}{p}+\int\frac{dy}{y}=\log c \Rightarrow \log p+\log y=\log c$$

$$\Rightarrow py=c \qquad \ldots(2)$$

Eliminating p between (1) and (2) we get the complete solution.

Now $\qquad (2) \Rightarrow p=\frac{c}{y}$

Thus, $\qquad (1) \Rightarrow y-2\frac{cx}{y}+y\frac{c^2}{y^2}=0 \Rightarrow \boldsymbol{y^2-2cx+c^2=0}$

This is the general solution.

Example 2. Solve : $p^3 - 4xyp + 8y^2 = 0$

Solution : The given equation is, $p^3 - 4xyp + 8y^2 = 0$ $\qquad \ldots(1)$

Solving for x, we get, $4x=\frac{p^2}{y}+\frac{8y}{p}$

Differentiating both sides w.r.t.y we get

$$4\frac{dx}{dy}=\frac{1}{y}2p\frac{dp}{dy}-\frac{1}{y^2}p^2+8y\left(-\frac{1}{p^2}\right)\frac{dp}{dy}+\frac{8}{p}$$

$$\Rightarrow \quad \frac{p^2}{y^2}-\frac{4}{p}=2\left(\frac{p}{y}-\frac{4y}{p^2}\right)\frac{dp}{dy} \Rightarrow \frac{p}{y}\left(\frac{p}{y}-\frac{4y}{p^2}\right)-2\left(\frac{p}{y}-\frac{4y}{p^2}\right)\frac{dp}{dy}=0$$

$$\Rightarrow \left(\frac{p}{y}-\frac{4y}{p^2}\right)\left(\frac{p}{y}-2\frac{dp}{dy}\right)=0$$

Ignoring the first factor, we get,

$$\frac{p}{y}-2\frac{dp}{dy}=0 \Rightarrow \frac{1}{2}\int\frac{dy}{y}-\int\frac{dp}{p}=\log c$$

$$\Rightarrow \frac{1}{2}\log y-\log p=\log c$$

$$\Rightarrow \log\left(\frac{p}{\sqrt{y}}\right)=\log c$$

$$\Rightarrow p=c\sqrt{y}$$

Putting $p=c\sqrt{y}$ in (1), we get

$$c^3(\sqrt{y})^3-4xyc\sqrt{y}+8y^2=0$$

$$\Rightarrow \quad c^3y(\sqrt{y})-4xy\sqrt{y}\,c+8y^2=0$$

$$\Rightarrow \quad 8y=c(4x-c^2)\sqrt{y} \Rightarrow 8\sqrt{y}=c(4x-c^2) \Rightarrow \sqrt{y}=\frac{c}{2}\left(x-\frac{c^2}{4}\right)$$

Thus, $\qquad \boldsymbol{y=k(x-k)^2,\left(\text{where } k=\frac{c}{2}\right),}$ **is the required solution.**

EXERCISES

Solve the following equations

1. $y = 2px + y^2 p^3$
2. $yp^2 - 2xp + y = 0$
3. $y = 3px + 6p^2 y^2$
4. $y^2 \log y = xpy + p^2$
5. $y = p^2 y + 2px$
6. $p^3 y^2 - 2xp + y = 0$
7. $p = \tan\left(x - \frac{p}{1+p^2}\right)$
8. $x = y + p^2$

Answers

1. $y^2 = 2cx + c^3$
2. $y^2 = 2cx - c^2$
3. $y^3 = 3cx + 6c^2$
4. $\log y = cx + c^2$
5. $y^2 = 2cy + c^2$
6. $2cx = c^3 + y^2$
7. $x = \frac{1}{1+p^2} + \tan^{-1} p;\quad y = -\frac{1}{1+p^2} + c$
8. $y = -\frac{1}{1+p^2} + c;\quad x = \frac{1}{1+p^2} + \tan^{-1} p$

3. Equations Solvable for y

Supposing that the equation, $f(x, y, p) = 0$ can be solved for y, in terms of x and p say,

$$y = f(x, p) \qquad \ldots(1)$$

Differentiating both sides w.r.t. x we get,

$$\frac{dy}{dx} = \varphi\left(x, p, \frac{dp}{dx}\right) \Rightarrow p = \varphi\left(x, p, \frac{dp}{dx}\right)$$

This is a first order differential equation in the variables x and p. If it is possible to solve the equation, let its solution be,

$$F(p, x, c) = 0 \qquad \ldots(2)$$

Eliminating p between (1) and (2) we get the complete solution of the given equation.

Note : If it is not possible to eliminate p between (1) and (2), then solve (1) and (2) for x and y in terms of p and these two expressions form the complete solution; here p is regarded as a parameter.

Example 1. Solve : $y = x + 2\tan^{-1} p$.

Solution : The given equation is, $y = x + 2\tan^{-1}p$

Differentiating both sides w.r.t. x we get,

$$\frac{dy}{dx} = 1 + \frac{2}{1+p^2}\frac{dp}{dx}$$

$$\Rightarrow \quad p = 1 + \frac{2}{1+p^2}\frac{dp}{dx} \Rightarrow \frac{2}{1+p^2}\frac{dp}{dx} + (1-p) = 0$$

$$\Rightarrow \frac{2\,dp}{(1-p)(1+p^2)} + dx = 0$$

$$\Rightarrow \int \frac{2\,dp}{(1-p)(1+p^2)} + \int dx = c \qquad \ldots(1)$$

Consider,

$$\frac{1}{(1-p)(1+p^2)} = \frac{A}{1-p} + \frac{Bp+C}{1+p^2}$$

$$\Rightarrow \quad 1 = A(1+p^2) + (Bp + C)(1-p)$$

Put, $p = 1$, we get $2A = 1 \Rightarrow A = \frac{1}{2}$

Equating the coefficient of p^2 and constant terms separately, we get

$$A - B = 0 \Rightarrow B = \frac{1}{2} \qquad A + C = 1 \Rightarrow C = \frac{1}{2}$$

$$\therefore \quad \frac{1}{(1-p)(1+p^2)} = \frac{1}{2}\cdot\frac{1}{1-p} + \frac{1}{2}\cdot\frac{p+1}{1+p^2}$$

$$\therefore \quad 2\int\frac{dp}{(1-p)(1+p^2)} = \int\frac{1}{1-p}dp + \int\frac{p}{1+p^2}dp + \int\frac{1}{1+p^2}dp$$

$$= -\log(1-p) + \frac{1}{2}\log(1+p^2) + \tan^{-1}p$$

$$= \log\sqrt{1+p^2} - \log(1-p) + \tan^{-1}p$$

$\therefore$ (1) becomes

$$\log\sqrt{1+p^2} - \log(1-p) + \tan^{-1}p + x = c$$

i.e. $\boldsymbol{x = c + \log(1-p) - \log\sqrt{1+p^2} - \tan^{-1}p}$, **and** $\boldsymbol{y = x + 2\tan^{-1}p}$ form the solution.

Example 2. Solve : $16x^2 + 2p^2y - p^3x = 0$.

Solution : The given equation is

$$16x^2 + 2p^2y - p^3x = 0$$

Solving for y, we get

$$y = \frac{p^3x - 16x^2}{2p^2}$$

Differentiating w.r.t.x, we get

$$\frac{dy}{dx} = \frac{2p^2\left[p^3 + x\,3p^2\frac{dp}{dx} - 32x\right] - (p^3x - 16x^2)4p\frac{dp}{dx}}{4p^4}$$

$$\Rightarrow \quad 2p^3\frac{dy}{dx} = p^4 + 3x\,p^3\frac{dp}{dx} - 32px - 2p^3x\frac{dp}{dx} + 32x^2\frac{dp}{dx}$$

$$\Rightarrow \quad 2p^4 = p^4 + p^3x\frac{dp}{dx} - 32px + 32x^2\frac{dp}{dx}$$

$$\Rightarrow \quad p^4 + 32px = \frac{dp}{dx}(p^3 + 32x)x$$

$$\Rightarrow \quad (p^3 + 32x)p - \frac{dp}{dx}x(p^3 + 32x) = 0$$

$$\Rightarrow \quad (p^3 + 32x)\left(p - x\frac{dp}{dx}\right) = 0$$

Ignoring the first factor, we get, $p - x\dfrac{dp}{dx} = 0$.

This is a differential equation in x and p. Separating the variables, we have

$$\frac{dp}{p} - \frac{dx}{x} = 0 \Rightarrow \int\frac{dp}{p} - \int\frac{dx}{x} = \log c \Rightarrow \log p - \log x = \log c \Rightarrow p = xc$$

Putting this value of p in the given equation, we get

$$16x^2 + 2x^2c^2y - x^3c^3x = 0 \Rightarrow \mathbf{16 + 2c^2y - c^3x^2 = 0}$$

This is the required solution.

EXERCISES

Solve the following equations

1. $y = 2px + y^2 p^3$ **2.** $y^2 (1 - p^2) = a$ **3.** $x p^2 - 2 yp + ax = 0$

4. $y = p \sin p + \cos p$ **5.** $y = xp^2 + p$ **6.** $4p^3 + 3xp = y$

Answers

1. $y^2 = 2cx + c^3$ 2. $y^2 - a = (x + c)^2$ 3. $2y = cx^2 + \frac{a}{c}$
4. $y = p \sin p + \cos p, x = \sin p + c$
5. $x = (\log p - p + c)(p - 1)^2, y = xp^2 + p$
6. $y = -\frac{8}{7}p^3 + c\,p^{-1/2}, x = -\frac{12}{7}p^2 + \frac{c}{3}p^{-3/2}$

2.11 Clairaut's Equations

The equation of the form, $y = px + f(p)$ ...(1)
is called the Clairaut's Equation. This is an equation solvable for y. Differentiating both sides w.r.t.x, we get

$$\frac{dy}{dx} = p + x\frac{dp}{dx} + f'(p)\frac{dp}{dx}$$

$$\Rightarrow \qquad \frac{dp}{dx}[x + f'(p)] = 0 \Rightarrow \frac{dp}{dx} = 0 \text{ or } x + f'(p) = 0$$

Now, $\frac{dp}{dx} = 0 \Rightarrow p = c$

Putting this value of p in (1), we get, $y = cx + f(c)$, as the general solution of the equation. Now consider the equation,

$$x + f'(p) = 0 \qquad ...(2)$$

By eliminating p between the equations (1) and (2) we obtain an equation of the form $\varphi(x, y) = 0$.

This is a solution of the equation (1). Further this solution does not contain the arbitrary constants and this cannot also be obtained from the general solution by giving any particular value for c. A solution of this type is called a **singular solution** of the given equation.

Geometrically the singular solution of the differential equation

$$y = px + f(p)$$

is the envelope of the family of curves represented by the general solution,

$$y = cx + f(c)$$

of the given differential equation.

Differentiating the given equation (1) partially w.r.t.p, we get

$$x + f'(p) = 0.$$

This equation is same as (2). Thus the p-eliminent of (1) and the equation obtained by differentiating (1) partially w.r.t.p, gives the singular solution.

Working rule of solving the Clairaut's equation.

1. To obtain the general solution replace p in the given equation by the arbitrary constant c.
2. To obtain the singular solution, differentiate the given equation partially w.r.t.p, and eliminate p between the given equation and the equation obtained.

Example 1. Find the general and singular solution of

$$\boldsymbol{y = px + \sin^{-1} p.}$$

Solution : The given equation is

$$y = px + \sin^{-1} p \qquad \ldots(1)$$

The given equation is of Clairaut's form. Thus the general solution is

$$\boldsymbol{y = cx + \sin^{-1} c}$$

To obtain the singular solution, differentiate the given equation partially w.r.t.p.

i.e. $$x + \frac{1}{\sqrt{1-p^2}} = 0 \qquad \ldots(2)$$

Now eliminate p between (1) and (2). To this end we have

$$x\sqrt{1-p^2} = -1 \;\Rightarrow\; 1-p^2 = \frac{1}{x^2} \Rightarrow p^2 = 1-\frac{1}{x^2} \Rightarrow p = \frac{\sqrt{x^2-1}}{x}$$

Putting this value of p in (1), we get

$$\boldsymbol{y = \sqrt{x^2-1} + \sin^{-1}\left(\frac{\sqrt{x^2-1}}{x}\right)}$$

This is the singular solution.

Example 2. Find the general and singular solution of

$$\boldsymbol{y = px + \frac{a}{p}.}$$

Solution : The given equation is $y = px + \dfrac{a}{p}$

The given equation is of Clairaut's form. Thus the general solution is

$$\boldsymbol{y = cx + \frac{a}{c}.}$$

To obtain the singular solution, differentiate the given equation partially w.r.t.p.

i.e. $$x - \frac{a}{p^2} = 0 \Rightarrow \frac{a}{p^2} = x \Rightarrow p^2 = \frac{a}{x} \Rightarrow p = \frac{\sqrt{a}}{\sqrt{x}}$$

Putting this value of p in the given equation, we get

$$y = x\frac{\sqrt{a}}{\sqrt{x}} + a\frac{\sqrt{x}}{\sqrt{a}} \;\Rightarrow\; y = \sqrt{ax}+\sqrt{ax} \;\Rightarrow\; \boldsymbol{y = 2\sqrt{ax}}.$$

This is the required singular solution.

Example 3. Find the general and singular solution of

$$\boldsymbol{(a^2 - x^2)\, p^2 + 2xy\, p + b^2 - y^2 = 0.}$$

Solution : The given equation is

$$(a^2 - x^2)\, p^2 + 2xy\, p + b^2 - y^2 = 0$$

$$\Rightarrow \quad y^2 - 2xy\, p + x^2 p^2 = a^2 p^2 + b^2$$

$$\Rightarrow \quad (y - xp)^2 = a^2p^2 + b^2 \Rightarrow y - xp = \pm\sqrt{a^2p^2+b^2} \Rightarrow y = xp \pm \sqrt{a^2p^2+b^2}$$

This is a Clairaut's equation. Thus the general solution is

$$y = xc \pm \sqrt{a^2c^2+b^2} \;\Rightarrow\; (a^2 - x^2)\, c^2 + 2xyc + b^2 - y^2 = 0$$

To find the singular solution, differentiate the given equation partially w.r.t.p.

i.e. $$2(a^2 - x^2)\, p + 2xy = 0 \;\Rightarrow\; p = \frac{xy}{x^2 - a^2}$$

Putting this value of p in the given equation, we get

$$(a^2-x^2)\frac{x^2y^2}{(x^2-a^2)^2}+2xy\frac{xy}{(x^2-a^2)}+b^2-y^2=0$$

$$\Rightarrow \quad \frac{x^2y^2}{a^2-x^2}-\frac{2x^2y^2}{a^2-x^2}+b^2-y^2=0 \Rightarrow -\frac{x^2y^2}{a^2-x^2}+b^2-y^2=0$$

$$\Rightarrow x^2y^2-b^2(a^2-x^2)+y^2(a^2-x^2)=0$$

$$\Rightarrow x^2b^2+a^2y^2=a^2b^2$$

$$\Rightarrow \frac{x^2}{a^2}+\frac{y^2}{b^2}=1$$

This is the singular solution.

Example 4. Find the general and singular solution of

$$(x^2-1)p^2-2xy\,p+y^2-1=0.$$

Solution : The given equation is

$$(x^2-1)p^2-2xy\,p+y^2-1=0$$

$$(x^2-1)p^2-2xy\,p+y^2-1=0$$

$$\Rightarrow \quad y^2-2xy\,p+x^2p^2=1+p^2 \Rightarrow (y-xp)^2=1+p^2$$

$$\Rightarrow y-xp=\pm\sqrt{1+p^2} \Rightarrow y=xp\pm\sqrt{1+p^2}$$

The is a Clairaut's equation. Thus the general solution is

$$y=xc+\pm\sqrt{1+c^2} \Rightarrow (x^2-1)c^2-2xyc+y^2-1=0.$$

To find the singular solution, differentiate the given equation partially w.r.t.p.

i.e. $$2(x^2-1)p-2xy=0 \Rightarrow (x^2-1)p=xy \Rightarrow p=\frac{xp}{x^2-1}.$$

Putting this value of p in the given equation, we get

i.e. $$(x^2-1)\frac{x^2y^2}{(x^2-1)^2}-2xy\frac{xy}{(x^2-1)}+y^2-1=0$$

$$\Rightarrow \quad \frac{x^2y^2}{x^2-1}-\frac{2x^2y^2}{x^2-1}+y^2-1=0 \Rightarrow -\frac{x^2y^2}{x^2-1}+y^2-1=0$$

$$\Rightarrow -x^2y^2+x^2y^2-y^2-x^2+1=0$$

$$\Rightarrow x^2+y^2=1$$

This is the singular solution.

Note : Some of the equations, which are not in the Clairaut's form, can be transformed into it by suitable change of variables.

Example 5. Find the general and singular solution of

$$x^2(y-px)=yp^2$$

by using the substitution $x^2=u$ and $y^2=v$.

Solution : The given equation is

$$x^2(y-px)=yp^2$$

Put, $x^2 = u$ and $y^2 = v \Rightarrow 2x\,dx = du$ and $2y\,dy = dv$

$$\Rightarrow \quad \frac{y}{x}\frac{dy}{dx} = \frac{dv}{du} \Rightarrow \frac{y}{x}p = P \quad \text{where} \quad p = \frac{dy}{dx} \text{ and } P = \frac{dv}{du} \Rightarrow p = \frac{x}{y}P$$

The given equation becomes

$$u\left(y - \frac{x^2}{y}P\right) = y\left(\frac{x^2}{y^2}\right)P^2$$

$$\Rightarrow \quad u\,(y^2 - x^2 P) = y^2\left(\frac{x^2}{y^2}\right)P^2$$

$$\Rightarrow \quad u\,(v - u\,P) = u\,P^2 \Rightarrow v - u\,p = P^2 \Rightarrow v = u\,p + P^2 \qquad \ldots(1)$$

This is a Clairaut's equation. Thus the general solution is,

$$v = u\,c + c^2 \Rightarrow y^2 = x^2\,c + c^2$$

To find the singular solution, differentiate (1) w.r.t.P partially.

i.e. $$u + 2\,P = 0 \Rightarrow P = -\frac{u}{2}$$

Putting this value of P in (1), we get

$$v = -\frac{u^2}{2} + \frac{u^2}{4} \Rightarrow v = -\frac{u^2}{2} \Rightarrow 2v + u^2 = 0 \Rightarrow \mathbf{2y^2 + x^4 = 0}$$

This is the singular solution.

Example 6. Find the general and singular solution of

$$\mathbf{(p - 1)\,e^{3x} + p^3\,e^{2y} = 0}$$

by using the substitution $u = e^x$ and $v = e^y$.

Solution : The given equation is

$$(p - 1)\,e^{3x} + p^3\,e^{2y} = 0$$

Now put $u = e^x$ and $v = e^y \Rightarrow du = e^x\,dx$ and $dv = e^y\,dy$

$$\Rightarrow \quad \frac{dv}{du} = \frac{e^y}{e^x}\frac{dy}{dx} \Rightarrow \frac{dy}{dx} = \frac{e^x}{e^y}\frac{dv}{du}$$

$$\Rightarrow \quad p = \frac{e^x}{e^y}P = \frac{u}{v}P \quad \text{where} \quad p = \frac{dy}{dx} \text{ and } P = \frac{dv}{du}$$

The given equation becomes

$$\left(\frac{u}{v}P - 1\right)u^3 + \frac{u^3}{v^3}P^3 v^2 = 0$$

$$\Rightarrow \quad \left(\frac{uP - v}{v}\right) + \frac{P^3}{v} = 0 \Rightarrow uP - v + P^3 = 0 \Rightarrow v = uP + P^3 \qquad \ldots(1)$$

This is a Clairaut's equation. Thus the general solution is,

$$v = uc + c^3 \Rightarrow e^x = e^y\,.\,c + c^3$$

To find the singular solution, differentiate (1) w.r.t.P. partially, we get

$$u + 3\,P^2 = 0 \Rightarrow P^2 = -\frac{u}{3} \Rightarrow P = \sqrt{-\frac{u}{3}}$$

Putting this value of P in (1), we get

$$v = u\sqrt{-\frac{u}{3}} + \left(-\frac{u}{3}\right)\sqrt{-\frac{u}{3}}$$

$$\Rightarrow \qquad v = \sqrt{-\frac{u}{3}\left(u - \frac{u}{3}\right)} \Rightarrow v = \sqrt{-\frac{u}{3}\left(\frac{2u}{3}\right)}$$

Squaring both sides we get

$$v^2 = -\frac{u}{3}\left(\frac{4u^2}{9}\right) \Rightarrow v^2 = -\frac{4u^2}{27} \Rightarrow 4u^2 + 27v^2 = 0 \Rightarrow \mathbf{4\,e^3x + 27\,e^2x = 0}$$

This is the singular solution.

Example 7. Find the general and singular solution of

$$\cos^2 y\, p^2 + \sin x \cos x \cos y\, p = \sin y \cos^2 x$$

using the transformation $u = \sin x$ and $v = \sin y$.

Solution : The given equation is

$$\cos^2 y\, p^2 + \sin x \cos x \cos y\, p = \sin y \cos^2 x$$

Put $\quad u = \sin x$ and $v = \sin y \Rightarrow du = \cos x\, dx$ and $d\,v = \cos y\, dy$

$$\Rightarrow \frac{dy}{dx} = \frac{\cos x}{\cos y} \cdot \frac{dv}{du} \Rightarrow p = \frac{\cos x}{\cos y} P \text{ where } P = \frac{dv}{du}$$

The given equation becomes

$$\cos^2 y \left(\frac{\cos^2 x}{\cos^2 y}\right) P^2 + \sin x \cos x \cos y \left(\frac{\cos x}{\cos y}\right) P = \sin y \cos^2 x$$

i.e. $\quad \cos^2 x\, P^2 + \sin x \cos^2 x\, P = \sin y \cos^2 x \Rightarrow P^2 + Pu = v \Rightarrow v = Pu + P^2 \quad \ldots(1)$

This is a Clairaut's equation. Thus the general solution is,

$$v = cu + c^2 \Rightarrow \mathbf{\sin y = c \sin x + c^2}$$

To find the singular solution, differentiate (1) partially w.r.t.P., we get

$$u + 2P = 0 \Rightarrow P = -\frac{u}{2}$$

Putting this value of p in (1), we get,

$$v = -\frac{u^2}{2} + \frac{u^2}{4} \Rightarrow v = -\frac{u^2}{4} \Rightarrow u^2 + 4v = 0 \Rightarrow \mathbf{\sin^2 x + 4 \sin y = 0}$$

This is the singular solution.

EXERCISES

I. Find the general and singular solutions of the following equations

1. $y = px + p^2$ **2.** $y = px + \frac{-ap}{\sqrt{1+p^2}}$ **3.** $xp^2 - yp + p^2 + 1 = 0$

4. $y = px - e^p$ **5.** $y = px + 2p^2$ **6.** $y + p^2 = xp + 1$

7. $y^2 - 2\,pxy + p^2\,(x^2 - 1) = m^2$ **8.** $y^2 + x^2 p^2 - 2xyp = \frac{4}{p^2}$ **9.** $p^2\,(x^2 - a^2) - 2xyp + y^2 + a^4 = 0$

II. Find the general and singular solutions of the following equations using the given transformation.

1. $y = 3\,px + 6\,y^2 p^2$; Put $y^3 = v$

2. $y = 2\,px + y^2 p^3$; Put $y^2 = v$

3. $(px - y)\,(x - py) = 2p$; Put $x^2 = u, y^2 = v$

4. $y^2(x - xp) = x^4 p^2$; Put $x = \frac{1}{u}, y = \frac{1}{v}$

5. $x(y - px) = x + py$; Put $x^2 = u, y^2 = v$

6. $(px - y)(py + x) = a^2 p$; Put $x^2 = u, y^2 = v$

7. $y^2 \log y = xpy \log x + x^2 p^2$; Put $\log y = v, \log x = u$

8. $(p x^2 + y^2)(px + y) = (p + 1)^2$; Put $x + y = u, xy = v$

Answers

1. $y = cx + c^2$; $x^2 + 4y = 0$

2. $y = cx + \frac{ac}{\sqrt{1+c^2}}$; $y^2 = \frac{(x^{2/3} - a^{2/3})^2}{x^{1/3}}$

3. $y = cx + c + \frac{1}{c}$; $y^2 = 4(x + 1)$

4. $y = cx - c^6$; $x + y - x \log x = 0$

5. $y = cx + 2c^2$; $x^2 + 8y = 0$

6. $y = cx + (1 - c^2)$; $4y - x^2 - 4 = 0$

7. $y^2 - 2cxy + c^2(x^2 - 1) = m^2$; $y^2 + m^2 x^2 = m^2$

8. $y^2 + x^2 c^2 - 2xyc = 4$; $y^2 = 8x$

9. $c^2(x^2 - a^2) - 2xyc + y^2 + a^4 = 0$ $y^2 = a^2(x^2 - a^2)$

II.

1. $3y^2 = 3cx + 2c^2$; $8y^3 + 3x^2 = 0$

2. $y^2 = cx + \frac{1}{8}c^3$; No singular solution.

3. $y^2 = cx^2 - \frac{2c}{1-c}$; $x - y + \sqrt{2} = 0$

4. $x = cy + c^2 xy$; $y + 4x^2 = 0$

5. $y^2 = cx^2 + 1 + c$; No singular solution.

6. $y^2 = x^2 c + \frac{a^2 c}{c+1}$; $y^2 + (x - a)^2 = 0$

7. $\log y = c \log x + c^2$; $(\log x)^2 + 4(\log y) = 0$

8. $c^2(x + y) - c\,xy - 1 = 0$; $x^2 y^2 - 4(x + y) = 0$

2.12 Linear Equations with constant coefficients

A general linear differential equation with constant coefficients of order n is of the form

$$a_0 \frac{d^n y}{dx^n} + a_1 \frac{d^{n-1} y}{dx^{n-1}} + a_2 \frac{d^{n-2} y}{dx^{n-2}} + \ldots\ldots + a_n y = f(x) \qquad \ldots\ldots (1)$$

where $a_0, a_1, a_2, \ldots\ldots a_n$ are constants and $f(x)$ is a function of x.

In the above equation it may be noted that the dependent variable and its derivatives appear only in the first degree.

We have already seen the method of solving the linear differential equation of first order. Now we shall consider the method of solving linear differential equations with constant coefficients of degree ≥ 2.

2.13 The Operator D

We shall introduce an operator denoted by D, which stands for, $\frac{d}{dx}$. i.e., $Dy = \frac{dy}{dx}$

D^2 is denoted for $\frac{d^2}{dx^2}$; D^3 is denoted for $\frac{d^3}{dx^3}$ and so on.

Using these notations, the equation (1) can be written as

$$(a_0 D^n + a_1 D^{n-1} + a_2 D^{n-2} + \ldots\ldots + a_n)\, y = f(x) \qquad \ldots\ldots (2)$$

In what follows we shall write

$$f(D) = a_0 D^n + a_1 D^{n-1} + a_2 D^{n-2} + \ldots\ldots + a_n$$

Thus with this notation, (2) becomes

$$F(C)\, y = f(x) \qquad \ldots\ldots (3)$$

The operator D satisfies the following laws similar to that of the fundamental laws of algebra

(i) $D(u + v) = Du + Dv$

(ii) $D^m . D^n u = D^{m+n} u$ [where m and n are +ve integers]

(iii) $D(cu) = c . Du$ [C is a constant]

(iv) $D^m (D^n u) = D^n (D^m u)$ [where m and n are +ve integers]

In the above laws u and v are functions of x.

Note :

1. The operator D is not commutative with variables - i.e. $D^m u . D^n v \neq D^m v . D^n u$.
2. The index laws ceases to hold when negative powers of D occur.

From now onwards we have every right in factorising $F(D)$ or performing any other operation, depending on the laws stated above.

The following are few useful and important results. We shall establish these for $n = 3$. However the results are true for any positive integer n involved.

I. $F(D)\, e^{ax} = e^{ax} F(a)$

Now, $F(D)e^{ax} = (a_0 D^3 + a_1 D^2 + a_2 D + a_3)\, e^{ax}$

$= a_0 D^3 e^{ax} + a_1 D^2 e^{ax} + a_2 D e^{ax} + a_3 e^{ax}$

$= a_0 a^3 e^{ax} + a_1 a^2 e^{ax} + a_2 a e^{ax} + a_3 e^{ax}$ $[\because D^3 e^{ax} = a^3 e^{ax}; D^2 e^{ax} = a^2 e^{ax}]$

$= (a_0 a^3 + a_1 a^2 + a_2 a + a_3)\, e^{ax}$

$= e^{ax} . F(a)$. [$F(a)$ is the expression obtained by putting $D = a$ in $F(D)$]

$\therefore$ $\mathbf{F(D)e^{ax} = e^{ax} F(a)}$

II. $F(D)(e^{ax}V) = e^{ax}F(D+a)V$, where V is any function of x.

Now, $F(D)[e^{ax}V] = (a_0 D^3 + a_1 D^2 + a_2 D + a_3)[e^{ax}V]$

$$= a_0 D^3[e^{ax}V] + a_1 D^2[e^{ax}V] + a_2 D[e^{ax}V] + a_3[e^{ax}V]$$

Now applying the Leibnitz's theorem for nth derivatives of a product of two functions, we have

$$D^3(e^{ax}V) = a^3 e^{ax}V + 3a^2 e^{ax}DV + 3ae^{ax}D^2V + e^{ax}D^3V$$

$$= e^{ax}[a^3 + 3a^2D + 3aD^2 + D^3]V = e^{ax}(D+a)^3 V$$

Similarly, $D^2(e^{ax}V) = e^{ax}(D+a)^2 V$ and $D(e^{ax}V) = e^{ax}(D+a)V$

$\therefore\ F(D)(e^{ax}V) = e^{ax}[a_0(D+a)^3 + a_1(D+a)^2 + a_2(D+a) + a_3]V = e^{ax}F(D+a)V$

Hence, $\mathbf{F(D)(e^{ax}V) = e^{ax}F(D+a)V}$

III. $F(D^2)\cos ax = F(-a^2)\cos ax$.

Here we shall take, $F(D^2) = a_0 D^4 + a_1 D^2 + a_2$ [i.e. of degree $2n$ where $n = 2$]

Now,

$$F(D^2)\cos ax = (a_0 D^4 + a_1 D^2 + a_2)\cos ax$$

$$= a_0 D^4 \cos ax + a_1 D^2 \cos ax + a_2 \cos ax$$

$$= [a_0(-a^2)^2 + a_1(-a^2) + a_2]\cos ax = F(-a^2)\cos ax$$

Similarly, $\mathbf{F(D^2)\sin ax = F(-a^2)\sin ax}$

First let us consider the method of solving the differential equation (1) when the right hand side function $f(x)$ is zero - i.e. $F(D) = 0$.

Consider the equation,

$$F(D)\,y = 0 \qquad \text{..... (4)}$$

Let, if possible, $y = c\,e^{mx}$, where c and m are some constants, be a solution of (4).

$\therefore$ for this value of y, (4) becomes,

$$c\,e^{mx}F(m) = 0$$

Here $F(m)$ is a polynomial in m of degree n.

$\therefore$ if m is a root of $F(m) = 0$, then $y = c\,e^{mx}$ is a solution of (4), whatever be the value of c.

Thus for each value of m which is a root of $F(m) = 0$, we have a solution $y = c\,e^{mx}$ of (4) and sum of all these independent solutions involving n arbitrary constants will give rise to the general solution of (4).

But there arise three cases, corresponding to real and distinct roots, repeated roots and imaginary roots, of $F(m) = 0$.

Case (i) : Let the roots of $F(m) = 0$ be all real and distinct - say $m_1, m_2, m_3, \dots\dots\dots, m_n$

Then,

$$c_1 e^{m_1 x}, c_2 e^{m_2 x}, c_3 e^{m_3 x}, \dots\dots\dots\dots\ c_n e^{m_n x}$$

are all solutions of (4).

Hence the general solution of (4) is

$$y = c_1 e^{m_1 x} + c_2 e^{m_2 x} + c_3 e^{m_3 x} + \dots\dots\dots + c_n e^{m_n x}$$

where $c_1, c_2, \ , \dots\dots\dots, c_n$ are arbitrary constants.

Case (ii) : Let $F(m) = 0$ has a root say m_1 repeated twice.

Then we know by theory of equations, $(D - m_1)^2$ is a factor of $F(m)$.

The solution corresponding to this factor is given by

$$(D - m_1)^2\,y = 0$$

i.e. $(D - m_1)(D - m_1)\, y = 0$

let $(D - m_1)\, y = v$

$\therefore$ $(D - m_1)\, v = 0 \Rightarrow \dfrac{dv}{dx} - m_1\, v = 0 \Rightarrow \dfrac{dv}{v} = m_1\, dx$

On integration we get,

$$\log v = \log c_1 + m_1 x \quad \Rightarrow \quad v = c_1\, e^{m_1 x}$$

i.e. $(D - m_1)\, y = c_1\, e^{m_1 x}$ $[\because (D - m_1)\, y = v]$

or $\dfrac{dy}{dx} - m_1\, y = c_1\, e^{m_1 x}$.

This is a linear equation of first degree, whose solution is given by,

$$y\, e^{-m_1 x} = c_1\, e^{m_1 x} \,.\, e^{-m_1 x}\, dx + c_2. \qquad [\text{Here } I.F = e^{-m_1 x}]$$

$$\Rightarrow \quad y\, e^{-m_1 x} = c_1\, x + c_2 \Rightarrow y = (c_1\, x + c_2)\, e^{m_1 x}$$

Hence the general solution of (4) is,

$$y = (c_1\, x + c_2)\, e^{m_1 x} + c_3\, e^{m_2 x} + \ldots\ldots\ldots + c_n\, e^{m_n x}$$

In general it can be shown that, if m_1 is a root of $F(m) = 0$, repeated r times, then the corresponding part of the solution is,

$$y = (c_1 + c_2\, x + \ldots\ldots\ldots + c_r\, x^{r-1})\, e^{m_1 x}$$

Case (iii) : Let the equation $F(m) = 0$ has a pair of imaginary (complex) roots say $m_1 = \alpha + i\,\beta$ and $m_2 = \alpha - i\,\beta$. Then, the solution corresponding to these roots will be

$$y = c_1\, e^{(\alpha + i\beta)x} + c_2\, e^{(\alpha - i\beta)x}$$

$$y = c_1\, e^{\alpha x} \,.\, e^{i\beta x} + c_2\, e^{\alpha x} \,.\, e^{-i\beta x}$$

$$y = c_1\, e^{\alpha x} \,.\, (\cos\beta x + i \sin\beta x) + c_2\, e^{\alpha x} \,.\, (\cos\beta x - i\sin\beta x)$$

$$[\because e^{i\theta} = \cos\theta + i\sin\theta]$$

$$y = (c_1 + c_2)\, e^{\alpha x} \cos\beta x + (c_1 - c_2)\, i\, e^{\alpha x} \sin\beta x$$

Let, $c_1 + c_2 = A$ and $i\,(c_1 - c_2) = B$, then, $y = e^{\alpha x}(A\cos\beta x + B\sin\beta x)$ which will be the solution corresponding to the roots $\alpha \pm i\,\beta$ of $F(m) = 0$.

Note : It looks at first sight as if the new arbitrary constant B is an imaginary number. But it is not so, for consider,

$$c_1 = 3 + 4\,i \text{ and } c_2 = 3 - 4\,i, \text{ then } A = 6 \text{ and } B = -8.$$

In general, it can be shown that if $F(m) = 0$ has imaginary roots $\alpha \pm i\,\beta$ repeated twice, then coresponding solutions of the equation are

$$y = e^{\alpha x}\,[(A_1 + A_2\, x)\cos\beta x + (B_1 + B_2\, x)\sin\beta x],$$

and so on.

The algebraic equation $F(m) = 0$ obtained by replacing D by m in $F(D)$ is called **Auxiliary equation.**

Example 1. Solve : $\dfrac{d^3y}{dx^3} - 13\dfrac{dy}{dx} + 12y = 0$

Solution : The given equation is,

$$\frac{d^3y}{dx^3} - 13\frac{dy}{dx} + 12y = 0$$

In the form of symbol D, the equation becomes

$$(D^3 - 13\,D + 12)\, y = 0$$

The auxiliary equation is, $m^3 - 13m + 12 = 0 \Rightarrow (m-1)(m+4)(m-3) = 0$.

$\therefore$ the roots are 1, – 4, 3 which are distinct. Hence the solution is

$$y = c_1 e^x + c_2 e^{-4x} + c_3 e^{3x}$$

Example 2. Solve : $(D^3 - 3D^2 + 4)y = 0$.

Solution : The given equation is, $(D^3 - 3D^2 + 4)y = 0$

The auxiliary equation is, $m^3 - 3m^2 + 4 = 0 \Rightarrow (m+1)(m-2)^2 = 0$

$\therefore$ the roots are – 1, 2, 2. Hence the solution is,

$$y = c_1 e^{-x} + (c_2 + c_3 x) e^{2x}.$$

Example 3. Solve : $(D^4 - D^3 - 9D^2 - 11D - 4)y = 0$.

Solution : The given equation is,

$$(D^4 - D^3 - 9D^2 - 11D - 4)y = 0$$

The auxiliary equation is,

$$m^4 - m^3 - 9m^2 - 11m - 4 = 0$$

By inspection – 1 is a root, thus $(m+1)$ is a factor. Thus

$$(m+1)(m^3 - 2m^2 - 7m - 4) = 0 \qquad \text{[By actual division]}$$

Again by inspection, – 1 is a root, thus we have

$$(m+1)^2(m^2 - 3m - 4) = 0$$

$$\Rightarrow \qquad (m+1)^2(m+1)(m-4) = 0 \Rightarrow (m+1)^3(m-4) = 0$$

$\therefore$ the roots of the auxiliary equation are, – 1, – 1, – 1, 4.

Hence the required solution is $y = c_1 e^{4x} + (c_2 + c_3 x + c_4 x^2) e^{-x}$.

Example 4. Solve : $\frac{d^3y}{dx^3} - 2\frac{d^2y}{dx^2} + 4\frac{dy}{dx} - 8y = 0$.

Solution : The given equation in the symbol D is,

$$(D^3 - 2D^2 + 4D - 8)y = 0$$

The auxiliary equation is, $m^3 - 2m^2 + 4m - 8 = 0$.

Clearly $m = 2$ is a root, thus the equation becomes,

$$(m-2)(m^2+4) = 0.$$

$\therefore$ the roots are 2, $\pm 2i$. Hence the required solution is

$$y = c_1 e^{2x} + c_2 \cos 2x + c_3 \sin 2x$$

Example 5. Solve : $(D^3 - D^2 - D - 2)y = 0$.

Solution : The given equation is, $(D^3 - D^2 - D - 2)y = 0$

The auxiliary equation is,

$$m^3 - m^2 - m - 2 = 0$$

$$\Rightarrow \qquad (m-2)(m^2 + m + 1) = 0$$

Now, $m - 2 = 0 \Rightarrow m = 2$, and $m^2 + m + 1 = 0 \Rightarrow m = -\frac{1}{2} \pm i\frac{\sqrt{3}}{2}$

Hence the required solution is,

$$y = c_1 e^{2x} + e^{-1/2\,x}\left(c_2 \cos\frac{\sqrt{3}}{2}x + c_3 \sin\frac{\sqrt{3}}{2}x\right)$$

EXERCISES

1. $\frac{d^2y}{dx^2}+3\frac{dy}{dx}+2y=0$
2. $\frac{d^2y}{dx^2}-7\frac{dy}{dx}+12y=0$
3. $(2D^2+D+2)\,y=0$
4. $(D^2-3D+2)\,y=0$, given when $x=0$, then $y=0$, $\frac{dy}{dx}=0$
5. $(D^3-6D^2+11D-6)\,y=0$
6. $(D^3-9D^2+23D-15)\,y=0$
7. $\frac{d^2y}{dx^2}-6\frac{dy}{dx}+9y=0$
8. $(D^3-D^2-D+1)\,y=0$
9. $(D^3-3D^2+4)\,y=0$
10. $(D^4+2D^3+D^2)\,y=0$
11. $(D^3+2D^2-D-2)\,y=0$
12. $(D^3-8)\,y=0$
13. $\frac{d^4y}{dx^4}+m^4y=0$
14. $(D^2+1)^2(D-1)^2\,y=0$
15. $(D^4-4D^3+8D^2-8D+4)\,y=0$

Answers

1. $y=c_1e^{-x}+c_2e^{-2x}$
2. $y=c_1e^{3x}+c_2e^{4x}$
3. $y=c_1e^{-2x}+c_2e^{-1/2\,x}$
4. $x=0$
5. $y=c_1e^{x}+c_2e^{2x}+c_3e^{3x}$
6. $y=c_1e^{x}+c_2e^{3x}+c_3e^{5x}$
7. $y=(c_1+c_2x)\,e^{2x}$
8. $y=(c_1+c_2x)\,e^{x}+c_3e^{-x}$
9. $y=c_1e^{-x}+(c_2+c_3x)\,e^{2x}$
10. $y=(c_1+c_2x)+(c_3+c_4x)\,e^{-x}$
11. $y=c_1e^{x}+c_2e^{-x}+c_3e^{-2x}$
12. $y=c_1e^{2x}+e^{-x}(c_2\cos x\sqrt{3}+c_3\sin x\sqrt{3})$
13. $y=c_1\,e^{\frac{mx}{\sqrt{2}}}+\cos\left(\frac{mx}{\sqrt{2}}+c_2\right)+c_3e^{\frac{mx}{\sqrt{2}}}\cos\left(\frac{mx}{\sqrt{2}}+c_4\right)$
14. $y=(c_1+c_2x)\sin x+(c_2+c_4x)\cos x+(c_5+c_6x)\,e^{x}$
15. $y=e^{x}(c_1+c_2x)\sin x+e^{x}(c_3+c_4x)\cos x$

2.14 The Complementary Function and the Particular Integral

In the previous section we have seen how to find the general solution of the linear equation when the right hand side function $f(x)$ is zero. Now we shall see a relation between the solution of the linear equation when $f(x)$ not equal to zero and the solution of the linear equation when $f(x)$ is zero.

Let, $y=u$ be a particular solution of

$$F(D)\,y=f(x) \qquad \text{..... (1)}$$

Then we have, $$F(D)\,u=f(x) \qquad \text{..... (2)}$$

Put, $y=u+v$ in (1), then, $\Rightarrow F(D)(u+v)=f(x)$

$\Rightarrow$ $$F(D)\,u+F(D)\,v=f(x) \qquad \text{..... (3)}$$

Consider (3) – (2), then we have,

$$F(D)\,v=0 \qquad \text{..... (4)}$$

If the solution of (4) be $v=F(x)$, containing n arbitrary constants, the general solution of the equation (1) is,

$$y=u+F(x)$$

The particular solution $y = u$ of (1) is called the **Particular integral** and the solution $y = F(x)$, containing n arbitrary constants, of (4) (which is same as (1) with right hand side function $f(x)$ as zero) is called the **complementary function.**

Hence the general solution of a linear equation with constant coefficients is the sum of the complementary function (C.F.) and a particular integral (P.I.), where the complementary function is the general solution of the equation when the right hand side function $f(x)$ is zero.

We have seen how to obtain C.F. - i.e. the solution of the linear equation when right hand side function $f(x)$ is zero. Now we shall see how to determine the P.I.

Let, $\dfrac{1}{F(D)} f(x)$ denote some function of x which when operated by $F(D)$ gives $f(x)$.

Clearly $\dfrac{1}{F(D)} f(x)$ is a particular integral of the equation (1). Thus, if we take

$$y = \frac{1}{F(D)} f(x)$$

the equation (1) is satisfied. So, from our definition,

$$F(D)\left\{\frac{1}{F(D)} f(x)\right\} = f(x)$$

Hence $\dfrac{1}{F(D)}$ is the inverse operator of $F(D)$. Hence

$$\textbf{P. I.} = \frac{1}{F(D)} f(x)$$

Note :

1. Naturally the above discussion suggests that $\dfrac{1}{D}$ is the operator inverse to D - i.e. the operator that integrates w.r.t.x. That is,

$$\frac{1}{D} f(x) = \int f(x)\, dx .$$

In general $\dfrac{1}{D^p}$ is the operator that integrates p times.

2. In the case of particular integral we do not add the constant of integration as we want only a particular solution.

We shall consider the method of finding P. I. when $f(x)$ takes few special forms.

1. Particular Integral when $f(x) = e^{mx}$

We know that,

$$F(D)\, e^{mx} = e^{mx} F(m)$$

This suggests, as long as $F(m) \neq 0$, $\dfrac{1}{F(m)} e^{mx}$ may be a value of $\dfrac{1}{F(D)} e^{mx}$.

This fact can be verified for,

$$F(D)\left\{\frac{1}{F(m)} e^{mx}\right\} = \frac{1}{F(m)} . F(D)\, e^{mx} . = \frac{1}{F(m)} . e^{mx} F(m) = e^{mx}$$

$$\therefore \quad \frac{1}{F(D)} e^{mx} = \frac{1}{F(m)} e^{mx} \textbf{ provided } F(m) \neq 0.$$

If $F(m) = 0$, then $(D - m)$ is a factor of $F(D)$. Let,

$$F(D) = (D - m)^r \varphi(D), \text{ where } \varphi(m) \neq 0$$

We know that, $F(D)(e^{mx}. V) = e^{mx} F(D + m) V$

This suggests that the following may be true if $V = 1$.

$$\frac{1}{F(D)}e^{mx} = \frac{1}{(D-m)^r\,\varphi(D)}e^{mx}$$

$$= \frac{1}{(D-m)^r}\left[\frac{1}{\varphi(D)}e^{mx}\right]$$

$$= \frac{1}{(D-m)^r}\cdot\frac{1}{\varphi(m)}\cdot e^{mx} \qquad [\because \varphi(m) \neq 0]$$

$$= \frac{1}{\varphi(m)}\cdot\frac{1}{(D-m)^r}\cdot e^{mx}.1$$

$$= \frac{e^{mx}}{\varphi(m)}\cdot\frac{1}{(D)^r}.1 \qquad \text{[replacing } D \text{ by } D+m]$$

$$= \frac{e^{mx}}{\varphi(m)}\cdot\frac{x^r}{r!}.1 \qquad \left[\because \frac{1}{D^r} \text{ integrates } 1, r \text{ times}\right]$$

Thus we have obtained a tentative result,

$$\frac{1}{F(D)}e^{mx} = \frac{e^{mx}}{\varphi(D)}\cdot\frac{x^r}{r!}$$

where m is a root of $F(D) = 0$, repeated r times.

The above result can be easily verified for its truthfulness - for, consider,

$$F(D)\left\{\frac{e^{mx}}{\varphi(m)}\cdot\frac{x^r}{r!}\right\} = (D-m)^r\,\varphi(D)\left\{\frac{e^{mx}}{\varphi(m)}\cdot\frac{x^r}{r!}\right\}$$

$$= \varphi(D)\left[(D-m)^r\left\{\frac{e^{mx}}{\varphi(m)}\cdot\frac{x^r}{r!}\right\}\right]$$

$$= \varphi(D)\left[\left\{\frac{e^{mx}}{\varphi(m)}D^r\cdot\frac{x^r}{r!}\right\}\right] \qquad \text{[replacing } D \text{ by } D+m]$$

$$= \varphi(D)\left[\frac{e^{mx}}{\varphi(m)}.1\right] \qquad \left[\because D^r \text{ differentiates } \frac{x^r}{r!}, r \text{ times}\right]$$

$$= \frac{1}{\varphi(m)}.e^{mx}\,\varphi(m) = e^{mx}$$

Hence, $$\frac{1}{F(D)}e^{mx} = \frac{e^{mx}}{\varphi(m)}\cdot\frac{x^r}{r!}$$

where m is a root of auxiliary equation $F(D) = 0$, repeated r times.

In particular, if m is a root of $F(D) = 0$ only once (i.e. $r = 1$), we have,

$$\frac{1}{F(D)}e^{mx} = \frac{e^{mx}}{\varphi(m)}\cdot\frac{x}{1!}$$

where $\varphi(m)$ is given by, $F(D) = (D-m)\,\varphi(m)$

If m is a root repeated twice, we have

$$\frac{1}{F(D)}e^{mx} = \frac{e^{mx}}{\varphi(m)}\cdot\frac{x^2}{2!}$$

where $\varphi(m)$ is given by, $F(D) = (D - m)^2 \varphi(m)$ and so on.

Note : 1. If $f(x)$ be a constant, say k, then,

$$\frac{1}{F(D)} k = \frac{1}{F(D)} k e^{0x} \qquad \text{[Here } m = 0]$$

$$\frac{1}{F(D)} k = k \cdot \frac{1}{F(0)} \qquad [D \text{ is replaced by } 0]$$

F (0) will be equal to the constant term that appears in $F(D)$.

2. If $f(x)$ is of the form $\sinh x$ or $\cosh x$, then we can proceed by writing,

$$\sinh x = \frac{e^x - e^{-x}}{2} \quad \text{and} \quad \cosh x = \frac{e^x + e^{-x}}{2}$$

2. Particular Integral when $f(x) = \cos mx$.

We know that, $F(D^2) \cos mx = F(-m^2) \cos mx$

Operating both sides with $\frac{1}{F(D^2)}$, we get,

$$\frac{1}{F(D^2)}\left[F(D^2) \cos mx\right] = \frac{1}{F(D^2)}\left[F(-m^2) \cos mx\right]$$

or

$$\cos mx = \frac{1}{F(D^2)} F(-m^2) \cos mx$$

Hence,

$$\frac{1}{F(D^2)} \cos mx = \frac{1}{F(-m^2)} \cos mx$$

Similarly,

$$\frac{1}{F(D^2)} \sin mx = \frac{1}{F(-m^2)} \sin mx$$

Thus to find the particular integral when the right hand side function is $\cos mx$ or $\sin mx$, we replace D^2 in $F(D)$ by $-m^2$ and proceed.

Note : The above method fails when $F(-m^2) = 0$. In such case we seek an alternate method which is explained in the example worked out latter.

3. Particular integral when $f(x) = x^m$ where m is a positive integer.

In this case to evaluate $\frac{1}{F(D^2)} x^m$. We write $\frac{1}{F(D)}$ as $[F(D)]^{-1}$ and expand this in ascending powers of D. Then we operate x^m with various terms of the expansion thus obtained. The result so obtained will be the particular integral.

Note : In the expansion of $[F(D)]^{-1}$ the term beyond the mth power of D need not be written, as the result of their operation on x^m would be zero.

i.e. $$D^r x^m = 0; \text{ if } r > m$$

4. Particular integral when $f(x) = e^{mx}$. V where V is any function of x.

We know that, $F(D)\, e^{mx} V = e^{mx} F(D + m) V$

This suggests that the following may be true,

$$\frac{1}{F(D)} e^{mx} V = e^{mx} \frac{1}{F(D+m)} V$$

This can be easily verified - for consider,

$$F(D)\left\{e^{mx} \cdot \frac{1}{F(D+m)} V\right\} = e^{mx} F(D+m) \frac{1}{F(D+m)} V = e^{mx} V$$

Hence,

$$\frac{1}{F(D)} e^{mx} V = e^{mx} \frac{1}{F(D+m)} V$$

Example 1. Solve : $\frac{d^2y}{dx^2} + 2\frac{dy}{dx} + y = 2e^{2x}$.

Solution : The given differential equation can be written as

$$(D^2 + 2D + 1)\, y = 2\, e^{2x}$$

The auxiliary equation is, $m^2 + 2m + 1 = 0 \Rightarrow (m+1)^2 = 0$

The roots are $-1, -1$. Thus, C.F. is $(c_1 + c_2 x)\, e^{-x}$

Now, P.I. $= \frac{1}{D^2 + 2D + 1} 2e^{2x} = \frac{1}{2^2 + 2.2 + 1} 2e^{2x} = \frac{2}{9} e^{2x}$

Hence the solution is, $\boldsymbol{y = (c_1 + c_2 x)\, e^{-x} + \frac{2}{9} e^{2x}}$

Example 2. Solve : $\frac{d^3y}{dx^3} - 3\frac{d^2y}{dx^2} + 4\frac{dy}{dx} - 2y = e^x$.

Solution : The given differential equation can be written as,

$$(D^3 - 3D^2 + 4D - 2)\, y = e^x$$

The auxiliary equation is, $m^3 - 3m^2 + 4m - 2 = 0$

Clearly $m = 1$ is a root.

$\Rightarrow$ $(m-1)(m^2 - 2m + 2) = 0 \Rightarrow m = 1$ or $m = 1 \pm i$

$\therefore$ C.F. is $c_1 e^x + e^x (c_2 \cos x + c_3 \sin x)$

Now, P.I. $= \frac{1}{D^3 - 3D^2 + 4D - 2} e^x$

Here, $F(1) = 0$, as 1 is a root of the equation $F(D) = 0$

$\therefore$ P.I. $= \frac{1}{(D-1)(D^2 - 2D + 2)} e^x = \frac{1}{1^2 - 2.1 + 2} \frac{x}{1!} e^x = x\, e^x$

Hence the solution is,

$$y = c_1 e^x + e^x (c_2 \cos x + c_3 \sin x) + xe^x$$

$$\Rightarrow \quad \boldsymbol{y = e^x [c_1 + c_2 \cos x + c_3 \sin x + x]}$$

Example 3. Solve : $(D^3 + 1)\, y = 3 + e^{-x} + 5\, e^{2x}$.

Solution : The given differential equation can be written as

$$(D^3 + 1)\, y = 3 + e^{-x} + 5\, e^{2x}$$

The auxiliary equation is

$$m^3 + 1 = 0 \Rightarrow (m+1)(m^2 - m + 1) = 0 \Rightarrow m = -1 \text{ and } m = \frac{1}{2} \pm \frac{\sqrt{3}}{2} i$$

Thus, C.F. is $c_1 e^{-x} + e^{\frac{1}{2}x} \left(c_2 \cos \frac{\sqrt{3}}{2} x + c_3 \sin \frac{\sqrt{3}}{2} x \right)$

Now, P.I. $= \frac{1}{D^3 + 1} (3 + e^{-x} + 5\, e^{2x})$

$$\Rightarrow \quad = \frac{1}{D^3 + 1} 3 + \frac{1}{D^3 + 1} e^{-x} + 5 \frac{1}{D^3 + 1} e^{2x}$$

Now consider,

$$\frac{1}{D^3+1}3 = \frac{1}{D^3+1}3 \,.\, e^{0x} = 3 \,.\, \frac{1}{0+1}e^{0x} = 3$$

$$\frac{1}{D^3+1}e^{-x} = \frac{1}{(D+1)(D^2-D+1)}e^{-x} = \frac{1}{1+1+1} \,.\, \frac{x}{1!} \,.\, e^{-x} = \frac{x}{3}e^{-x}$$

$$\frac{1}{D^3+1}e^{2x} = \frac{1}{8+1}e^{2x} = \frac{1}{9}e^{2x}$$

$\therefore$ $$\text{P.I.} = 3 + \frac{x}{3}e^{-x} + \frac{5}{9}e^{2x}.$$

Hence the solution is,

$$y = c_1 e^{-x} + e^{1/2\,x}\left(c_2 \cos\frac{\sqrt{3}}{2}x + c_3 \sin\frac{\sqrt{3}}{2}x\right) + 3 + \frac{x}{3}e^{-x} + \frac{5}{9}e^{2x}.$$

Example 4. Solve : $\frac{d^3y}{dx^3} + \frac{d^2y}{dx^2} - \frac{dy}{dx} - y = \cos 2x$ **.**

Solution : The given differential equation is, $(D^3 + D^2 - D - 1)\, y = \cos\, 2x$

The auxiliary equation is, $m^3 + m^2 - m - 1 = 0$

Clearly $m = 1$ is a root.

$\therefore$ $(m-1)(m^2 + 2m + 1) = 0 \Rightarrow (m-1)(m+1)^2 = 0 \Rightarrow$ the roots are $1, -1, -1$

$\therefore$ $\text{C.F.} = c_1 e^x + (c_2 + c_3 x)\, e^{-x}$

Now,
$$\text{P.I.} = \frac{1}{D^3 + D^2 - D - 1}\cos 2x$$

$$= \frac{1}{D \,.\, D^2 + D^2 - D - 1}\cos 2x$$

$$= \frac{1}{D \,.\, (-4) + (-4) - D - 1}\cos 2x \qquad \text{[replacing } D \text{ by } (-2^2)]$$

$$= \frac{1}{-5\,(D+1)}\cos 2x$$

$$= \frac{1}{-5}\frac{D-1}{D^2-1}\cos 2x \qquad \text{[Note this step]}$$

$$= \frac{1}{-5}(D-1)\frac{1}{-4-1}\cos 2x$$

$$= \frac{1}{25}(D-1)\cos 2x$$

$$= \frac{1}{25}(-2\sin 2x - \cos 2x) = -\frac{1}{25}(2\sin 2x + \cos 2x)$$

Hence the solution is,

$$y = c_1 e^x + (c_2 + c_3 x)\, e^{-x} - \frac{1}{25}(2\sin 2x + \cos 2x).$$

Example 5. Solve : $(D^2 - 2D + 5)\, y = \sin 3x$.

Solution : The given differential equation can be written as

$$(D^2 - 2D + 5)\, y = \sin 3x$$

The auxiliary equation is

$$m^2 - 2m + 5 = 0 \Rightarrow m = 1 \pm 2i$$

$\therefore$ C.F. $= e^x (c_1 \cos 2x + c_2 \sin 2x)$.

Now, $$\text{P.I.} = \frac{1}{D^2 - 2D + 5} \sin 3x$$

$$= \frac{1}{-9 - 2D + 5} \sin 3x \quad \text{[replacing } D^2 \text{ by } -(3^2)]$$

$$= \frac{1}{-2(D+2)} \sin 3x$$

$$= -\frac{1}{2}(D-2) \cdot \frac{1}{D^2 - 4} \sin 3x \quad \text{(Note this step)}$$

$$= -\frac{1}{2}(D-2) \cdot \frac{1}{-9-4} \sin 3x$$

$$= \frac{1}{26}(D-2) \sin 3x = \frac{1}{26}(3 \cos 3x - 2 \sin 2x)$$

Hence the solution is,

$$\mathbf{y = e^x(c_1 \cos 2x + c_2 \sin 2x) + \frac{1}{26}(3 \cos 3x - 2 \sin 2x).}$$

Example 6. Solve : $\dfrac{d^2y}{dx^2} - y = 2 + 5x$.

Solution : The given differential equation can be written as,

$$(D^2 - 1)\, y = 2 + 5x$$

The auxiliary equation is, $m^2 - 1 = 0 \Rightarrow (m-1)(m+1) = 0 \Rightarrow$ roots are $1, -1$.

$\therefore$ C.F. is $c_1 e^x + c_2 e^{-x}$

Now, $$\text{P.I.} = \frac{1}{D^2 - 1}(2 + 5x)$$

$$= -(1 - D^2)^{-1}(2 + 5x)$$

$$= -(1 + D^2 + D^4 + \ldots\ldots)(2 + 5x) = -(2 + 5x)$$

Hence the solution is, $\mathbf{y = c_1 e^x + c_2 e^{-x} - (2 + 5x).}$

Example 7. Solve : $(D^3 - D^2 - 6D)\, y = 1 + x^2$.

Solution : The given equation is, $(D^3 - D^2 - 6D)\, y = 1 + x^2$

The auxiliary equation is,

$$m^3 - m^2 - 6m = 0 \Rightarrow m(m^2 - m - 6) = 0 \Rightarrow m(m-3)(m+2) = 0$$

$\therefore$ the roots are $0, 3, -2$.

$\therefore$ C.F. $= c_1 + c_2 e^{3x} + c_3 e^{-2x}$

Now, $$\text{P.I.} = \frac{1}{D^3 - D^2 - 6D}(1 + x^2)$$

$$= \frac{1}{D(D^2 - D - 6)}(1 + x^2)$$

$$= \left(\frac{1}{-6D}\right) \frac{1}{1 - \left(\dfrac{D^2 - D}{6}\right)}(1 + x^2)$$

$$= \frac{1}{-6D}\left[1-\left(\frac{D^2-D}{6}\right)\right]^{-1}(1+x^2)$$

$$= \frac{1}{-6D}\left[1+\frac{D^2-D}{6}+\frac{(D^2-D)^2}{36}+.....\right](1+x^2)$$

$$= \frac{1}{-6D}\left[1+\frac{7D^2}{36}-\frac{D}{6}+......\right](1+x^2)$$

$$= -\frac{1}{6D}\left[1+x^2+\frac{7}{18}-\frac{x}{3}\right]$$

$$= -\frac{1}{6D}\left[x^2-\frac{x}{3}+\frac{25}{18}\right] = -\frac{1}{6}\left[\frac{x^3}{3}-\frac{x^2}{6}+\frac{25}{18}x\right] = -\frac{1}{108}\left[6x^3-3x^2+25x\right]$$

Hence the solution is,

$$y = c_1 + c_2 e^{3x} + c_3 e^{-2x} - \frac{1}{108}\left[6x^3 - 3x^2 + 25x\right].$$

Example 8. Solve : $\frac{d^2y}{dx^2} + 3\frac{dy}{dx} + 2y = e^{2x} . \sin x$ **.**

Solution : The given equation is, $(D^2 + 3D + 2)\, y = e^{2x} \sin x$

The auxiliary equation is, $m^2 + 3m + 2 = 0 \Rightarrow (m+1)(m+2) = 0$

$\therefore$ the roots are $-1, -2$

$\therefore$ C.F. $= c_1 e^{-x} + c_2 e^{-2x}$

Now,

$$\text{P.I.} = \frac{1}{D^2+3D+2} e^{2x} \sin x$$

$$= e^{2x} \frac{1}{(D+2)^2 + 3(D+2) + 2} \sin x$$

$$= e^{2x} \frac{1}{D^2+7D+12} \sin x$$

$$= e^{2x} \frac{1}{-(1)^2+7D+12} \sin x$$

$$= e^{2x} \frac{1}{7D+11} \sin x$$

$$= e^{2x} (7D-11) \frac{1}{49D^2-121} \sin x$$

$$= e^{2x} (7D-11) \frac{1}{49(-1)-121} \sin x$$

$$= \frac{e^{2x}}{-170} (7\cos x - 11 \sin x)$$

Hence the solution is,

$$y = c_1 e^{-x} + c_2 e^{-2x} - \frac{e^{2x}}{170}(7\cos x - 11\sin x).$$

Example 9. Solve : $\frac{d^3y}{dx^3} + y = 3 + e^{-x} + 5\,e^x x^2.$

Solution : The given equation is, $(D^3 + 1)\,y = 3 + e^{-x} + 5\,e^x x^2$

The auxiliary equation is, $m^3 + 1 = 0 \Rightarrow (m+1)(m^2 - m + 1) = 0 \Rightarrow m = -1, \frac{1}{2} \pm \frac{i\sqrt{3}}{2}$

$$\therefore \quad \text{C.F.} = c_1 e^{-x} + e^{\left(\frac{x}{2}\right)}\left(c_2 \cos\frac{\sqrt{3}}{2}x + c_3 \sin\frac{\sqrt{3}}{2}x\right)$$

Now,
$$\text{P.I.} = \frac{1}{D^3+1}\,(3 + e^{-x} + 5\,e^x x^2)$$

$$= \frac{1}{D^3+1}3 + \frac{1}{D^3+1}e^{-x} + 5\frac{1}{D^3+1}e^x x^2$$

Consider,

$$\frac{1}{D^3+1}3 = 3\frac{1}{D^3+1}e^{0x} = 3$$

$$\frac{1}{D^3+1}e^{-x} = \frac{1}{(D+1)(D^2-D+1)}e^{-x} = \frac{1}{3}\frac{x}{1!}e^{-x} = \frac{x}{3}e^{-x}$$

$$\frac{1}{D^3+1}e^x x^2 = e^x\frac{1}{(D+1)^3+1}x^2$$

$$= e^x\frac{1}{D^3+3D^2+3D+2}x^2$$

$$= \frac{e^x}{2}\frac{1}{\left(1+\frac{D^3+3D^2+3D}{2}\right)}x^2$$

$$= \frac{e^x}{2}\left(1+\frac{D^3+3D^2+3D}{2}\right)^{-1}x^2$$

$$= \frac{e^x}{2}\left(1-\frac{D^3+3D^2+3D}{2}+\frac{(D^3+3D^2+3D)^2}{4}\ldots\ldots\right)x^2$$

$$= \frac{e^2}{2}\left(x^2-3-3x+\frac{9}{2}\right)$$

$$\therefore \quad \text{P.I.} = 3 + \frac{x}{3}e^{-x} + \frac{e^x}{2}\left(x^2-3-3x+\frac{9}{2}\right) = \frac{e^x}{4}(2x^2-6x+3)+\frac{x}{3}e^{-x}+3.$$

Hence the solution is, y = C.F. + P.I.

Two Important Results.

I. $\frac{1}{D^2+a^2}\sin ax = \frac{-x}{2a}\cos ax$

We cannot evaluate this by replacing D^2 by $-a^2$, as this gives zero in the denominator. Therefore we shall adopt a different method by using the formula $e^{i\theta} = \cos\theta + i\sin\theta$

Now, $\sin ax$ = Imaginary part of e^{iax}

Now,
$$\frac{1}{D^2+a^2}\sin ax = \text{Imaginary part of } \frac{1}{D^2+a^2}e^{iax}$$

$$= \text{Imaginary part of } \frac{1}{(D-ia)(D+ia)}e^{iax}$$

$$= \text{Imaginary part of } \frac{1}{2ai} \cdot \frac{x}{1!} e^{iax}$$

$$= \text{Imaginary part of } \frac{x}{2ai} (\cos ax + i \sin ax)$$

$$= \text{Imaginary part of } \frac{x}{2a} (-i \cos ax + \sin ax)$$

$$= -\frac{x}{2a} \cos ax$$

$$\therefore \quad \frac{1}{D^2 + a^2} \sin ax = -\frac{x}{2a} \cos ax$$

II. $\frac{1}{D^2 + a^2} \cos ax = \frac{x}{2a} \sin ax$

As above, $\frac{1}{D^2 + a^2} \cos ax = \text{Real part of } \frac{1}{D^2 + a^2} e^{iax}$

$$\therefore \quad \frac{1}{D^2 + a^2} \cos ax = \frac{x}{2a} \sin ax$$

Example 10. Solve : $(D^2 + 4)\, y = \sin^2 x$.

Solution : The given differential equation can be written as, $(D^2 + 4)\, y = \sin^2 x$

The auxiliary equation is, $m^2 + 4 = 0 \Rightarrow m = \pm 2i$

$\therefore$ C.F. $= c_1 \cos 2x + c_2 \sin 2x$

Now, $\quad \text{P.I.} = \frac{1}{D^2+4}\left[\frac{1}{2}(1 - \cos 2x\right]$

$$= \frac{1}{2}\frac{1}{D^2+4}.1 - \frac{1}{2}\frac{1}{D^2+4}.\cos 2x$$

Consider, $\quad \frac{1}{D^2+4}.1 = \frac{1}{D^2+4} e^{0x} = \frac{1}{4}$

Again, $\quad \frac{1}{D^2+4}.1 = \frac{1}{D^2+2^2}.\cos 2x = \frac{x}{4}\sin 2x \qquad \left[\because \frac{1}{D^2+a^2}\cos ax = \frac{x}{2a}\sin ax\right]$

$\therefore \quad \text{P.I.} = \frac{1}{8} - \frac{x}{8}\sin 2x.$

Hence the solution is, $\boldsymbol{y = c_1 \cos 2x + c_2 \sin 2x + \frac{1}{8} - \frac{x}{8} \sin 2x.}$

Example 11. Solve : $(D^3 - 3D^2 + 9D - 27)\, y = \cos 3x$.

Solution : The given equation is, $(D^3 - 3D^2 + 9D - 27)\, y = \cos 3x$

The auxiliary equation is, $m^3 - 3m^2 + 9m - 27 = 0 \Rightarrow (m-3)(m^2+9) = 0$

$\therefore$ the roots are $3, 3i, -3i$

$\therefore$ C.F. $= c_1 e^{3x} + c_2 \cos 3x + c_3 \sin 3x$

Now, $\quad \text{P.I.} = \frac{1}{D^3 - 3D^2 + 9D - 27} \cos 3x$

Here the denominator gives zero, when D^2 is replaced by -3^2.

Now, $\quad \text{P.I.} = \frac{1}{(D^2+9)(D-3)} \cos 3x$

$$= \frac{1}{(D^2+9)}(D+3)\frac{1}{(D^2-9)} \cos 3x$$

$$= -\frac{1}{18}\frac{1}{(D^2+9)}(-3 \sin 3x + 3 \cos 3x)$$

$$= -\frac{1}{6}\frac{1}{(D^2+9)}(\cos 3x - \sin 3x)$$

Now, $\dfrac{1}{D^2+9}\cos 3x = \dfrac{x}{6}\sin 3x, \quad \dfrac{1}{D^2+9}\sin 3x = -\dfrac{x}{6}\cos 3x$

$\therefore$ P.I. $= -\dfrac{x}{36}(\sin 3x + \cos 3x)$.

Hence the solution is, $y = c_1 e^{3x} + c_2 \cos 3x + c_3 \sin 3x - \dfrac{x}{36}(\sin 3x + \cos 3x)$.

Example 12. Solve : $(D^2 - 6D + 9)\, y = e^{3x}(x^2 + 7x + 5)$.

Solution : The given equation is, $(D^2 - 6D + 9)\, y = e^{3x}(x^2 + 7x + 5)$

The auxiliary equation is, $m^2 - 6m + 9 = 0 \Rightarrow (m-3)^2 = 0 \Rightarrow$ the roots are 3, 3.

$\therefore$ C.F. $= (c_1 + c_2 x)\, e^{3x}$

Now, P.I. $= \dfrac{1}{D^2-6D+9} e^{3x}(x^2+7x+5)$

$$= e^{3x}\frac{1}{(D+3)^2-6(D+3)+9}(x^2+7x+5)$$

$$= e^{3x}\frac{1}{D^2}(x^2+7x+5)$$

$$= e^{3x}\frac{1}{D}\frac{1}{D}(x^2+7x+5) = e^{3x}\frac{1}{D}\left(\frac{x^3}{3}+\frac{7x^2}{2}+5x\right)$$

$$= e^{3x}\left(\frac{x^4}{12}+\frac{7x^3}{6}+\frac{5x^2}{2}\right)$$

$$= \frac{e^{3x}}{12}(x^4 + 14x^3 + 30x^2)$$

Hence the solution is, $y = (c_1 + c_2 x)\, e^{3x} + \dfrac{e^{3x}}{12}(x^4 + 14x^3 + 30x^2)$.

EXERCISES

Solve the following differential equations

1. $\dfrac{d^2y}{dx^2} + \dfrac{dy}{dx} + y = e^{-x}$
2. $(D^2 + 3D - 4)\, y = 12\, e^{2x}$
3. $\dfrac{d^3y}{dx^3} + y = 3 + e^{-x} + 5\, e^{2x}$
4. $(D^3 - 1)\, y = (e^x + 1)^2$
5. $(D^3 - 7D + 6)\, y = e^{2x}$ when $x = 0, y = 0$.
6. $(D^4 - 18D^2 + 81)\, y = 36\, e^{3x}$
7. $(D^2 - 3D + 2)\, y = e^x + e^{2x}$
8. $(4D^2 - 1)\, y = e^{x/2} + 12e^x$
9. $(D^2 + 16)\, y = 14 \cos 3x$
10. $(D^2 + 4)\, y = \sin 2x$
11. $(D^2 + 9)\, y = \cos 3x$
12. $(D^2 + 1)\, y = 12 \cos 2x - \sin x$

13. $(D^2 - 2D + 5)\,y = \sin 3x$

14. $(D^2 - 4D + 4)\,y = e^{-4x} + 5\cos 3x$

15. $(D^2 - 3D + 2)\,y = 6e^{3x} + \sin 2x$

16. $(D^2 - 4D + 4)\,y = \sin 2x$

17. $(D^3 - 3D^2 + 9D - 27)\,y = \cos 3x$

18. $(D^2 - 4)\,y = x^2$

19. $\dfrac{d^2y}{dx^2} - 5\dfrac{dy}{dx} + 6y = x$

20. $\dfrac{d^2y}{dx^2} + 4y = x^2 + 3$

21. $\dfrac{d^3y}{dx^3} + 8y = x^4 + 2x + 1$

22. $\dfrac{d^2y}{dx^2} - 2\dfrac{dy}{dx} + y = x - 1$

23. $(D^3 - 2D^2 + D)\,y = e^{2x} + x^2 + x$

24. $(D^2 + D - 6)\,y = x$

25. $\dfrac{d^2y}{dx^2} - 2\dfrac{dy}{dx} + 5y = e^{2x}\sin x$

26. $\dfrac{d^3y}{dx^3} - 3\dfrac{d^2y}{dx^2} + 3\dfrac{dy}{dx} - y = x\,e^x + e^x$

27. $\dfrac{d^2y}{dx^2} - 3\dfrac{dy}{dx} + 2y = x^2 e^{3x}$

28. $\dfrac{d^3y}{dx^3} - \dfrac{d^2y}{dx^2} - \dfrac{dy}{dx} + y = e^x(1 + \sin x)$

29. $\dfrac{d^2y}{dx^2} - 10\dfrac{dy}{dx} + 16y = e^{4x}\sin 2x$

30. $\dfrac{d^2y}{dx^2} - y = e^x(1 + x^2)$

31. $(D^2 + 2D + 2)\,y = 2e^{-x}\sin x$

32. $(D^2 - 2D + 1)\,y = x^2 e^{3x}$

Answers

1. $y = e^{-1/2\,x}\left(A\cos\dfrac{\sqrt{3}}{2}x + B\sin\dfrac{\sqrt{3}}{2}x\right) + e^{-x}$

2. $y = c_1 e^{-4x} + c_2 e^x + 2e^{2x}$

3. $y = c_1 e^{-x} + e^{x/2}\left(c_2\cos\dfrac{x\sqrt{3}}{2} + c_3\sin\dfrac{x\sqrt{3}}{2}\right) + 3 + \dfrac{5}{9}e^{2x} + \dfrac{x}{3}e^{-x}$

4. $y = e^{-x/2}\left(c_1\cos\dfrac{x\sqrt{3}}{2}x + c_2\sin\dfrac{x\sqrt{3}}{2}\right) + e^x\left(c_3 + \dfrac{2}{3}x\right) + \dfrac{1}{7}e^{2x} - 1$

5. $y = c\,(e^{-x} - e^{6x}) + \dfrac{1}{4}e^x(1 - e^x)$

6. $y = (c_1 + c_2)\,e^x + (c_3 + c_4)\,e^{-1/3\,.\,x} + \dfrac{x^2}{2}e^{3x}$

7. $y = (c_1 - x)\,e^x + (c_2 + x)\,e^{2x}$

8. $y = c_1 e^{x/2} + c_2 e^{-x/2} + \dfrac{x}{2}e^{x/2} + 4e^x$

9. $y = c_1\cos 4x + c_2\sin 4x + 2\cos 3x$

10. $y = c_1\cos 2x + c_2\sin 2x - \dfrac{x}{4}\cos 2x$

11. $y = c_1\cos 3x + c_2\sin 3x + \dfrac{x}{6}\sin 3x$

12. $y = \left(c_1 + \dfrac{1}{2}\right)\cos x + c_2\sin x - 4\cos 2x$

13. $y = c_1 e^{3x} + c_2 e^{2x} - \dfrac{7(\cos 3x + 9\sin 2x)}{130}$

14. $y = (c_1 + c_2 x)\,e^{2x} + \dfrac{1}{36}e^{-4x} - \dfrac{5}{169}(12\sin x + 5\cos 3x)$

15. $y = c_1 e^x + c_2 e^{2x} + 3e^{3x} + \dfrac{1}{40}(6\cos 2x - 2\sin 2x)$

16. $y = (c_1 + c_2 x)\,e^{2x} + \dfrac{1}{8}\cos 2x$

17. $y = c_1 e^{3x} + c_2\cos 3x + c_3\sin 3x - \dfrac{x}{36}(\sin 3x + \cos 3x)$

18. $y = c_1 e^{2x} + c_2 e^{-2x} - \frac{1}{4}x^2 - \frac{1}{8}$

19. $y = c_1 e^{2x} + c_2 e^{3x} + \frac{1}{6}x + \frac{5}{36}$

20. $y = c_1 \cos 2x + c_2 \sin 2x + \frac{1}{4}\left(x^2 + \frac{5}{2}\right)$

21. $y = c_1 e^{-2x} + e^x (c_2 \cos \sqrt{3}\, x + c_3 \sin \sqrt{3}\, x) + \frac{1}{8}(x^4 - x + 1)$

22. $y = (c_1 + c_2 x) e^x + x + 1$

23. $y = c_1 + (c_2 x + c_3) x e^{-x} + \frac{x}{6}(2x^2 - 9x + 24) + \frac{e^{2x}}{19}$

24. $y = c_1 e^{-3x} + c_2 e^{2x} - \frac{x}{6} - \frac{1}{30}$

25. $y = e^x (c_1 \sin 2x + c_2 \cos 2x) - \frac{1}{10} e^{2x} (\cos x - 2 \sin x)$

26. $y = e^x \left(c_1 + c_2 x + c_3 x^2 + \frac{1}{6}x^3 + \frac{1}{24}x^4 \right)$

27. $y = c_1 e^{2x} + c_2 e^x - \frac{3e^{3x}}{4}(2x + 3)$

28. $y = (c_1 + c_2 x) e^x + c_3 e^x + e^x \left[\frac{x^2}{4} + \frac{1}{5}(\cos 3x - 2 \sin x) \right]$

29. $y = c_1 e^{8x} + c_2 e^{2x} + \frac{e^{4x}}{40}(\cos 2x - 3 \sin x)$

30. $y = c_1 e^x + c_2 e^{-x} + \frac{xe^x}{3}(9 - 3x + x^2)$

31. $y = e^{-x}(c_1 \cos x + c_2 \sin x) - x e^{-x} \cos x$

32. $y = (c_1 + c_2 x) e^x + \frac{e^{3x}}{4}\left[x^2 - 2x + \frac{3}{2} \right]$

2.15 Homogeneous Linear Differential Equations

A Linear Differential Equation of the type,

$$x^n \frac{d^n y}{dx^n} + a_1 x^{n-1} \frac{d^{n-1} y}{dx^{n-1}} + a_2 x^{n-2} \frac{d^{n-2} y}{dx^{n-2}} + \ldots\ldots + a_n = f(x) \quad \ldots (1)$$

where $a_1, a_2, a_3, \ldots\ldots, a_n$ are constants and $f(x)$ is a function of x, is called a **Cauchy's Homogeneous Differential Equation of order n.**

This differential equation (1) can be transferred into a linear differential equation with constant coefficients by a suitable substitution.

The equation (1) can be written as

$$(x^n D^n + a_1 x^{n-1} D^{n-1} + a_2 x^{n-2} D^{n-2} + \ldots\ldots + a_n) y = f(x) \quad \ldots (2)$$

Put, $z = \log_e x$, i.e. $x = e^z \Rightarrow \frac{dz}{dx} = \frac{1}{x}$

Now, $\frac{dy}{dx} = \frac{dy}{dz} \cdot \frac{dz}{dx} \Rightarrow \frac{dy}{dx} = \frac{1}{x} \cdot \frac{dy}{dz} \Rightarrow x \cdot \frac{dy}{dx} = \frac{dy}{dz} \Rightarrow x . \frac{d}{dx} = \frac{d}{dz}$

i.e. $x D = D_1$, where $D_1 = \frac{d}{dz}$.

Again, $\frac{d^2y}{dx^2} = \frac{d}{dx}\left(\frac{dy}{dx}\right)$

$= \frac{1}{x}\cdot\frac{d}{dz}\left(\frac{1}{x}\cdot\frac{dy}{dz}\right)$ $\left(\because \frac{d}{dx} = \frac{1}{x}\frac{d}{dz}\right)$

$= \frac{1}{x^2}\cdot\frac{d^2z}{dz^2} - \frac{1}{x^2}\cdot\frac{dy}{dz} = \frac{1}{x^2}\left(\frac{d^2y}{dz^2} - \frac{dy}{dz}\right)$

$\therefore$ $x^2\frac{d^2y}{dx^2} = \frac{d^2y}{dz^2} - \frac{dy}{dz} \Rightarrow x^2 D^2 y = (D_1^2 - D_1)\,y$ $\left(\text{where } D_1^2 = \frac{d^2}{dz^2}\right)$

i.e. $x^2 D^2 = D_1 (D_1 - 1)$.

Similarly, $x^3 D^3 = D_1 (D_1 - 1)(D_1 - 2)$ and so on.

Hence the equation becomes,

$$F(D_1)\,y = f_1(z) \qquad \text{... (3)}$$

where $f_1(z)$ is the value of $f(x)$ in terms of z.

The equation (3) is now clearly is a linear differential equation with constant coefficients and therefore can be solved by the methods seen earlier.

Example 1. Solve : $4x^2\frac{d^2y}{dx^2} + 4x\frac{dy}{dx} - y = 4x^2$.

Solution : The given equation is,

$$4x^2\frac{d^2y}{dx^2} + 4x\frac{dy}{dx} - y = 4x^2 \Rightarrow (4x^2 D^2 + 4x D - 1)\,y = 4x^2$$

Put $x = e^z$; i.e. $z = \log x$, then

$$x^2 D^2 = D_1 (D_1 - 1) \text{ and } x D = D_1 \qquad \left(\text{where } D_1 = \frac{d}{dz}\right)$$

$\therefore$ the equation becomes,

$$[4D_1 (D_1 - 1) + 4 D_1 - 1]\,y = 4\,e^{2z} \Rightarrow (4D_1^2 - 1)\,x = 4e^{2z}$$

which is a linear equation with constant coefficients.

The auxiliary equation is, $4m^2 - 1 = 0 \Rightarrow (2m - 1)(2m + 1) = 0 \Rightarrow$ the roots are $\frac{1}{2}, -\frac{1}{2}$.

$\therefore$ C.F. $= c_1 e^{1/2z} + c_2 e^{-1/2z}$

$\therefore$ P.I. $= \frac{1}{4D_1^2 - 1}4e^{2z} = 4\cdot\frac{1}{15}\cdot e^{2z} = \frac{4}{15}e^{2z}$

$\therefore$ the solution is

$$y = c_1 e^{1/2z} + c_2 e^{-1/2z} + \frac{4}{15}e^{2z}$$

$$\Rightarrow \quad y = c_1\sqrt{x} + c_2\frac{1}{\sqrt{x}} + \frac{4}{15}x^2 \qquad [\because x = e^z]$$

Example 2. Solve : $x^2\frac{d^2y}{dx^2} + x\frac{dy}{dx} + 9y = 3x^2 + \sin(3\log x)$.

Solution : The given equation is,

$$x^2\frac{d^2y}{dx^2} + x\frac{dy}{dx} + 9y = 3x^2 + \sin(3\log x)$$

$\Rightarrow$ $(x^2 D^2 + x D + 9)\, y = 3x^2 + \sin(3 \log x)$

Put, $x = e^z$; i.e. $z = \log x$

$\therefore$ $x^2 D^2 = D_1 (D_1 - 1)$ and $x D = D_1$ $\left(\text{where } D_1 = \dfrac{d}{dz}\right)$

$\therefore$ the equation becomes,

$$[D_1 (D_1 - 1) + D_1 + 9]\, y = 3 e^{2z} + \sin(3z) \Rightarrow (D_1^2 + 9)\, y = 3 e^{2z} + \sin(3z)$$

$\therefore$ the auxiliary equation is, $m^2 + 9 = 0 \Rightarrow m = 3i, -3i$

$\therefore$ C.F. $= c_1 \cos 3z + c_2 \sin 3z$

Now, P. I. $= \dfrac{1}{D_1^2 + 9}(3 e^{2z} + \sin 3z) = \dfrac{1}{D_1^2 + 9} 3e^{2z} + \dfrac{1}{D_1^2 + 9} \sin 3z$

Consider, $\dfrac{1}{D_1^2 + 9} 3e^{2z} = 3 \cdot \dfrac{1}{4 + 9} e^{2z} = \dfrac{3}{13} e^{2z}$

Consider, $\dfrac{1}{D_1^2 + 9} \sin 3z$. Here $D_1^2 + 9 = 0$, when D_1^2 is replaced by $-(3)^2$

$\therefore$ we shall consider,

$$\frac{1}{D_1^2 + 9} \sin 3z = \text{Imaginary part of } \frac{1}{D_1^2 + 9} e^{3iz} \quad [\because e^{3iz} = \cos 3z + i \sin 3z]$$

Now, $\dfrac{1}{D_1^2 + 9} e^{3iz} = \dfrac{1}{(D_1 - 3i)(D_1 + 3i)} e^{3iz}$

$$= \frac{1}{6!} \cdot \frac{x}{1!} \cdot e^{3iz} = \frac{z}{6!} (\cos 3z + i \sin 3z)$$

$\therefore$ $\dfrac{1}{D_1^2 + 9} \sin 3z = -\dfrac{z}{6} \cos 3z$. Thus, P. I. $= \dfrac{3}{13} e^{2z} - \dfrac{z}{6} \cos 3z$

Hence the solution is,

$$y = c_1 \cos 3z + c_2 \sin 3z + \frac{3}{13} e^{2z} - \frac{z}{6} \cos 3z$$

$$\mathbf{y = c_1 \cos(\log x^3) + c_2 \sin(\log x^3) + \frac{3}{13} x^2 - \frac{\log x}{6} \cos(\log x^3).}$$

Example 3. Solve : $\mathbf{x^3 \dfrac{d^2y}{dx^3} + 2x \dfrac{dy}{dx} - 2y = x^2 \log x + 3x}$.

Solution : The given equation is,

$$x^3 \frac{d^3y}{dx^3} + 2x \frac{dy}{dx} - 2y = x^2 \log x + 3x$$

$\Rightarrow$ $(x^3 D^3 + 2 x D - 2)\, y = x^2 \log x + 3x$

Put, $x = e^z$; i.e. $z = \log x$, then

$x^3 D^3 = D_1 (D_1 - 1)(D_1 - 2)$ and $x D = D_1$ $\left(\text{where } D_1 = \dfrac{d}{dz}\right)$

$\therefore$ the equation becomes,

$$[D_1 (D_1 - 1)(D_1 - 2) + 2 D_1 - 2]\, y = e^{2z} \cdot z + 3 e^z$$

$\Rightarrow$ $[D_1^3 - 3 D_1^2 + 4 D_1 - 2]\, y = e^{2z} \cdot z + 3 e^z$

The auxiliary equation is, $m^3 - 3m^2 + 4m - 2 = 0$

Clearly $m = 1$ is a root of the equation.

$\therefore \quad (m-1)(m^2-2m+2)=0 \Rightarrow m=1, 1\pm i$

$\therefore \quad \text{C.F.} = c_1 e^z + e^z (c_2 \cos z + c_3 \sin z)$

$$\therefore \quad \text{P. I.} = \frac{1}{D_1^3 - 3D_1^2 + 4D_1 - 2}(e^{2z}.z + 3e^z)$$

$$= \frac{1}{D_1^3 - 3D_1^2 + 4D_1 - 2} e^{2z}.z + \frac{1}{D_1^3 - 3D_1^2 + 4D_1 - 2} 3e^z$$

Now consider,

$$\frac{1}{D_1^3 - 3D_1^2 + 4D_1 - 2} e^{2z}(z) = e^{2z}\frac{1}{(D_1+2)^3 - 3(D_1+2)^2 + 4(D_1+2) - 2}(z)$$

$$= e^{2z}\frac{1}{D_1^3 + 3D_1^2 + 4D_1 + 2}(z)$$

$$= \frac{e^{2z}}{2}\frac{1}{\left(1 + \frac{D_1^3 + 3D_1^2 + 4D_1}{2}\right)}(z)$$

$$= \frac{e^{2z}}{2}\left(1 + \frac{D_1^3 + 3D_1^2 + 4D_1}{2}\right)^{-1}(z)$$

$$= \frac{e^{2z}}{2}\left(1 - \frac{D_1^3 + 3D_1^2 + 4D_1}{2} + \ldots\ldots\right)(z)$$

$$= \frac{e^{2z}}{2}\left(z - \frac{4}{2}\right) = \frac{e^{2z}}{2}(z-2)$$

Now, $$\frac{1}{D_1^3 - 3D_1^2 + 4D_1 - 2} 3e^z = 3\frac{1}{(D_1-1)(D^2-2D+2)}.e^z = \frac{3}{1}.\frac{z}{1!}.e^z = 3\,ze^z$$

Hence the solution is,

$$y = c_1 e^z + e^z (c_2 \cos z + c_3 \sin z) + \frac{e^{2z}}{2}(z-2) + 3z e^z$$

$$\Rightarrow \quad \mathbf{y = c_1 x + x [c_2 \cos (\log x) + c_3 \sin (\log x)] - \frac{1}{2} x^2 (\log x - 2) + 3x \log x}$$

Note : The differential equation of the form

$$(a+bx)^n D^n y + a_1 (a+bx)^{n-1} D^{n-1} y + \ldots\ldots + a_n y = f(x)$$

can be reduced to the homogeneous linear form by putting $z = a + bx$.

Example 4. Solve : $(1+x)^2 \frac{d^2y}{dx^2} + (1+x)\frac{dy}{dx} + y = 4 \cos \log (1+x)$.

Solution : The given equation is

$$(1+x)^2 \frac{d^2y}{dx^2} + (1+x)\frac{dy}{dx} + y = 4 \cos \log (1+x)$$

Put, $1 + x = z; \Rightarrow \frac{dz}{dx} = 1$

Now, $$\frac{dy}{dx} = \frac{dy}{dz}.\frac{dz}{dx} \quad \Rightarrow \quad \frac{dy}{dx} = \frac{dy}{dz}.1$$

Now, $$\frac{d^2y}{dx^2} = \frac{d}{dx}\left(\frac{dy}{dx}\right) \Rightarrow \frac{d^2y}{dx^2} = \frac{d}{dz}\left(\frac{dy}{dz}\right) = \frac{d^2y}{dz^2} \qquad \left(\text{since } \frac{dz}{dx} = 1\right)$$

$\therefore$ equation becomes, $z^2 \dfrac{d^2y}{dz^2} + z\dfrac{dy}{dz} + y = 4 \cos \log z$

This is a homogeneous linear equation

Put, $z = e^u$; i.e. $u = \log z$

$$\therefore \quad z^2 D^2 = D_1 (D_1 - 1) \text{ and } z D = D_1 \qquad \left(\text{Here } D = \frac{d}{dz} \text{ and } D_1 = \frac{d}{du}\right)$$

$\therefore$ equation further takes the form,

$$[D_1 (D_1 - 1) + D_1 + 1]\, y = 4 \cos u \Rightarrow (D_1^2 + 1)\, y = 4 \cos u$$

The auxiliary equation is, $m^2 + 1 = 0 \Rightarrow m = \pm i$

$\therefore$ C.F. $= c_1 \cos u + c_2 \sin u$

$$\therefore \quad \text{P. I.} = \frac{1}{D_1^2 + 1} 4 \cos u$$

$$= 4 . \frac{1}{D_1^2 + 1} \cos u = 4 . \frac{u}{2} \sin u = 2\, u \sin u \qquad \left(\because \frac{1}{D^2 + a^2} \cos ax = \frac{x}{2a} \sin ax\right)$$

$\therefore$ the solution is,

$$y = c_1 \cos u + c_2 \sin u + 2\, u \sin u$$

$$\Rightarrow \quad y = c_1 \cos (\log z) + c_2 \sin (\log z) + 2 (\log z) \sin (\log z)$$

$$\Rightarrow \quad \mathbf{y = c_1 \cos [\log (1 + x)] + \sin [\log (1 + x)] [c_2 + 2 \log (1 + x)]}$$

Example 5. Solve : $\mathbf{(2x - 1)^3 \dfrac{d^3y}{dx^3} + (2x - 1) \dfrac{dy}{dx} - 2y = 0}$**.**

Solution : The given equation is

$$(2x - 1)^3 \frac{d^3y}{dx^3} + (2x - 1) \frac{dy}{dx} - 2y = 0$$

Put, $2x - 1 = z; \Rightarrow \dfrac{dz}{dx} = 2$

$$\therefore \quad \frac{dy}{dx} = 2\frac{dy}{dz}; \quad \frac{d^2y}{dx^2} = 2^2 \frac{d^2y}{dz^2}; \quad \frac{d^3y}{dx^3} = 2^3 \frac{d^3y}{dz^3}$$

$\therefore$ Equation becomes, $8\, z^3 \dfrac{d^3y}{dz^3} + 2z \dfrac{dy}{dz} - 2y = 0$

Put, $z = e^u \Rightarrow u = \log z \Rightarrow z^3 D^3 = D_1 (D_1 - 1) (D_1 - 2)$ and $z D = D_1$

Here, $D = \dfrac{d}{dz}$ and $D_1 = \dfrac{d}{du}$.

$\therefore$ we have,

$$[8 D_1 (D_1 - 1) (D_1 - 2) + 2D_1 - 2]\, y = 0 \Rightarrow (8 D_1^3 - 24 D_1^2 + 18 D_1 - 2)\, y = 0$$

The auxiliary equation is,

$8m^3 - 24 m^2 + 18m - 2 = 0 \Rightarrow 4m^3 - 12 m^2 + 9m - 1 = 0 \Rightarrow (m - 1) (4m^2 - 8m + 1) = 0$

$\therefore$ the roots are, $m = 1, 1 \pm \dfrac{\sqrt{3}}{2}$

$$\therefore \quad \text{C.F.} = c_1 e^u + c_2\, e^{(1+\sqrt{3/2})u} + c_3\, e^{(1-\sqrt{3/2})u}$$

Since the right hand side is function (of x) in the given equation is zero, the solution is

$$y = c_1 e^u + c_2\, e^{(1+\sqrt{3/2})u} + c_3\, e^{(1-\sqrt{3/2})u}$$

$$\Rightarrow \qquad y = c_1 z + c_2 z^{(1+\sqrt{3/2})} + c_3 z^{(1-\sqrt{3/2})}$$

$$\Rightarrow \qquad y = z\,[c_1 + c_2 z^{\sqrt{3/2}} + c_3 z^{\sqrt{-3/2}}]$$

$$\Rightarrow \qquad \mathbf{y = (2x-1)\,[c_1 + c_2 (2x-1)^{\sqrt{3/2}} + c_3 (2x-1)^{\sqrt{3/2}}]}$$

EXERCISES

Solve the following differential equations

1. $x^2 \dfrac{d^2y}{dx^2} + x\dfrac{dy}{dx} - 4y = x^2$

2. $x^2 \dfrac{d^2y}{dx^2} - 3x\dfrac{dy}{dx} + 4y = 2x^2$

3. $x^2 \dfrac{d^2y}{dx^2} + 2x\dfrac{dy}{dx} - 20y = (x+1)^2$

4. $x^2 \dfrac{d^2y}{dx^2} - 2x\dfrac{dy}{dx} + 2y = 2x^3$

5. $x^3 \dfrac{d^3y}{dx^3} + x^2 \dfrac{d^2y}{dx^2} = x$

6. $x^3 \dfrac{d^3y}{dx^3} + x\dfrac{dy}{dx} - y = 3x^4$

7. $(x^2 D^2 - 2x D + 2)y = \dfrac{1}{x}$

8. $x^3 \dfrac{d^3y}{dx^3} + 2x^2 \dfrac{d^2y}{dx^2} + 2y = 10\left(x + \dfrac{1}{x}\right)$

9. $x^2 \dfrac{d^2y}{dx^2} + 3x\dfrac{dy}{dx} + y = x + \dfrac{1}{x}$

10. $x^2 \dfrac{d^2y}{dx^2} - x\dfrac{dy}{dx} + 2y = x \log x$

11. $x^2 \dfrac{d^2y}{dx^2} + 5x\dfrac{dy}{dx} + 4y = x \log x$

12. $x^2 \dfrac{d^2y}{dx^2} + 2x\dfrac{dy}{dx} = \log x$

13. $(x^2 D^2 + 1)\, y = 3x^2$

14. $x^3 \dfrac{d^3y}{dx^3} + 3x^2 \dfrac{d^2y}{dx^2} + x\dfrac{dy}{dx} + y = x + \log x$

15. $\dfrac{d^3y}{dx^3} - \dfrac{4}{x}\dfrac{d^2y}{dx^2} + \dfrac{5}{x^2}\dfrac{dy}{dx} - \dfrac{2y}{x^3} = 1$ **[Hint :** Multiply throughout by x^3**]**

16. $(x^2 D^2 - 3xD + 4)y = x^3$

17. $x^2 \dfrac{d^2y}{dx^2} - x\dfrac{dy}{dx} + y = 2x \log x$, given that $y = 1$ and $\dfrac{dy}{dx} = 0$ when $x = 1$.

18. $(x^2 D^2 - 3x D + 5)\, y = x^2 \sin \log x$

19. $(x^2 D^2 - x D + 4)\, y = \cos(\log x) + x \sin(\log x)$

20. $(3x+2)^2 \dfrac{d^2y}{dx^2} + 3(3x+2)\dfrac{dy}{dx} - 36y = 3x^2 + 4x + 1$

21. $(5+2x)\dfrac{d^3y}{dx^3} - 6(5+2x)\dfrac{dy}{dx} + 8y = 0$

22. $(x+a)^2 \dfrac{d^3y}{dx^3} - 4(x+a)\dfrac{dy}{dx} + 6y = x$

23. $(1+2x)^2 \dfrac{d^3y}{dx^3} - 6(1+2x)\dfrac{dy}{dx} + 16y = 8(1+2x)^2$

Answers

1. $y = c_1 x^2 + c_2 x^{-2} + \dfrac{1}{4} x^2 \log x$

2. $y = x^2 [c_1 + c_2 \log x + (\log x)^2]$

3. $y = c_1 x^{-5} + c_2 x^4 - \left(\dfrac{x^2}{14} + \dfrac{x}{9} + \dfrac{1}{20}\right)$

4. $y = c_1 x + c_2 x^2 + x^3$

5. $y = c_1 + (c_2 + c_3 \log x)\, x + \dfrac{(\log x)^2}{2} x$

6. $y = x\,(c_1 + c_2 \log x) + c_3\, x^{-2}$

7. $y = c_1\, x + c_2\, x^2 + x^{-\frac{1}{6}}$

8. $y = c_1\, x^{-1} + x\,[c_2 \cos(\log x) + c_3 \sin(\log x)] + 5x + \dfrac{2(\log x)}{x}$

9. $y = (c_1 + c_2 \log x)\, x^{-1} + \dfrac{x}{4} + \dfrac{(\log x)^2}{2} \cdot \dfrac{1}{x}$

10. $y = x\,[c_1 \cos(\log x) + c_2 \sin(\log x) + x \log x]$

11. $y = x^{-2}(c_1 + c_2 \log x) + \dfrac{1}{9} x\,(\log x^{-\frac{2}{3}})$

12. $y = c_1 + c_2\, x^{-1} + \dfrac{1}{2}(\log x)^2 - \log x$

13. $y = \sqrt{x}\left[A\cos\left(\dfrac{\sqrt{3}}{2}\log x\right) + B\sin\left(\dfrac{\sqrt{3}}{2}\log x\right)\right] + x^2$

14. $y = c_1\, x^{-1} + c_2\, \sqrt{x}\cos\left(\dfrac{\sqrt{3}}{2}\log x + c_3\right) + \dfrac{1}{2}x + \log x$

15. $y = c_1\, x^2 + x^{\frac{5}{2}}\left(c_2 x^{\sqrt{21/2}} + c_2 x^{\sqrt{21/2}}\right) - \dfrac{1}{5}x^3$

16. $y = (c_1 + c_2 \log x)x^2 + \dfrac{1}{(m-2)} x^m$

17. $y = (1 - \log x)\, x + \dfrac{1}{3}\, x\,(\log x)^2$

18. $y = x^2\,[c_1 \cos(\log x) + c_2 \sin(\log x)] - \dfrac{x^2}{2} \log x \,.\, \cos(\log x)$

19. $y = x\,[c_1 \cos(\sqrt{3}\, \log x) + c_2 \sin(\sqrt{3}\, \log x)] + \dfrac{3}{5}\cos(\log x)$
$- \dfrac{2}{5}\sin(\log x) + \dfrac{1}{2} x \sin(\log x)$

20. $y = c_1\,(3x+2)^2 + c_2\,(3x+2)^{-2} + \dfrac{1}{100}\left[(3x+2)^2 \log(3x+2) + 1\right]$

21. $y = (5+2x)^2\left[c_1(5+2x)\sqrt{2} + c_2(5+2x) - \sqrt{2}\right]$

22. $y = c_1\,(x+a)^3 + c_2\,(x+a)^2 + \dfrac{x+a}{2} - \dfrac{a}{b}$

23. $y = \left[c_1 + c_2 \log(1+2x)\right](1+2x)^2 + (1+2x)^2\left[\log(1+2x)\right]^2$

2.16 Simultaneous Linear Differential Equations with constant coefficients

If the variables x and y are functions of `$t$' then $x$ and $y$ can be obtained in terms of `t', when two differential equations involving x, y and t are known. Such equations are called ordinary simultaneous differential equations. We shall consider only those equations which are linear with constant coefficients.

If $D = \dfrac{d}{dt}$ then such equations are given by

$$f_1(D)\, x + f_2(D)\, y = \varphi(t) \quad \text{... (1)}$$

$$g_1(D)\, x + g_2(D)\, y = \Psi(t) \quad \text{... (2)}$$

where $f_1(D), f_2(D), g_1(D)$ and $g_2(D)$ are functions of D with constant coefficients $\varphi(t)$ and $\Psi(t)$ functions of independent variable t.

Working rule for solving the set of simultaneous equation

1. Operate (1) with $g_2(D)$ and (2) with $f_2(D)$ and subtract.
2. Result of the first step would be
$[g_2(D) f_1(D) - f_2(D) g_1(D)]\, x = g_2(D)\, \varphi(t) - f_2(D)\, \Psi(t)$
This is clearly a linear differential equation with constant coefficients.
3. Solution of the equation obtained in step (3) be $x = f(t)$.
4. Substitute $x = f(t)$ in (1) or (2) to obtain y in terms of t, say $y = g(t)$
5. $x = f(t)$ and $y = g(t)$ constitute the complete solution.

Example 1. Solve the simultaneous equations

$$\frac{dx}{dt} + 7x - y = 0 \quad \text{and} \quad \frac{dy}{dt} + 2x + 5y = 0.$$

Solution : The given equations can be written as

$$(D+7)\, x - y = 0 \qquad \text{... (1)}$$

$$(D+5)\, y + 2x = 0 \qquad \text{... (2)}$$

Operate (1) with $(D + 5)$, we get

$$(D+7)(D+5)x - (D+5)\, y = 0 \qquad \text{... (3)}$$

Consider (2) + (3), then we get,

$$(D+7)(D+5)\, x + 2x = 0 \Rightarrow (D^2 + 12\, D + 37)\, x = 0$$

This is clearly a linear differential equation with constant coefficient.

The auxiliary equation is, $m^2 + 12\, m + 37 = 0 \Rightarrow m = \dfrac{-12 \pm \sqrt{-4}}{2} = -6 \pm i$

$\therefore$ the general solution is, $\boldsymbol{x = e^{-6t}(c_1 \cos t + c_2 \sin t)}$ ... (4)

Now consider,

$$Dx = e^{-6t}(-c_1 \sin t + c_2 \cos t) - 6\, e^{-6t}(c_1 \cos t + c_2 \sin t)$$

(1) $\Rightarrow$ $Dx + 7x - y = 0$

$\Rightarrow$ $y = 7x + D\,x$

$\Rightarrow$ $y = 7\,[e^{-6t}(c_1 \cos t + c_2 \sin t)] + e^{-6t}(-c_1 \sin t + c_2 \cos t)$
$\qquad - 6\, e^{-t}(c_1 \cos t + c_2 \sin t)$

$\therefore$ $\boldsymbol{y = e^{-6t}[(c_1 + c_2) \cos t - (c_1 - c_2) \sin t]}$ ... (5)

The relations (4) and (5) constitute the complete solution of the given equation.

Example 2. Solve the simultaneous equations

$$D^2 x - 3x - y = e^t \quad \text{and} \quad Dy - 2x = 0.$$

Solution : The given equations can be written as

$$(D^2 - 3)\, x - y = e^t \qquad \text{... (1)}$$

$$D\, y - 2x = 0 \qquad \text{... (2)}$$

Operate (1) with D, we get

$$(D^2 - 3)\, Dx - Dy = De^t \qquad \text{... (3)}$$

Consider (2) + (3), then we get,

i.e. $\quad (D^2 - 3)\, D\, x - 2x = e^t$ ... (4)

This is clearly a linear differential equation with constant coefficient.

The auxiliary equation is, $m^3 - 3\, m - 2 = 0 \Rightarrow (m-2)(m+1)^2 = 0 \Rightarrow m = 2, -1, -1$

$\therefore$ C.F. is $(c_1 + c_2\, t)\, e^{-t} + c_3\, e^{2t}$

Now, P.I. $= \frac{1}{D^3 - 3D - 2} e^t = -\frac{1}{4} e^t$

$\therefore$ the solution of (4) is, $x = (c_1 + c_2 t) e^{-t} + c_3 e^{2t} - \frac{1}{4} e^t$ ... (5)

Now, (2) $\Rightarrow Dy = 2x$

$$\Rightarrow \quad Dy = 2\left[(c_1 + c_2 t) e^{-t} + c_3 e^{2t} - \frac{1}{4} e^t\right]$$

$$\Rightarrow \quad y = \frac{1}{D}\left\{2\left[(c_1 + c_2 t) e^{-t} + c_3 e^{2t} - \frac{1}{4} e^t\right]\right\}$$

$$y = 2\left[-(c_1 + c_2 t) e^{-t} - c_2 e^{-t} + \frac{c_3}{2} e^{2t} - \frac{1}{4} e^t\right]$$

$$y = -2\left[(c_1 + c_2 + c_2 t) e^{-t} - \frac{c_3}{2} e^{2t} - \frac{1}{4} e^t\right] \quad \text{... (6)}$$

The relations (5) and (6) constitute the complete solution of the given equation.

EXERCISES

Solve the following differential equations; $D = \frac{d}{dt}$

1. $\frac{dx}{dt} = 3x - y; \quad \frac{dy}{dt} = x + y$
2. $\frac{dx}{dt} - 2x + 2\frac{dy}{dt} = 2 - 4e^{2x}; \quad 2\frac{dx}{dt} - 3x + 3\frac{dy}{dt} = y$
3. $(3D + 2)\, x + (D - 6)\, y = 5\, e^x; \ (4D + 2)\, x + (D - 8)\, y = 5\, e^x + 2x - 2$
4. $Dx = 3x + 2y; \quad Dy + 5x + 3y = 0$
5. $\frac{d^2x}{dt^2} - 3x - 4y = 0; \quad \frac{d^2x}{dt^2} + x + y = 0$
6. $(D^2 + 6)\, y + D\, x = 0; \ (D + 2)\, y + (D - 2)\, x = 2$
7. $\frac{dx}{dt} + x = y + e^t; \quad \frac{dy}{dt} + y = x + e^t$
8. $D^2 y - (2D - 1)\, x = 1; \ (2D + 1)\, y + (D^2 - 4)\, x = 0$
9. $(D^2 - 3D)\, x - (D - 2)\, y = 14x + 7; \ (D - 3)\, x + D\, y = 1$

Answers

1. $x = (c_1 + c_2 t)e^{2t}; \quad y = (c_1 - c_2 + c_2 t)\, e^{2t}$
2. $x = c_1 e^t + c_2 e^{-2t} + 5\, e^{2t} - 1, \ y = \frac{1}{2} c_1 e^t - c_2 e^{-2t} - e^{2t} + 3$
3. $x = c_1 \cos 2t + c_2 \sin 2t + 2\, e^t - 3t + 5, \ y = c_2 \sin 2t + c_1 \cos 2t + e^t - x$
4. $x = c_1 \cos t + c_2 \sin t, \ y = \frac{1}{2}[(c_2 - 3c_1) \cos t - (c_1 + 3c_2) \sin t]$
5. $x = (c_1 + c_2 t)\, e^t + (c_3 + c_4 t)\, e^{-t}, \ y = \frac{1}{2}\ (c_2 - c_1 - c_2 t)e^t - \frac{1}{2}\ (c_4 + c_3 + c_4 t)e^{-t}$
6. $x = -1 - 5c_1 e^{3t} + c_3 \cos 2t - c_2 \sin 2t, \ y = c_1 e^{3t} + c_2 \cos 2t + c_3 \sin 2t$
7. $x = c_1 + c_2 e^{-2t} + e^t; \quad y = c_1 - c_2 e^{-2t} + e^t$

8. $x = 1 + c_1 e^t + c_2 e^{-t} + c_3 \cos t + c_4 \sin t,$

$y = 4 + c_1 e^t - 3c_2 e^{-t} + (c_3 - 2c_4) \cos t + (2c_3 + c_4) \sin t$

9. $x = 2 + c_1 e^t + c_2 e^{3t} + c_3 e^{-2t}, \quad y = 7 + 7t + 2c_1 e^t - \frac{5}{2} c_3 e^{-2t}$

2.17 Linear Equations of Second order with Variable Coefficients

The readers are familiar with some methods of solving the linear differential equations with constant coefficients. In this section we shall see few methods of solving the linear differential equations of second order with variable coefficients.

The general form of the linear equation of second order with variable coefficients is of the form

$$\frac{d^2y}{dx^2} + P\frac{dy}{dx} + Qy = R \quad \text{or} \quad y_2 + P y_1 + Qy = R$$

where P, Q and R functions of x.

In this text we discuss the following methods of solving the linear differential equations of second order with variable coefficients.

1. **Complete solution in terms of known integral belonging to complementary function.**
2. **Method of changing Independent Variable.**
3. **Method of changing Dependent Variable. (Reduction to normal form).**
4. **Method of variation of Parameters.**
5. **Method by finding a First Integral. (Exact equation)**

2.18 Complete Solution in terms of Known Integral belonging to complementary Functions.

Consider the equation,

$$\frac{d^2y}{dx^2} + P\frac{dy}{dx} + Qy = R \qquad \dots (1)$$

Let $y = u$ be a known integral in the complementary function (C.F.) of (1). i.e. $y = u$ is a solution of (1), when its right side is taken to be zero.

Thus

$$\frac{d^2u}{dx^2} + P\frac{du}{dx} + Qu = 0 \qquad \dots (2)$$

Let the complete solution of (1) be $y = u \,.\, v$ where v has to be determined which is a function of x.

Consider $\quad y = u \,.\, v \Rightarrow \dfrac{dy}{dx} = u\dfrac{dv}{dx} + v\dfrac{du}{dx} \Rightarrow \dfrac{d^2y}{dx^2} = u\dfrac{d^2v}{dx^2} + 2\dfrac{du}{dx}\,.\,\dfrac{dv}{dx} + v\dfrac{d^2u}{dx^2}$

Now putting, $y = u \,.\, v$ in (1), we get,

$$\frac{d^2}{dx^2}(u\,v) + P\frac{d}{dx}(u\,v) + Q\,(u\,.\,v) = R$$

$$\Rightarrow \quad \left(u\frac{d^2v}{dx^2} + 2\frac{du}{dx}\,.\,\frac{dv}{dx} + v\frac{d^2u}{dx^2}\right) + P\left(u\frac{dv}{dx} + v\frac{du}{dx}\right) + Q\,u\,.\,v = R$$

$$\Rightarrow \quad u\frac{d^2v}{dx^2} + \frac{dv}{dx}\left(2\frac{du}{dx} + Pu\right) = R \qquad \dots (3)$$

$$\left[\because \frac{d^2u}{dx^2} + p\frac{du}{dx} + Qu = 0 \text{ from (2)}\right]$$

Put, $\frac{dv}{dx} = z \quad \therefore \quad \frac{d^2v}{dx^2} = \frac{dz}{dx}$

Thus (3) becomes,

$$u\frac{dz}{dx} + \left(2\frac{du}{dx} + Pu\right)z = R \Rightarrow \frac{dz}{dx} + \left(\frac{2}{u}\frac{du}{dx} + P\right)z = \frac{R}{u} \quad \text{... (4)}$$

This is a first order linear differential equation. Its integrating factor is given by,

$$\text{I.F.} = e^{\int\left(\frac{2}{u}\cdot\frac{du}{dx} + P\right)dx} = e^{2\log u + \int P\,dx} = u^2 e^{\int P\,dx}$$

$\therefore$ Solution of (4) is

$$z\,.\,u^2 e^{\int P\,dx} = \int \frac{R}{u}\,.\,u^2 e^{\int P\,dx}\,dx + C_1$$

$$\Rightarrow \quad z = \frac{e^{-\int P\,dx}}{u^2}\left[\int R\,.\,u\,e^{\int P\,dx}\,dx + C_1\right]$$

$$\Rightarrow \quad \frac{dv}{dx} = \frac{e^{-\int P\,dx}}{u^2}\left[\int R\,u\,e^{\int P\,dx}\,dx + C_1\right] \quad \text{... (5)}$$

Integrating both sides w.r.t. x, we get v as a function of x. Putting this value of v in $y = uv$ we get the complete solution of (1).

Note 1. Since the given differential equation is of second order, it will have two arbitrary constants.

2. If in the given equation the r.h.s. function is zero. i.e. $R = 0$, the equation (5) becomes,

$$\frac{dv}{dx} = C_1\left[\frac{e^{-\int P\,dx}}{u^2}\right] \Rightarrow v = C_1\int\left[\frac{1}{u^2}\,.\,e^{-\int P\,dx}\right]dx + C_2$$

Thus in this case the complete solution will be

$$\boldsymbol{y = u\,v = C_2\,u + C_1\,u\int\left[\frac{1}{u^2}\,.\,e^{-\int P\,dx}\right]dx\,.}$$

Thus in the case of differential equation, when r.h.s. function is zero, the other part of the C.F. can be obtained interms of the given part of C.F. by using the following formula.

$$\boldsymbol{u\int\left[\frac{1}{u^2}\,.\,e^{-\int P\,dx}\right]dx}$$

where u is the given part of C.F. The usefulness of this can be seen in solving the equations using the variation of parameters.

Working Rule for finding Complete Solution when an Integral of C.F. is known.

(a) Write the given equation in the standard form

$$\frac{d^2y}{dx^2} + P\frac{dy}{dx} + Qy = R$$

(i.e. making the coefficients of $\frac{d^2y}{dx^2}$ as unity)

(b) If an integral u of C.F. is given, assume $y = uv$ to be complete solution of the given equation.

(c) Put $y = u\,v$ in the given equation, then the equation reduces to the form

$$\frac{d^2v}{dx^2} + \left(P + \frac{2}{u}\frac{du}{dx}\right)\frac{dv}{dx} = \frac{R}{u}.$$

(d) Put $\frac{dv}{dx} = z$ and solve the linear differential equation obtained. (if $R = 0$ the equation reduces to the "variable separable" equation and hence solve.)

(e) In the obtained solution put $\frac{dv}{dx} = z$ and integrate w.r.t. x, then we get v. Thus $y = u\,v$ is the required complete solution of (1).

Note A. If an integral of C.F. is not given, one can obtain it in certain cases by using the following table.

(i) If $1 + P + Q = 0$, then an integral of C.F. is e^x.

(ii) If $1 - P + Q = 0$, then an integral of C.F. is e^{-x}.

(iii) If $a^2 + a\,P + Q = 0$, then an integral of C.F. is e^{ax}.

(iv) If $P + Q\,x = 0$, then an integral of C.F. is x.

(v) If $2 + 2\,P\,x + Q\,x^2 = 0$, then an integral of C.F. is x^2.

Example 1. Solve : $x\frac{d^2y}{dx^2} - 2(x+1)\frac{dy}{dx} + (x+2)\,y = (x-2)\,e^{2x};\quad (x > 0)$

Solution : Since $x > 0$, dividing throughout by x, we have,

$$\frac{d^2y}{dx^2} - 2\left(1+\frac{1}{x}\right)\frac{dy}{dx} + \left(1+\frac{2}{x}\right)y = \left(1-\frac{2}{x}\right)e^{2x} \qquad \text{... (1)}$$

Here, $P = -2\left(1+\frac{1}{x}\right);\; Q = \left(1+\frac{2}{x}\right);\; R = \left(1-\frac{2}{x}\right)e^{2x}$

Consider, $1 + P + Q = 1 - 2 - \frac{2}{x} + 1 + \frac{2}{x} = 0 \Rightarrow y = e^x$ is an integral of (1)

Let the complete solution of (1) be $y = e^x\,v$, where v is a function of x, which has to be determined.

Putting, $y = e^x\,v$ in (1) we get, $\frac{d^2v}{dx^2} + P_1\frac{dv}{dx} = R_1 \qquad \text{... (2)}$

where, $P_1 = P + \frac{2}{u}\frac{du}{dx}, \quad = -2\left(1+\frac{1}{x}\right) + 2e^{-x}.e^{x} = -\frac{2}{x} \qquad (\because u = e^x)$

and, $R_1 = \frac{R}{u} = \left(1-\frac{2}{x}\right)\frac{e^{2x}}{e^x} = \left(1-\frac{2}{x}\right)e^x$

$\therefore$ (2) becomes, $\frac{d^2v}{dx^2} - \frac{2}{x}\frac{dv}{dx} = \left(1-\frac{2}{x}\right)e^x$

Put, $\frac{dv}{dx} = z$, so that $\frac{d^2v}{dx^2} = \frac{dz}{dx}$. Thus (2) becomes,

$$\frac{dz}{dx} - \frac{2}{x}z = \left(1-\frac{2}{x}\right)e^x \qquad \text{... (3)}$$

which is a linear differential equation of order one.

Now, I.F. $= e^{\int -\frac{2}{x}dx} = e^{-2\log x} = \frac{1}{x^2}$. Thus solution of (3) is

$$z \cdot \frac{1}{x^2} = \int \frac{1}{x^2}\left(1 - \frac{2}{x}\right) e^x \, dx + C_1$$

$$z \cdot \frac{1}{x^2} = \int (x^{-2} \cdot e^x - 2\, x^{-3}\, e^x)\, dx + C_1$$

$$z \cdot \frac{1}{x^2} = \int x^{-2} \cdot e^x \, dx + x^{-2}\, e^x - \int x^{-2} \cdot e^x \, dx + C_1 \qquad \text{[Integrating } - 2x^{-3} e^x \text{ by parts]}$$

$$z \cdot \frac{1}{x^2} = x^{-2} \cdot e^x + C_1 \Rightarrow z = e^x + C_1 x^2 \Rightarrow \frac{dv}{dx} = e^x + C_1 x^2$$

On integration, we have

$$v = \int (e^x + C_1 x^2)\, dx + C_2 \Rightarrow v = e^x + C_1 \frac{x^3}{3} + C_2$$

Hence the complete solution of (1) is

$$y = e^x \left[e^x + \frac{C_1 x^2}{3} + C_2 \right] \Rightarrow \boldsymbol{y = e^{2x} + \frac{C_1}{3} x^2 e^x + C_2 e^x}$$

Example 2. Solve : $x^2 y'' + xy' - y = 2x^2$ $(x > 0)$, given that $\frac{1}{x}$ is a part of the complementary function and $y(1) = y'(1) = 0$.

Solution : Since $x > 0$, dividing throughout by x^2, we have

$$y'' + \frac{1}{x} y' - \frac{1}{x^2} = 2 \qquad \dots (1)$$

Here it is given $y = \frac{1}{x}$ is a part of the complementary function.

Putting $y = \frac{1}{x}\, v$ in (1), the equation becomes,

$$\frac{d^2v}{dx^2} + \left(P + \frac{2}{u}\frac{du}{dx}\right)\frac{dv}{dx} = \frac{R}{u}, \qquad \left(\text{where } u = \frac{1}{x}\right)$$

Now, $\quad P + \frac{2}{u}\frac{du}{dx} = \frac{1}{x} + 2x\left(-\frac{1}{x^2}\right) = -\frac{1}{x}$ and $\frac{R}{u} = 2x$

$\therefore$ we have

$$\frac{d^2v}{dx^2} - \frac{1}{x}\frac{dv}{dx} = 2x \qquad \dots (2)$$

Put, $\quad \frac{dv}{dx} = z. \Rightarrow \frac{d^2v}{dx^2} = \frac{dz}{dx}.$

Thus, equation (2) becomes

$$\frac{dz}{dx} - \frac{1}{x} z = 2x$$

Now, I.F. $= e^{\int -\left(\frac{1}{x}\right) dx} = e^{-\log x} = \frac{1}{x}$

$\therefore$ Solution of (2) is

$$z \cdot \frac{1}{x} = \int 2x \cdot \frac{1}{x}\, dx + C_1 \Rightarrow z \cdot \frac{1}{x} = 2x + C_1 \Rightarrow z = 2x^2 + C_1 x$$

i.e. $\quad \frac{dv}{dx} = 2x^2 + C_1 x$

On integration, we have,

$$v = \int (2x^2 + C_1 x)\, dx + C_2 \Rightarrow v = \frac{2}{3}x^3 - C_1\frac{x^2}{2} + C_2$$

Hence the complete solution is,

$$y = \frac{1}{x}\left(\frac{2}{3}x^3 - C_1\frac{x^2}{2} + C_2\right)$$

$$y = \frac{2}{3}x^2 - \frac{C_1}{2}x + C_2\frac{1}{x} \Rightarrow y = \frac{2}{3}x^2 - C_1^* x + C_2\frac{1}{x} \qquad \left(\text{where } C_1^* = \frac{C_1}{2}\right)$$

By data, $y(1) = 0 \Rightarrow$ i.e. $\frac{2}{3} - C_1^* + C_2 = 0$ ... (3)

also by data $y'(1) = 0$,

Now, $y' = \frac{4}{3}x - C_1^* - C_2\frac{1}{x^2} \Rightarrow \frac{4}{3} - C_1^* - C_2 = 0$ ... (4)

Solving (3) and (4) we get $C_1^* = \frac{3}{2}$ and $C_2 = \frac{1}{2}$

$\therefore$ **The required particular solution is, $y = x^2 - \frac{3}{2}x + \frac{1}{2x}$.**

Example 3. Solve $\frac{d^2y}{dx^2} + \frac{1}{x}\frac{dy}{dx} - \frac{1}{x^2}y = 0$, given that $x + \frac{1}{x}$ is a solution.

Solution : The given equation is, $\frac{d^2y}{dx^2} + \frac{1}{x}\frac{dy}{dx} - \frac{1}{x^2}y = 0$ ... (1)

By data $x + \frac{1}{x}$ is an integral. Putting $y = v\left(x + \frac{1}{x}\right)$ in (1), we get

$$\frac{d^2v}{dx^2} + \left(P + \frac{2}{u}\frac{du}{dx}\right)\frac{dv}{dx} = 0, \qquad \left(\text{where } u = x + \frac{1}{x}\right)$$

Now, $P + \frac{2}{u}\frac{du}{dx} = \frac{1}{x} + \frac{2x}{x^2+1}\left(1 - \frac{1}{x^2}\right) = \frac{3x^2 - 1}{x(x^2+1)}$

Thus we have,

$$\therefore \quad \frac{d^2v}{dx^2} + \frac{3x^2-1}{x(x^2+1)} \cdot \frac{dv}{dx} = 0 \qquad \text{... (2)}$$

Put, $\frac{dv}{dx} = z \Rightarrow \frac{d^2v}{dx^2} = \frac{dz}{dx}$. Thus equation (2) becomes

$$\frac{dz}{dx} + \frac{3x^2-1}{x(x^2+1)} \cdot z = 0$$

$$\frac{dz}{z} + \frac{3x^2-1}{x(x^2+1)} \cdot dx = 0 \Rightarrow \frac{dz}{z} + \left[-\frac{1}{x} + \frac{4x}{(x^2+1)}\right] dx = 0$$

Integrating we get,

$$\log z - \log x + 2\log(x^2+1) = \log C_1$$

$$\Rightarrow \quad z(x^2+1)^2 = C_1 x \Rightarrow z = \frac{C_1 x}{(x^2+1)^2}$$

Now, $\frac{dv}{dx} = \frac{C_1 x}{(x^2+1)^2} \Rightarrow dv = C_1\frac{x\,dx}{(x^2+1)^2} \Rightarrow \int dv = C_1\int\frac{x\,dx}{(x^2+1)^2} + C_2$

Putting, $x^2 + 1 = t$, we have $x\,dx = \frac{1}{2}\,dt$. Thus we have,

$$\int dv = \frac{C_1}{2}\int \frac{dt}{t^2} + C_2 \Rightarrow v = -\frac{C_1}{2}\frac{1}{t} + C_2 \Rightarrow v = -\frac{C_1}{2(x^2+1)} + C_2$$

∴ The complete solution is

$$y = v\left(x + \frac{1}{x}\right) = \left(x + \frac{1}{x}\right)\left(C_2 - \frac{C_1}{2(x^2+1)}\right)$$

$$\Rightarrow \quad y = C_2\left(\frac{x^2+1}{x}\right) - \frac{C_1}{2x}$$

$$\Rightarrow \quad y = \frac{1}{x}\left(C_2 - \frac{C_1}{2}\right) + C_2 x \Rightarrow \mathbf{y = C_1^* \frac{1}{x} + C_2 x} \quad \left(\text{where } C_1^* = C_2 - \frac{C_1}{2}\right)$$

EXERCISES

1. Solve $y'' + y' \cot x - y \operatorname{cosec}^2 x = 0$, given that $\cot x$ is a solution.
2. Solve $x y'' - (1 + x) y' + y = 0, x > 0$, given that $(x + 1)$ is a solution.
3. Solve $x \frac{d^2y}{dx^2} - (2x-1)\frac{dy}{dx} + (x - 1) y = 0, x > 0$, given that e^x is a solution.
4. Verify that $y = e^{2x}$ is a solution of $x y'' - (1 + x) y' + 2(1 - x) y = 0$, and obtain the general solution.
5. Solve $(x^2 + 1) y'' - 2 x y' + 2 y = 0$, given that $\cot x$ is a solution and $y = 1$ and $y' = 0$, when $x = 0$.
6. Solve $\frac{d^2y}{dx^2} - \cot x \frac{dy}{dx} - (1 - \cot x) y = e^x \sin x$, given that e^x is a part of complementary function and $y = (0) = y'(0) = -\frac{1}{2}$.
7. Solve the following equations by finding a part of the C.F. (using the rules given in Note A)

(i) $(1 - x) y'' + x y' - y = 0, x \neq 1$

(ii) $(x + 1) \frac{d^2y}{dx^2} - 2(x+3)\frac{dy}{dx} + (x + 5) y = e^x, x \neq 1$

(iii) $\frac{d^2y}{dx^2} - \cot x \frac{dy}{dx} - (1 - \cot x) y = e^x \sin x$

(iv) $x^2 \frac{d^2y}{dx^2} - 2x(1+x)\frac{dy}{dx} + 2(1 + x) y = x^3, x > 0$

(v) $x^2 \frac{d^2y}{dx^2} - (x^2 - 2x)\frac{dy}{dx} + (x - 2) y = x^2 e^x, x > 0$

(vi) $(x + 2) \frac{d^2y}{dx^2} - (2x+5)\frac{dy}{dx} + 2y = (x + 1) e^x$

(vii) $x y'' - 2(x + 1) x y'' + (x + 2)y = (x - 2) e^x, x \neq 0$ given that $y = e^2$ and $y' = 2e^2$ when $x = 1$

Answers

1. $y = C_1 \operatorname{cosec} x + C_2 \cot x$

2. $y = C_1 e^x + C_2 (1 + x)$

3. $y = C_1 e^x \log x + C_2 e^x$

4. $y = -\dfrac{C_1}{9}(3x+1)e^{-x} + C_2 e^{2x}$

5. $y = 1 - x^2$

6. $y = -\dfrac{1}{2}e^x \cos x + \dfrac{C_1}{\sqrt{5}} e^{-x}\left(x + \tan^{-1}\dfrac{1}{2}\right) + C_2 e^x$ and $C_1 : C_2 = -5 : 1$

7. **(i)** $y = C_1 e^x + C_2 x$

(ii) $y = \dfrac{1}{5} C_1 e^x (x+1)^5 + C_2 e^x - \dfrac{1}{4} x e^x$

(iii) $y = -\dfrac{C_1}{5} e^{-x}(2\sin x + \cos x) + C_2 e^x - \dfrac{1}{2} e^x \cos x$

(iv) $y = \dfrac{1}{2} C_1 x e^{2x} + C_2 x - \dfrac{1}{2} x^2$ **(v)** $y = C_1 x e^x + C_2 x (x-1) e^x$

(vi) $y = -\dfrac{C_1}{4}(2x+5) + C_2 e^{2x} - e^x$ **(vii)** $y = e^{2x}$

2.19 Method of Changing the Independent Variable

Consider the equation

$$\frac{d^2y}{dx^2} + P\frac{dy}{dx} + Qy = R \qquad \text{... (1)}$$

Now let us change the independent variable x to z, where z is a function of x, i.e. $z = f(x)$ (say).

We have,

$$\frac{dy}{dx} = \frac{dy}{dz}\cdot\frac{dz}{dx}$$

and,

$$\frac{d^2y}{dx^2} = \frac{d}{dx}\left(\frac{dy}{dz}\cdot\frac{dz}{dx}\right) = \frac{dy}{dz}\cdot\frac{d^2z}{dx^2} + \frac{d^2y}{dz^2}\left(\frac{dz}{dx}\cdot\frac{dz}{dx}\right) = \frac{d^2y}{dx^2}\left(\frac{dz}{dx}\right)^2 + \frac{dy}{dz}\left(\frac{d^2z}{dx^2}\right)$$

The equation (1) becomes,

$$\frac{d^2y}{dx^2}\left(\frac{dz}{dx}\right)^2 + \frac{dy}{dz}\cdot\frac{d^2z}{dx^2} + P\frac{dy}{dz}\cdot\frac{dz}{dx} + Q\,y = R$$

$$\Rightarrow \quad \frac{d^2y}{dx^2}\left(\frac{dz}{dx}\right)^2 + \left(\frac{d^2z}{dx^2} + P\cdot\frac{dz}{dx}\right)\frac{dy}{dx} + Q\,y = R$$

Dividing throughout by $\left(\dfrac{dz}{dx}\right)^2$, we get,

$$\frac{d^2y}{dx^2} + P_1\frac{dy}{dx} + Q_1 y = R_1 \qquad \text{... (2)}$$

$$P_1 = \frac{\dfrac{d^2z}{dx^2} + P\dfrac{dz}{dx}}{\left(\dfrac{dz}{dx}\right)^2} \qquad Q_1 = \frac{Q}{\left(\dfrac{dz}{dx}\right)^2} \qquad R_1 = \frac{R}{\left(\dfrac{dz}{dx}\right)^2}$$

Here P_1, Q_1 and R_1 are functions of x. However, these can be converted to functions of z by using the relation $z = f(x)$. Thus the equation (2) is now an equation with z as independent variable.

We choose z as a function of x in one of the following ways, so that the converted equation (2) can be solved by a method known to us already.

1. Suppose $Q_1 = \pm kf(x)$, then we choose the new independent variable z, such that $\left(\frac{dz}{dx}\right)^2 = kf(x)$. Observe that we omit the negative sign.

Since, $\left(\frac{dz}{dx}\right)^2 = kf(x)$, we have $\left(\frac{dz}{dx}\right) = \sqrt{[kf(x)]}$. $\Rightarrow Q_1 = \frac{\pm kf(x)}{kf(x)} = \pm k$, a constant.

By choosing z, such that $\left(\frac{dz}{dx}\right)^2 = kf(x)$, if P_1 also turns out to be a constant, then (2) will be a linear differential equation with constant coefficients, which can be solved.

2. Suppose we choose z such that, $\frac{d^2z}{dx^2} + P\frac{dz}{dx} = 0$ then, $P_1 = 0$. Solving we have

$$z = \int x \, . \, I \, . \, dx, \qquad \text{where, } I = e^{-\int P\,dx}$$

If for this value of z, Q_1 also turns out to be a constant, the equation (2) can be solved easily.

Example 1. Solve : $\cos x\frac{d^2y}{dx^2} + \sin x\frac{dy}{dx} - 2y\cos^3 x = 2\cos^5 x$ by changing the independent variable.

Solution : Dividing throughout by $\cos x$, we have

$$\frac{d^2y}{dx^2} + \tan x\frac{dy}{dx} - 2y\cos^2 x = 2\cos^4 x \qquad \dots (1)$$

Here, $P = \tan x$, $Q = -2\cos^2 x$, $R = 2\cos^4 x$

Let us choose z such that,

$$\left(\frac{dz}{dx}\right)^2 = 2\cos^2 x \Rightarrow \left(\frac{dz}{dx}\right) = \sqrt{2}\cos x \Rightarrow z = \sqrt{2}\sin x$$

With this change (1) becomes,

$$\frac{d^2y}{dz^2} + P_1\frac{dy}{dz} + Q_1 y = R_1$$

where, $$P = \frac{\frac{d^2z}{dx^2} + P\frac{dz}{dx}}{\left(\frac{dz}{dx}\right)^2} = \frac{-\sqrt{2}\sin x + \sqrt{2}\tan x\cos x}{2\cos^2 x} = 0$$

$$Q_1 = \frac{Q}{\left(\frac{dz}{dx}\right)^2} = -1 \qquad R_1 = \frac{R}{\left(\frac{dz}{dx}\right)^2} = \frac{2\cos^4 x}{2\cos^2 x} = \cos^2 x$$

Thus (2) becomes, $\frac{d^2y}{dz^2} - y = 1 - \frac{z^2}{2}$ $\qquad \left(\because \cos^2 x = 1 - \sin^2 x = 1 - \frac{z^2}{2}\right)$

$$\Rightarrow \qquad (D^2 - 1)\,y = 1 - \frac{z^2}{2}.$$

The auxiliary equation is, $m^2 - 1 = 0$. The roots are ± 1. Thus, C.F. $= C_1 e^z + C_2 e^{-z}$

Now, $$\text{P.I.} = \frac{1}{D^2 - 1}\left[1 - \frac{1}{2}z^2\right]$$

$$= -(1 - D^2)^{-1}\left[1 - \frac{1}{2}z^2\right]$$

$$= -[1 + D^2 + D^4 + ...]\left[1 - \frac{1}{2}z^2\right] = -\left[1 - \frac{1}{2}z^2 - 1\right] = \frac{1}{2}z^2$$

$\therefore$ The solution is,

$$y = C_1 e^z + C_2 e^{-z} + \frac{1}{2}z^2$$

$\Rightarrow$ $$y = C_1 e^{\sqrt{2}\sin x} + C_2 e^{-\sqrt{2}\sin x} + \sin^2 x \qquad [\because z = \sqrt{2}\sin x]$$

Example 2. Solve :

$$\frac{d^2y}{dx^2} + (2\cos x + \tan x)\frac{dy}{dx} + y\cos^2 x = \cos^4 x$$

by change of independent variable.

Solution : Here $P = 2\cos x + \tan x$, $Q = \cos^2 x$, $R = \cos^4 x$

Let us choose the new independent variable z such that,

$$\left(\frac{dz}{dx}\right)^2 = \cos^2 x \Rightarrow \frac{dz}{dx} = \cos x \Rightarrow z = \sin x$$

In terms of z the given equation becomes

$$\frac{d^2y}{dz^2} + P_1\frac{dy}{dz} + Q_1 y = R_1$$

where, $$P_1 = \frac{\dfrac{d^2z}{dx^2} + P\dfrac{dz}{dx}}{\left(\dfrac{dz}{dx}\right)^2} = \frac{-\sin x + (2\cos x + \tan x)\cos x}{\cos^2 x} = \frac{2\cos^2 x}{\cos^2 x} = 2$$

$$Q_1 = \frac{Q}{\left(\dfrac{dz}{dx}\right)^2} = \frac{\cos^2 x}{\cos^2 x} = 1$$

$$R_1 = \frac{R}{\left(\dfrac{dz}{dx}\right)^2} = \frac{\cos^4 x}{\cos^2 x} = \cos^2 x = 1 - \sin^2 x = 1 - z^2$$

Thus we have,

$$\frac{d^2y}{dz^2} + 2\frac{dy}{dz} + y = 1 - z^2 \Rightarrow (D^2 + 2D + 1)\,y = 1 - z^2$$

The auxiliary equation is, $m^2 + 2m + 1 = 0 \Rightarrow (m+1)^2 = 0 \Rightarrow m = -1, -1$

$\therefore$ $\quad$ C.F. $= (C_1 + C_2 z)\,e^{-z}$

Now, $\quad$ P.I. $= \dfrac{1}{D^2 + 2D + 1}\left[1 - z^2\right]$

$$= (1 + D)^{-2}\left[1 - z^2\right]$$

$$= [1 - 2D + 3D^2 - 4D^3 + ...]\left[1 - z^2\right] = 1 - z^2 + 4z - 6 = -z^2 + 4z - 5$$

$\therefore$ The solution is,

$$y = (C_1 + C_2 z)\,e^{-z} + (4z - z^2 - 5)$$

$\Rightarrow$ $$\mathbf{y = (C_1 + C_2 \sin x)\, e^{-\sin x} + (4\sin x - \sin^2 x - 5)} \qquad (\because z = \sin x)$$

Example 3. Solve : $(1+x^2)^2 \frac{d^2y}{dx^2}+2x(1+x^2)\frac{dy}{dx}+y=0$, using the transformation $z=\tan^{-1}x$.

Solution : The given equation is,

$$(1+x^2)^2 \frac{d^2y}{dx^2}+2x(1+x^2)\frac{dy}{dx}+y=0$$

Dividing throughout by $(1+x^2)^2$, we get

$$\frac{d^2y}{dx^2}+\frac{2x}{(1+x^2)}\frac{dy}{dx}+\frac{1}{(1+x^2)^2}y=0$$

Here, $P=\frac{2x}{1+x^2},\ Q=\frac{1}{(1+x^2)^2},\ R=0$

The transformation is given by $z=\tan^{-1}x$

$$\therefore \quad \frac{dz}{dx}=\frac{1}{1+x^2} \quad \text{and} \quad \frac{d^2z}{dx^2}=-\frac{2x}{(1+x^2)^2}$$

In terms of z, the given equation becomes

$$\frac{d^2y}{dz^2}+P_1\frac{dy}{dz}+Q_1\,y=R_1$$

where, $$P_1=\frac{\frac{d^2z}{dx^2}+P\frac{dz}{dx}}{\left(\frac{dz}{dx}\right)^2}=\frac{-\frac{2x}{(1+x^2)^2}+\frac{2x}{(1+x^2)^2}}{\left(\frac{1}{1+x^2}\right)^2}=0,\ Q_1=\frac{Q}{\left(\frac{dz}{dx}\right)^2}=1$$

$\therefore$ The equation becomes,

$$\frac{d^2y}{dz^2}+y=0 \Rightarrow (D^2+1)\,y=0.$$

The auxiliary equation is, $m^2+1=0 \Rightarrow m=\pm i$. This general solution is,

$y=C_1\cos z+C_2\sin z \quad \Rightarrow \quad$ **$y=C_1\cos(\tan^{-1}x)+C_2\sin(\tan^{-1}x)$.**

Example 4. Solve : $x\frac{d^2y}{dx^2}-\frac{dy}{dx}-4x^3y=8x^3\sin x^2$ $(x>0)$ using the transformation $z=x^2$.

Solution : The given equation is

$$x\frac{d^2y}{dx^2}-\frac{dy}{dx}-4x^3y=8x^3\sin x^2;\quad (x>0)$$

Dividing throughout by x, we get

$$\frac{d^2y}{dx^2}-\frac{1}{x}\frac{dy}{dx}-4x^2y=8x^2\sin x^2$$

The transformation is given by,

$$z=x^2 \quad \Rightarrow \quad \frac{dz}{dx}=2x \quad \text{and} \quad \frac{d^2z}{dx^2}=2.$$

Under this transformation, the given equation becomes

$$\frac{d^2y}{dz^2}+P_1\frac{dy}{dz}+Q_1\,y=R_1$$

where, $$P_1 = \frac{\frac{d^2z}{dx^2} + P\frac{dz}{dx}}{\left(\frac{dz}{dx}\right)^2} = \frac{2 + \left(-\frac{1}{x}\right)2x}{4x^2} = 0$$

$$Q_1 = \frac{Q}{\left(\frac{dz}{dx}\right)^2} = \frac{-4x^2}{4x^2} = -1, \; R_1 = \frac{R}{\left(\frac{dz}{dx}\right)^2} = \frac{8x^2 \sin x^2}{4x^2} = 2 \sin x^2 = 2 \sin x^2$$

∴ We have,

$$\frac{d^2y}{dz^2} - y = 2 \sin z \Rightarrow (D^2 - 1)\, y = 2 \sin z$$

The auxiliary equation is, $m^2 - 1 = 0 \Rightarrow m = \pm 1$

∴ C.F. is $C_1 e^z + C_2 e^{-z}$

Now, $$\text{P.I.} = \frac{1}{D^2 - 1} 2 \sin z = \frac{1}{-1-1} 2 \sin z = -\sin z$$

∴ The solution is, $y = C_1 e^z + C_2 e^{-z} - \sin z \Rightarrow \mathbf{y = C_1 e^{x^2} + C_2 e^{-x^2} + \sin x^2}$

EXERCISES

1. Solve the following differential equations by changing the independent variable.

 (i) $(1 + x^2)^2 \frac{d^2y}{dx^2} + 2(1+x^2)\frac{dy}{dx} + 4y = 0$

 (ii) $\frac{d^2y}{dx^2} + \tan x \frac{dy}{dx} + 4(\cos^2 x)\, y = 0$

 (iii) $\sin^2 x \frac{d^2y}{dx^2} + \sin x \cos x \frac{dy}{dx} + 4y = 0$

 (iv) $x^6 \frac{d^2y}{dx^2} + 3x^5 \frac{dy}{dx} + a^2 y = \frac{1}{x^2}$

 (v) $x \frac{d^2y}{dx^2} - \frac{dy}{dx} + 4x^3 y = x^5$

 (vi) $\cos x \frac{d^2y}{dx^2} + \sin x \frac{dy}{dx} - 2y \cos^3 x = 2 \cos^5 x$

 (vii) $\frac{d^2y}{dx^2} + \cot x \frac{dy}{dx} - (\sin^2 x)\, y = \cos x - \cos^2 x$

 (viii) $\frac{d^2y}{dx^2} + (3 \sin x - \cot x)\frac{dy}{dx} + 2y \sin^2 x = e^{-\cos x} \sin^2 x$

2. Solve the following equations by using the given transformation.

 (i) $x \frac{d^2y}{dx^2} + (2x^2 - 1)\frac{dy}{dx} - 24x^3 y = 4x^3 \sin x^2;\ x > 0,\ z = x^2.$

 (ii) $(1 - x^2)\frac{d^2y}{dx^2} - x\frac{dy}{dx} - a^2 y = 0;\ z = \sin^{-1} x$, given that $y(0) = -\frac{1}{2}$ and $y'(0) = a$

 (iii) $\frac{d^2y}{dx^2} - (1 + 4e^x)\frac{dy}{dx} + 3e^{2x} y = e^{2(x + e^x)};\ z = e^x.$

Answers

1. (i) $y(1+x^2) = C_1(1-x^2) + 2C_2 x$

(ii) $y = C_1 \cos(2\sin x) + C_2 \sin(2\sin x)$

(iii) $y = C_1 \cos\left(2\log\tan\frac{x}{2}\right) + C_2 \sin\left(2\log\tan\frac{x}{2}\right)$

(iv) $y = C_1 \cos\left(\frac{a}{2x^2}\right) - C_2 \sin\left(\frac{a}{2x^2}\right) + \frac{1}{a^2x^2}$

(v) $y = C_1 \cos x^2 + C_2 \sin x^2 + \frac{x^2}{4}$ (vi) $y = C_1 e^{\sqrt{2}\sin x} + C_2 e^{-\sqrt{2}\sin x} + \sin^2 x$

(vii) $y = C_1 e^{-\cos x} + C_2 e^{\cos x} - \cos x$ (viii) $y = C_1 e^{\cos x} + C_2 e^{2\cos x} + \frac{1}{6}e^{-\cos x}$

2. (i) $y = C_1 e^{2x^2} + C_2 e^{-3x^2} - \frac{1}{50}(\cos x^2 + 7\sin x^2)$

(ii) $4y = C_1 e^{a\sin^{-1}x} - C_2 e^{-\sin^{-1}x}$ (iii) $y = C_1 e^z + C_2 e^{3z} - e^{2z}$; where $z = e^x$

2.20 Method of Changing Dependent Variable (Reduction to Normal Form)

In this method, we would like to reduce the differential equation

$$\frac{d^2y}{dx^2} + P\frac{dy}{dx} + Qy = R$$

by a suitable substitution, to the form

$$\frac{d^2v}{dx^2} + Iv = Sqq$$

which is known as the **normal form** i.e. a form in which the first derivative is abscent. The process of bringing into normal form is called the **method of changing dependent variable,** or removing the first derivative.

Consider the equation,

$$\frac{d^2y}{dx^2} + P\frac{dy}{dx} + Qy = R \qquad \text{... (1)}$$

Let $y = uv$, where u and v are functions of x. Then we have,

$$\frac{dy}{dx} = u\frac{dv}{dx} + v\frac{du}{dx} \text{ and } \frac{d^2y}{dx^2} = u\frac{d^2v}{dx^2} + 2\frac{du}{dx}\cdot\frac{dv}{dx} + v\frac{d^2u}{dx^2}$$

Now putting $y = uv$ in (1), we have

$$u\frac{d^2v}{dx^2} + 2\frac{du}{dx}\cdot\frac{dv}{dx} + v\frac{d^2u}{dx^2} + P\left(u\frac{dv}{dx} + v\frac{du}{dx}\right) + Qu\,.\,v = R$$

On rearrangement we have

$$u\frac{d^2v}{dx^2} + \left(Pu + 2\frac{du}{dx}\right)\cdot\frac{dv}{dx} + v\left(\frac{d^2u}{dx^2} + P\frac{du}{dx} + Qu\right) = R$$

$$\Rightarrow \frac{d^2v}{dx^2} + \left(P + \frac{2}{u}\frac{du}{dx}\right)\cdot\frac{dv}{dx} + \frac{1}{u}\left(\frac{d^2u}{dx^2} + P\frac{du}{dx} + Qu\right)v = \frac{R}{u} \qquad \text{... (2)}$$

To remove the first order derivative $\frac{dv}{dx}$ from (2), we must choose u to satisfy the equation

$$P+\frac{2}{u}\frac{du}{dx}=0 \quad\Rightarrow\quad \frac{du}{u}=-\frac{1}{2}P\,dx \quad\Rightarrow\quad \int\frac{du}{u}=-\frac{1}{2}\int P\,dx$$

That is, $\log u=-\frac{1}{2}\int P\,dx \quad\Rightarrow\quad u=e^{-\left(\frac{1}{2}\right)\int P\,dx}$

Thus, the suitable substitution for the dependent variable to remove the first derivative is $y=u\,v$, where $u=e^{-\left(\frac{1}{2}\right)\int P\,dx}$.

Further, for this u, we have

$$\frac{1}{u}\left(\frac{d^2u}{dx^2}+P\frac{du}{dx}+Q\,u\right)=Q-\frac{1}{4}P^2-\frac{1}{2}\frac{dP}{dx} \qquad \text{... (3)}$$

Hence the equation (2) becomes,

$$\frac{d^2v}{dx^2}+I\,v=S \qquad \text{... (4)}$$

where, $I=Q-\frac{1}{4}P^2-\frac{1}{2}\frac{dP}{dx}$ and $S=\frac{R}{u}$

The equation (4) is called the **Normal form.**

Working rule for the method of changing the dependent Variable (bringing into Normal form)

1. Bring the differential equation to standard form

$$\frac{d^2y}{dx^2}+P\frac{dy}{dx}+Q\,y=R$$

i.e. keeping the coefficient of $\frac{d^2y}{dx^2}$ as unity.

2. To bring it to normal form (i.e. to remove the first order derivative) choose $u=e^{-\left(\frac{1}{2}\right)\int P\,dx}$.

3. Put $y=u\,v$ in the equation, which reduces to the normal form $\frac{d^2v}{dx^2}+I\,v=S$, where

$$I=Q-\frac{1}{4}P^2-\frac{1}{2}\frac{dP}{dx} \quad\text{and}\quad S=\frac{R}{u}.$$

4. Solving this equation, we obtain v, then the complete solution is given by $y=u\,v$.

Note : The equation (Normal form) obtained in the step (3) above can be solved if it takes two special forms.

(i) when I = constant. In this case the equation can be solved as it will be a linear equation with constant coefficients.

(ii) when $I=\frac{\text{constant}}{x^2}$, then the resulting equation reduces to homogeneous form which can be reduced to equation with constant coefficients by putting $x=e^z$

Example 1. Solve by changing the dependent variable.

$$x^2\frac{d^2y}{dx^2}-2x(1+x)\frac{dy}{dx}+2(1+x)y=x^3\ (x>0)$$

given that $y=\frac{1}{2}$ and $\frac{dy}{dx}=1$, when $x=1$.

Solution : The given equation is,

$$x^2 \frac{d^2y}{dx^2} - 2x(1+x)\frac{dy}{dx} + 2(1+x)y = x^3 \qquad \text{... (1)}$$

Since $x > 0$, dividing throughout by x^2, we get,

$$\frac{d^2y}{dx^2} - 2\left(\frac{1+x}{x}\right)\frac{dy}{dx} + 2\left(\frac{1+x}{x^2}\right)y = x$$

Here, $P = -2\left(\frac{1+x}{x}\right);\ Q = 2\left(\frac{1+x}{x^2}\right);\ R = x$

To remove the first derivative, we shall choose

$$u = e^{-\frac{1}{2}\int P\,dx} = e^{-\frac{1}{2}\int -2\left(\frac{1+x}{x}\right)dx} = e^{\int\left(\frac{1}{x}+1\right)dx} = e^{\log x + x} = x e^x$$

Putting $y = uv$, the equation reduces to, $\frac{d^2v}{dx^2} + Iv = S$,

where,

$$I = Q - \frac{1}{4}P^2 - \frac{1}{2}\frac{dP}{dx} = 2\left(\frac{1+x}{x^2}\right) - \frac{1}{4}\times 4\left(\frac{1+x}{x}\right)^2 - \frac{1}{2}\cdot\frac{2}{x^2} = -1$$

$$S = \frac{R}{u} = \frac{x}{x e^x} = e^{-x},$$

Thus, we have, $\frac{d^2v}{dx^2} - v = e^{-x} \Rightarrow (D^2 - 1)v = e^{-x}$

The auxiliary equation is, $m^2 - 1 = 0 \Rightarrow m = \pm 1$. Thus C.F. is $C_1 e^x + C_2 e^{-x}$,

Now, $$\text{P.I.} = \frac{1}{D^2-1}e^{-x} = \frac{1}{(D+1)(D-1)}e^{-x} = x \cdot \frac{1}{-2}e^{-x} = -\frac{x}{2}e^{-x}$$

$\therefore$ The complete solution of the normal form is,

$$v = C_1 e^x + C_2 e^{-x} - \frac{x}{2}e^{-x}.$$

Hence the general solution of the given equation is

$$y = x e^x \left(C_1 e^x + C_2 e^{-x} - \frac{x}{2}e^{-x}\right).$$

By data, $y = \frac{1}{2}$ when $x = 1 \Rightarrow \frac{1}{2} = e\left(C_1 e + C_2 e^{-1} - \frac{1}{2}e^{-1}\right) \Rightarrow e^2 C_1 + C_2 = 1$ ... (2)

Again by data $\frac{dy}{dx} = 1$, when $x = 1 \Rightarrow 3e^2 C_1 + C_2 = 2$ ... (3)

Solving (3) and (4) we get, $C_1 = \frac{1}{2}e^{-2}$ and $C_2 = \frac{1}{2}$

Hence the required general solution is, $\mathbf{2y = x(e^{2x-2} + 1 - x)}$.

Example 2. Solve by changing the dependent variable.

$$\frac{d^2y}{dx^2} - 2\tan x\frac{dy}{dx} - (a^2+1)y = e^x \sec x$$

by reducing it to normal form.

Solution : The given equation is

$$\frac{d^2y}{dx^2} - 2\tan x\frac{dy}{dx} - (a^2+1)y = e^x \sec x$$

Here, $P = -2\tan x;\quad Q = -(a^2+1);\quad R = e^x \sec x$

To remove the first derivative, we shall choose,

$$u = e^{-\frac{1}{2}\int P\,dx} = e^{-\frac{1}{2}\int -2\tan x\,dx} = e^{\int \tan x\,dx} = e^{\log \sec x} = \sec x$$

Putting $y = u\,v$, the equation reduces to

$$\frac{d^2v}{dx^2} + I\,v = S,$$

where,

$$I = Q - \frac{1}{4}P^2 - \frac{1}{2}\frac{dP}{dx}$$

$$= -(a^2+1) - \frac{1}{4}\,.\,4\tan^2 x - \frac{1}{2}(-2\sec^2 x)$$

$$= -a^2 - 1 - \tan^2 x + \sec^2 x = -a^2 \qquad [\because 1 + \tan^2 x = \sec^2 x]$$

$$S = \frac{R}{u} = \frac{e^x \sec x}{\sec x} = e^x$$

$\therefore$ We have,

$$\frac{d^2v}{dx^2} - a^2 v = e^x \Rightarrow (D^2 - a^2)\,v = e^x$$

The auxiliary equation is, $m^2 - a^2 = 0 \Rightarrow m = \pm a$. Thus C.F. is $C_1 e^{ax} + C_2 e^{-ax}$

Now, $\quad$ P.I. $= \dfrac{1}{D^2 - a^2}e^x = \dfrac{1}{1-a^2}e^x$

$\therefore$ The complete solution of the normal form is, $v = C_1 e^{ax} + C_2 e^{-ax} + \dfrac{1}{1-a^2}e^x$

Hence the general solution of the given equation is

$$\mathbf{y = u\,v = \sec x\left[C_1 e^{ax} + C_2 e^{-ax} + \frac{1}{1-a^2}e^x\right].}$$

Example 3. Solve by changing the dependent variable

$$\mathbf{x^2\frac{d^2y}{dx^2} - 4x^3\frac{dy}{dx} + (4x^4 - 2x^2 + 5)\,y = x^4 e^{x^2},\ (x > 0)}$$

by removing the first derivative.

Solution : The given equation is

$$x^2\frac{d^2y}{dx^2} - 4x^3\frac{dy}{dx} + (4x^4 - 2x^2 + 5)\,y = x^4 e^{x^2}$$

Since $x > 0$, dividing throughout by x^2, we get

$$\frac{d^2y}{dx^2} - 4x\frac{dy}{dx} + \left(4x^2 - 2x + \frac{5}{x^2}\right)y = x^2 e^{x^2}$$

Here, $\quad P = -4x;\quad Q = 4x^2 - 2 + \dfrac{5}{x^2};\quad R = x^2 e^{x^2}$

To remove the first derivative, we shall choose,

$$u = e^{-\frac{1}{2}\int P\,dx} = e^{x^2}$$

Putting $y = u\,v$, the equation reduces to

$$\frac{d^2 v}{dx^2} + I\,v = S,$$

where,

$$I = Q - \frac{1}{4}P^2 - \frac{1}{2}\frac{dP}{dx} = 4x^2 - 2 + \frac{5}{x^2} - 4x^2 + 2 = \frac{5}{x^2}$$

and, $$S = \frac{R}{u} = x^2$$

$\therefore$ We have,

$$\frac{d^2 v}{dx^2} + \frac{5}{x^2}v = x^2 \quad \Rightarrow \quad x^2\frac{d^2 v}{dx^2} + 5\,v = x^4$$

This is a homogeneous differential equation which can be reduced to linear equation with constant coefficients. To this end, we shall put,

$$x = e^z \Rightarrow z = \log x$$

Then, $x^2 D^2 = D_1(D_1 - 1)$, where $D = \dfrac{d}{dx}$ and $D_1 = \dfrac{d}{dz}$

$\therefore$ the equation becomes, $(D_1^2 - D_1 + 5)\,v = e^{4z}$

The auxiliary equation is, $m^2 - m + 5 = 0 \Rightarrow m = \dfrac{1}{2} \pm i\dfrac{\sqrt{19}}{2}$

$$\therefore \quad \text{C.F.} = e^{\frac{1}{2}z}\left[C_1\cos\frac{\sqrt{19}}{2}.z + C_2\sin\frac{\sqrt{19}}{2}.z\right]$$

Now, $$\text{P.I.} = \frac{1}{D_1^2 - D_1 + 5}e^{4z} = \frac{1}{17}e^{4z}$$

$\therefore$ The complete solution of the normal form is,

$$v = x^{\frac{1}{2}}\left[C_1\cos\left(\frac{\sqrt{19}}{2}.\log x\right) + C_2\sin\left(\frac{\sqrt{19}}{2}.\log x\right)\right] + \frac{1}{17}x^4 \qquad [\because z = \log x]$$

Hence the general solution of the given equation is

$$\boldsymbol{y = u\,v = e^{x^2}.\sqrt{x}\left[C_1\cos\left(\frac{\sqrt{19}}{2}.\log x\right) + C_2\sin\left(\frac{\sqrt{19}}{2}.\log x\right)\right] + \frac{1}{17}x^4}$$

EXERCISES

1. Solve the following differential equations by changing the dependent variable.

(i) $\dfrac{d^2y}{dx^2} - 2\tan x\dfrac{dy}{dx} + 5\,y = 3$

(ii) $\dfrac{d^2y}{dx^2} + \dfrac{2}{x}\dfrac{dy}{dx} + n^2 y = 0$

(iii) $x\dfrac{d}{dx}\left(x\dfrac{dy}{dx} - y\right) - 2x\dfrac{dy}{dx} + 2\,y + x^2 y = 0, x > 0$

(iv) $\dfrac{d^2y}{dx^2} - 2\cot x\dfrac{dy}{dx} + (1 + 2\cot^2 x)\,y = 0$

(v) $\dfrac{d^2y}{dx^2} - 4x\dfrac{dy}{dx} + (4x^2 - 3)\,y = e^{x^2}$

(vi) $\frac{d^2y}{dx^2} - 2\tan x \frac{dy}{dx} + 5y = (\sec x)e^x$

(vii) $\frac{d^2y}{dx^2} - \frac{2}{x}\frac{dy}{dx} + \left(1+\frac{2}{x^2}\right)y = x e^x$

(viii) $\frac{d^2y}{dx^2} - 4x\frac{dy}{dx} + (4x^2-1)y = -3e^{x^2}\sin 2x$

(ix) $4x^2\frac{d^2y}{dx^2} + 4x^5\frac{dy}{dx} + (x^8+6x^4+4)y = 0$

(x) $\frac{d^2y}{dx^2} - \frac{1}{\sqrt{x}}\frac{dy}{dx} + \frac{1}{4x^2}(-8+\sqrt{x}+x)y = 0$

2. Solve the following equations by reducing them to normal form.

(i) $x\frac{d^2y}{dx^2} - 2(x+1)\frac{dy}{dx} + (x+2)y = (x-2)e^x;\ x>0$, given that $y = e^2$ and $y = 2e^2$, when $x = 1$

(ii) $\frac{d^2y}{dx^2} + 4\operatorname{cosec} 2x\frac{dy}{dx} + 2\tan^2 xy = e^x\cot x$

Answers

1. (i) $y = \sec x\left[C_1\cos\sqrt{6}\,.\,x + 2C_2\sin\sqrt{6}\,.\,x\right]$

(ii) $y = \frac{1}{x}(C_1\cos nx + C_2\sin nx)$

(iii) $y = x(C_1\cos x + C_2\sin x)$

(iv) $y = \sin x(C_1 + C_2 x)$

(v) $y = e^{x^2}(C_1e^x + C_2e^{-x} - 1)$

(vi) $y = \sec x\left[C_1\cos\sqrt{6}\,.\,x + 2C_2\sin\sqrt{6}\,.\,x + \frac{e^x}{7}\right]$

(vii) $y = x\left(C_1\cos x + C_2\sin x + \frac{e^x}{2}\right)$

(viii) $y = e^{x^2}(C_1\cos x + C_2\sin x + \sin 2x)$

(ix) $y = e^{-\left(\frac{x^4}{8}\right)}\sqrt{x}\left[C_1\cos\left(\frac{\sqrt{3}}{2}\log x\right) + C_2\sin\left(\frac{\sqrt{3}}{2}\log x\right)\right]$

(x) $y = e^{\sqrt{x}}\,x(C_1x^2 + C_2x^{-1})$

2. (i) $y = xe^x\left(C_1x^2 + \frac{C_2}{x} + 1 - \frac{x}{2}\right)$ where $C_1 = \frac{e}{2}$ and $C_2 = \frac{4e-3}{6}$

(ii) $y = \sec x\left[C_1e^{\sqrt{2}x} + C_2e^{-\sqrt{2}x} - e^x\right]\cot x$

2.21 Method of Variation of Parameters

In this method it will be convenient to use suffixes to denote differentiation, for example y_1 and y_2 stand for $\frac{dy}{dx}$ and $\frac{d^2y}{dx^2}$ respectively.

Consider the equation,

$$y_2 + P\,y_1 + Q\,y = R \qquad \text{... (1)}$$

where P, Q and R are functions of x.

If u is a solution of the equation,

$$y_2 + P\,y_1 + Q\,y = 0 \qquad \text{... (2)}$$

the function $C_1\,u$ is also a solution for any arbitrary constant C_1. In the method of solving the equation (1), when part of the complementary function is known, we replaced the constant C_1 by a function $v\,(x)$ and considered the possibility of the existence of a solution of the equation (1), of the form $y = u\,.\,v$. In the present method we would like to discuss the possibility of the existence of a solution in the form $y = A\,u + B\,v$, where A and B are functions of x, where $y = C_1\,u + C_2\,v$ is the general solution of (2) i.e. homogeneous equation of (1). In other words, we discuss the possibility of having a solution of (1) by replacing the arbitrary constants C_1 and C_2 by the functions of x, in the complementary function of (1).

Let the complementary function of (1) be

$$y = C_1\,u + C_2\,v$$

That is, it is the complete solution of the equation (2). Hence C_1 and C_2 are arbitrary constants and u and v are functions of x.

Consider,

$$y = A\,u + B\,v \qquad \text{... (3)}$$

where A and B are now functions of x.

We shall try to determine A and B so that $A\,u + B\,v$ is the complete solution of (1).

Since we have two unknown functions A and B, in terms of which y has been expressed, we require at least two conditions to determine A and B. We are having one condition that is, $y = A\,u + B\,v$ to be a solution to the equation (1). We are therefore free to impose a condition on A and B so that it will suit to our purpose. One such condition may be $B = 0$. But this will lead to the method that we have already discussed in § 5.7. Therefore, we shall consider a different condition.

Differentiating (3), we get,

$$y_1 = A\,u_1 + B\,v_1 + A_1\,u + B_1\,v$$

To avoid the involvement of the derivatives of A and B of higher order than first, we choose A and B such that

$$A_1\,u + B_1\,v = 0^* \qquad \text{... (4)}$$

∴ we have,

$$y_1 = A\,u_1 + B\,v_1 \qquad \text{... (5)}$$

Differentiating again, we get $y_2 = A\,u_2 + B\,v_2 + A_1\,u_1 + B_1 v_1$ ... (6)

Putting the values of y_1 and y_2, given by (3), (5) and (6) in (1), we get [this is done as (3) is assumed to be a solution of (1)].

---

*We could have selected A and B that $A_1\,u + B_1\,v = \sin x\,e^x$, or any other suitable functions, but for simplicity we have chosen in such a way that $A_1\,u + B_1\,v = 0$.

$$A\,u_2 + B\,v_2 + A_1 u_1 + B_1 v_1 + P\,(A\,u_1 + B\,v_1) + Q\,(A\,u + B\,v) = R$$

$$\Rightarrow \qquad A\,(u_2 + P\,u_1 + Q\,u) + B\,(v_2 + P\,v_1 + Q\,v) + (A_1 u_1 + B_1 v_1) = R$$

Since u and v are solutions of (2), we have,

$$u_2 + P\,u_1 + Q\,u = 0 \text{ and } v_2 + P\,v_1 + Q\,v = 0$$

Thus, we have,

$$A_1\,u_1 + B_1 v_1 = R \qquad \text{... (7)}$$

Now solving the equations (4) and (7) we can obtain A_1 and B_1. But the solution to this system [(4) and (7)] of equations exists if

$$W = \begin{vmatrix} u & v \\ u_1 & v_1 \end{vmatrix} \neq 0$$

The determinant in the right side is called **Wronskian** of the functions u and v. It can be shown that the Wronskian does not vanish, if u and v are linearly independent i.e. $u \neq kv$ for any k. Thus we get,

$$A_1 = \frac{-v\,R}{W} \text{ and } B_1 = \frac{uR}{W}$$

Integrating these we can find A and B as

$$A = f(x) + C_1 \text{ and } B = g(x) + C_2$$

Putting these in (3), the complete solution of (1) is

$$y = C_1 u + C_2 v + u\,f\,(x) + v\,g\,(x)$$

where C_1 and C_2 are arbitrary constants.

Note 1. By observing the equation $y = C_1 u + C_2 v$ and the equation (3), we can make out that the solutions for the equations (1) and (2) are of same form except that A and B are functions of x in (3). For this reason this method is called the **"Variation of Parameters"**.

2. This method is useful if the complete solution of (2) is known, i.e. C.F. of (1).

Working Rule for the method of Variation of Parameters.

1. Bring the given differential equation to the standard form

$$\frac{d^2y}{dx^2} + P\frac{dy}{dx} + Qy = R$$

2. If the C.F. is not given, find the C.F. of the given equation and let it be $C_1 u + C_2 v$
3. Regard $y = A\,u + B\,v$, where A and B are functions of x as the complete solution of the given equation.
4. Find the functions A and B, by integrating

$$A_1 = \frac{-v\,R}{W} \text{ and } B_1 = \frac{u\,R}{W}$$

where, $W = \begin{vmatrix} u & v \\ u_1 & v_1 \end{vmatrix}$.

5. Putting these values of A and B in $y = A\,u + B\,v$, we get the complete solution of the given equation.

Example 1. Solve $\dfrac{d^2y}{dx^2} - y = \dfrac{2}{1+e^x}$, **by the method of variation of parameters.**

Solution : The given equation is, $(D^2 - 1)\,y = \dfrac{2}{1+e^x}$

The auxiliary equation is, $m^2 - 1 = 0 \quad \Rightarrow \quad m = \pm 1$

$\therefore \qquad$ C.F. $= y = C_1 e^x + C_2 e^{-x}$

Let the complete solution of the equation be, $y = A e^x + B e^{-x}$, where A and B are functions of x.

Consider,

$$W = \begin{vmatrix} u & v \\ u_1 & v_1 \end{vmatrix} \qquad \left(\text{where } u = e^x,\ v = e^{-x}\right)$$

$$W = \begin{vmatrix} e^x & e^{-x} \\ e^x & -e^{-x} \end{vmatrix} = -1 - 1 = -2$$

$$\therefore \quad A_1 = \frac{-vR}{W} = \frac{-e^{-x}}{-2} \cdot \frac{2}{1+e^x} = \frac{e^{-x}}{1+e^x} \qquad \left(\because R = \frac{2}{1+e^x}\right)$$

$$B_1 = \frac{uR}{W} = \frac{e^x}{-2} \cdot \frac{2}{1+e^x} = -\frac{e^x}{1+e^x}$$

$$\therefore \quad A = \int \frac{e^{-x}}{1+e^x}\, dx + C_1$$

$$= \int \frac{dz}{z^2(1+z)} + C_1 \qquad \left(\text{Put } e^x = z \Rightarrow dx = \frac{dz}{z}\right)$$

$$= \int \left(\frac{1}{z^2} - \frac{1}{z} + \frac{1}{1+z}\right) dz + C_1$$

$$= -\frac{1}{z} - \log z + \log(1+z) + C_1$$

$$= -\frac{1}{z} + \log\left(\frac{1+z}{z}\right) + C_1 = -e^{-x} + \log(1+e^{-x}) + C_1$$

Again, $$B = -\int \frac{e^x}{1+e^x}\, dx + C_2 = -\log(1+e^x) + C_2$$

Hence the complete solution is,

$$y = e^x\,[-e^{-x} + \log(1+e^{-x}) + C_1] + e^{-x}\,[C_2 - \log(1+e^x)]$$

$$\Rightarrow \quad \mathbf{y = C_1 e^x + C_2 e^{-x} - 1 + e^x \log(1+e^{-x}) - e^{-x} \log(1+e^x)}$$

Example 2. Solve $x^2 y_2 + x y_1 - y = x^2 e^x,\ x > 0$, by the method of variation of parameters.

Solution : The given equation is, $x^2 y_2 + x y_1 - y = x^2 e^x$ ... (1)

Consider the equation

$$x^2 \frac{d^2 y}{dx^2} + x \frac{dy}{dx} - y = 0 \qquad \text{... (2)}$$

This is a homogeneous equation,

Put, $x = e^z, \Rightarrow z = \log x$

Now $x^2 D^2 = D_1(D_1 - 1)$ and $xD = D_1$ $\left(\text{where } D = \frac{d}{dx} \text{ and } D_1 = \frac{d}{dz}\right)$

$\therefore$ the equation becomes,

$$D_1(D_1 - 1) + (D_1 - 1)\, y = 0 \Rightarrow (D_1^2 - 1)\, y = 0$$

The auxiliary equation is, $D_1^2 - 1 = 0 \Rightarrow D_1 = \pm 1$

$\therefore$ the complete solution of the equation (2) is, $y = C_1 x + C_2 \frac{1}{x}$ $[\because x = e^z]$

Let the complete solution of the (1) be, $y = Ax + B\frac{1}{x}$, where A and B are functions of x.
Consider,

$$W = \begin{vmatrix} u & v \\ u_1 & v_1 \end{vmatrix} \qquad \left(\text{where } u = x,\ v = \frac{1}{x}\right)$$

$$W = \begin{vmatrix} x & \frac{1}{x} \\ 1 & -\frac{1}{x^2} \end{vmatrix} = -\frac{1}{x} - \frac{1}{x} = -\frac{2}{x}$$

$\therefore$ A_1 and B_1 are given by

$$A_1 = \frac{-vR}{W} = -\frac{\frac{1}{x}e^x}{\left(-\frac{2}{x}\right)} \Rightarrow A_1 = \frac{1}{2}e^x \qquad \text{[Note that } R = e^x\text{]}$$

$$B_1 = \frac{uR}{W} = \frac{x\,e^x}{\left(-\frac{2}{x}\right)} = -\frac{1}{2}x^2 e^x$$

Now, $\quad A = \int \frac{1}{2} e^x\, dx + C_1 \Rightarrow A = \frac{1}{2}e^x + C_1$

Again, $\quad B = \int -\frac{1}{2}x^2\, e^x\, dx + C_2$

$\Rightarrow \quad B = -\frac{1}{2}\left[x^2\, e^x - \int 2x\, e^x\, dx\right] + C_2$ (Integrating by parts)

$\Rightarrow \quad B = -\frac{1}{2}\left[x^2 e^x - 2x\, e^x - 2e^x\right] + C_2$

$\Rightarrow \quad B = -\frac{1}{2}e^x\left[x^2 - 2x + 2\right] + C_2$

The complete solution is,

$$y = \left(C_1 + \frac{1}{2}e^x\right)x + \left[C_2 - \frac{1}{2}e^x\,(x^2 - 2x + 2)\right]\frac{1}{x}$$

$$\Rightarrow \quad \boldsymbol{y = C_1 x + C_2\frac{1}{x} + e^x\left(1 - \frac{1}{x}\right).}$$

Example 3. Solve $(1-x)\frac{d^2y}{dx^2} + x\frac{dy}{dx} - y = (1-x)^2,\ x \neq 1$ by the method of variation of parameters.

Solution : Since $x \neq 1$, dividing throughout by $(1 - x)$, we get

$$\frac{d^2y}{dx^2} + \frac{x}{(1-x)}\frac{dy}{dx} - \frac{1}{(1-x)}y = (1-x)$$

Here, $\quad P = \frac{x}{1-x},\ Q = -\frac{1}{1-x},\ R = 1-x$

Consider, $\quad 1 + P + Q = 1 + \frac{x}{1-x} - \frac{1}{1-x} = \frac{1-x+x-1}{1-x} = 0 \Rightarrow e^x$ is a part of C.F.

Also consider, $P+Qx=\frac{x}{1-x}-\frac{x}{1-x}=0$. $\Rightarrow$ x is also a part of C.F.

$\therefore$ the C.F. of the given equation is, $y=C_1e^x+C_2x$

Let the complete solution of the given equation be, $y=A\,e^x+B\,x$ where A and B are functions of x, to be determined.

Now,

$$W=\begin{vmatrix} u & v \\ u_1 & v_1 \end{vmatrix}=W=\begin{vmatrix} e^x & x \\ e^x & 1 \end{vmatrix}=e^x-x\,e^x \qquad \text{(where, } u=e^x,\ v=x\text{)}$$

$\therefore$ A_1 and B_1 are given by,

$$A_1=\frac{-vR}{W}=-\frac{x\,(1-x)}{(1-x)\,e^x}=-x\,e^{-x} \text{ and } B_1=\frac{uR}{W}=\frac{e^x\,(1-x)}{(1-x)\,e^x}=1$$

$$\therefore \quad A=\int -x\,e^{-x}\,dx+C_1=x\,e^{-x}+e^{-x}+C_1=A=(x+1)\,e^{-x}+C_1$$

Again, $\quad B=\int 1\,dx+C_2=x+C_2$

Hence the complete solution is,

$$y=e^x\,[(x+1)\,e^{-x}+C_1]+x\,[x+C_2]$$

$$\Rightarrow \quad \boldsymbol{y=x^2+x+1+C_1e^x+C_2x}.$$

Example 4. Solve $(x^2+1)\frac{d^2y}{dx^2}-2x\frac{dy}{dx}+2y=6\,(x^2+1)^2$, given $y=-2$ when $x=-2$ and $y=4$ when $x=2$, by the method of variation of parameters.

Solution : Dividing throughout by (x^2+1), we get

$$\frac{d^2y}{dx^2}-\frac{2x}{(x^2+1)}\frac{dy}{dx}+\frac{2}{(x^2+1)}y=6\,(x^2+1) \quad \text{... (1)}$$

Here, $\quad P=-\frac{2x}{(x^2+1)},\ Q=\frac{2}{(x^2+1)},\ R=6\,(x^2+1)$

Consider $\quad P+xQ=-\frac{2x}{(x^2+1)}+\frac{2x}{(x^2+1)}=0 \Rightarrow x$ is a part of the C.F.

As it is not possible to obtain other part by inspection, we shall use the note 2 given under the method of solving the equation, when an integral of C.F. is given, i.e. the other part of C.F. is given by,

$$v=u\int\left[\frac{1}{u^2}e^{-\int P\,dx}\right]dx$$

$$=x\int\frac{1}{x^2}(x^2+1)\,dx \qquad \left[\because e^{-\int P\,dx}=e^{\int\frac{2x}{x^2+1}dx}=e^{\log(x^2+1)}=1+x^2\right]$$

$$=x\int\left(1+\frac{1}{x^2}\right)dx=x^2-1$$

Thus, the C.F. of the given equation is, $C_1\,(x^2-1)+C_2\,x$.

Let the complete solution of the (1) be, $y=A\,(x^2-1)+Bx$, where A and B are functions of x.

Now, $\quad W=\begin{vmatrix} u & v \\ u_1 & v_1 \end{vmatrix}=W=\begin{vmatrix} x^2-1 & x \\ 2x & 1 \end{vmatrix}=-(1+x^2) \qquad \text{(where } u=x^2-1, v=x\text{)}$

$\therefore A_1$ and B_1 are given by,

$$A_1 = \frac{-vR}{W} = \frac{-6(x^2+1)x}{-(1+x^2)} = 6x \text{ and } B_1 = \frac{uR}{W} = \frac{(x^2-1)6(x^2+1)}{-(1+x^2)} = -6(x^2-1)$$

Now, $\quad A = \int 6x\,dx + C_1 = 3x^2 + C_1, \quad B = \int -6(x^2-1)\,dx + C_2 = -2x^3 + 6x + C_2$

Hence, the complete solution of (1) is

$$y = (3x^2 + C_1)(x^2-1) + (-2x^3 + 6x + C_2)\,x$$

$$\Rightarrow \quad y = C_2 x + C_1(x^2-1) + x^4 + 3x^2$$

By data $y = -2$ when $x = -2$ and $y = 4$ when $x = 2$. Thus, we have,

$$3C_1 - 2C_2 = -30 \text{ and } 3C_1 + 2C_2 = -24$$

Solving we get, $C_2 = \frac{3}{2}$ and $C_1 = -9$

Hence, the required solution is,

$$y = \left(\frac{3}{2}\right)x - 9(x^2-1) + x^4 + 3x^2 \Rightarrow \mathbf{y = \left(\frac{3}{2}\right)x - 6x^2 + x^4 + 9}.$$

EXERCISES

1. Solve the following equations by the method of variation of parameters.

(i) $y_1 + 9y = \sec 3x$ **(ii)** $y_2 + y = \sec x$ **(iii)** $y_2 + y = \operatorname{cosec} x$

(iv) $\frac{d^2y}{dx^2} + 2\frac{dy}{dx} + 2y = 2e^x \cos x$

(v) $x^2\frac{d^2y}{dx^2} + x\frac{dy}{dx} - y = x^2 e^x$, given that $y = \frac{e}{2}$ and $\frac{dy}{dx} = 0$, when $x = 1$.

(vi) $x^2\frac{d^2y}{dx^2} - 2x\frac{dy}{dx} + 2y = \frac{1}{x^2}$ **(vii)** $\frac{d^2y}{dx^2} - \frac{4}{x}\frac{dy}{dx} + \frac{6}{x^2}y = x^2 \sin x$

(viii) $\frac{d^2y}{dx^2} - 6\frac{dy}{dx} + 9y = x^{-2}e^{3x}$ **(ix)** $x^2\frac{d^2y}{dx^2} - 2x(1+x)\frac{dy}{dx} + 2(1+x)y = x^3$

2. Solve $x^2 y_2 - x(x+2)y_1 + (x+2)y = x^3$, by the method of variation of parameters, if $y = x$ and $y = xe^x$ are solutions of

$$[x^2D^2 - x(x+2)D + (x+2)]\,y = 0,$$

given that $y = -\frac{1}{4}$ and $y_1 = -1$ when $x = 1$.

3. Solve $(x^2-1)\frac{d^2y}{dx^2} - 2x\frac{dy}{dx} + 2y = (x^2-1)^2$ by the method of variation of parameters, if $y = x$ and $y = x^2 + 1$ are solutions of

$$[(x^2-1)D^2 - 2xD + 2]\,y = 0,$$

given that $y = 1$ when $x = -1$ and $y = 0$ when $x = 1$.

Answers

1. (i) $y = C_1 \cos 3x + C_2 \sin 3x + \frac{x}{3} \cos 3x + \frac{1}{9} \sin 3x \log(\cos 3x)$

(ii) $y = C_1 \cos x + C_2 \sin x + x \cos x + \sin x \log(\cos x)$

(iii) $y = C_1 \cos x + C_2 \sin x - x \cos x + \sin x \log(\sin x)$

(iv) $y = e^{-x}(C_1 \cos x + C_2 \sin x) + \frac{e^x}{4}(\sin x + \cos x)$

(v) $12xy = 9e - 4e x^2 + 12(x-1)e^x$ (vi) $y = C_1 x^2 + C_2 x + \frac{1}{12x^2}$

(vii) $y = C_1 x^3 + C_2 x^2 - x^2 \sin x$ (viii) $y = (C_1 + C_2 x)x - (\log x + 1)e^{3x}$

(ix) $y = C_1 x e^x + C_2 x - \frac{1}{4} x - \frac{1}{2} x^2$

2. $y = \frac{3x}{2} + \frac{x}{4e} e^x - x(x+1)$ 3. $12y = 2x^4 - x^2 - 6x + 5$

2.22. Method by Finding a First Integral (Exact Equations)

Definition : A second degree linear equation

$$a_0(x)\frac{d^2y}{dx^2} + a_1(x)\frac{dy}{dx} + a_2(x)\,y = f(x) \qquad \text{... (1)}$$

is said to be exact equation, if there exists a first order linear equation

$$b_0(x)\frac{dy}{dx} + b_1(x)\,y = g(x) \qquad \text{... (2)}$$

such that (1) can be obtained from (2) by direct differentiation. In such case (2) is called the First Integral of the equation (1).

An equation of (1) can be solved if it is exact, by finding its first integral and solving it. The following theorem gives us a necessary and sufficient condition for an equation to be exact and further it gives a formula to write down the first integral, if the given equation is exact.

Theorem : A necessary and sufficient condition for the equation

$$a_0\frac{d^2y}{dx^2} + a_1\frac{dy}{dx} + a_2 y = f(x) \qquad \text{... (A)}$$

to be exact is $\frac{d^2a_0}{dx^2} - \frac{da_1}{dx} + a_2 = 0$. Here a_0, a_1 and a_2 are functions of x.

Proof : Let the equation (A) be exact. Then there exists a first order equation

$$b_0\frac{dy}{dx} + b_1 y = g(x) \qquad \text{... (B)}$$

where b_0, b_1 are functions of x, such that (A) can be obtained on direct differentiation of this equation.

Differentiating (B), we get,

$$b_0\frac{d^2y}{dx^2} + \left[\frac{db_0}{dx} + b_1\right]\frac{dy}{dx} + \frac{db_1}{dx}y = g'(x) \qquad \text{... (C)}$$

Comparing the equations (A) and (C), we get

$$a_0 = b_0; \quad a_1 = \frac{db_0}{dx} + b_1; \quad a_2 = \frac{db_1}{dx}; \ f = g'$$

This gives,

$$a_2 = \frac{db_1}{dx} = \frac{d}{dx}\left(a_1 - \frac{db_0}{dx}\right) = \frac{da_1}{dx} - \frac{d^2 b_0}{dx^2} = \frac{da_1}{dx} - \frac{d^2 a_0}{dx^2} \qquad [\because a_0 = b_0]$$

Hence, $$\frac{d^2 a_0}{dx^2} - \frac{da_1}{dx} + a_2 = 0.$$

Conversely suppose (A) is such that, $\dfrac{d^2 a_0}{dx^2} - \dfrac{da_1}{dx} + a_2 = 0$

$\therefore$ (A) can be written as

$$a_0 \frac{d^2 y}{dx^2} + a_1 \frac{dy}{dx} + \left(\frac{da_1}{dx} - \frac{d^2 a_0}{dx^2}\right) y = f(x)$$

$$\Rightarrow \qquad a_0 \frac{d^2 y}{dx^2} - \frac{d^2 a_0}{dx^2} y + a_1 \frac{dy}{dx} + \frac{da_1}{dx} y = f(x)$$

$$\Rightarrow \qquad \frac{d}{dx}\left\{a_0 \frac{dy}{dx} + \left(a_1 - \frac{da_0}{dx}\right) y\right\} = f(x)$$

This equation can be written as,

$$\frac{d}{dx}\left\{a_0 \frac{dy}{dx} + \left(a_1 - \frac{d\,a_0}{dx}\right) y - \int f(x)\,dx + C\right\} = 0$$

where C is any arbitrary constant.

This shows that the equation (A) can be obtained on differentiating

$$\boldsymbol{a_0 \frac{dy}{dx} + \left(a_1 - \frac{d\,a_0}{dx}\right) y - \int f(x)\,dx + C = 0}$$

Thus the equation (A) is exact and hence the theorem.

From the proof of the above theorem it follows that if (A) is exact, (i.e. if the condition for exactness is satisfied), then we can write the first integral of (A) as

$$a_0 \frac{dy}{dx} + \left(a_1 - \frac{d\,a_0}{dx}\right) y - \int f(x)\,dx + C = 0$$

Solving this equation the complete solution of (A) can be obtained.

Example 1. Show that the equation

$$\boldsymbol{x^2 (1+x) \frac{d^2 y}{dx^2} + 2x(2+3x)\frac{dy}{dx} + 2(1+3x)\,y = 0}$$

is exact and solve it.

Solution. In the given equation,

$$a_0 = x^2 (1+x); \quad a_1 = 2x(2+3x) \text{ and } a_2 = 2(1+3x)$$

Consider,

$$\frac{d^2 a_0}{dx^2} - \frac{da_1}{dx} + a_2 = 2 + 6x - (4 + 12x) + 2 + 6x$$
$$= 2 + 6x - 4 - 12x + 2 + 6x = 0$$

$\therefore$ the given equation is exact. The first integral is given by

$$a_0 \frac{dy}{dx} + \left(a_1 - \frac{d\,a_0}{dx}\right) y - \int f(x)\,dx + C_1 = 0$$

$$\Rightarrow \qquad x^2(1+x)\frac{dy}{dx} + [2x(2+3x) - (2x + 3x^2)]\,y + C_1 = 0 \qquad [\because f(x) = 0]$$

$$\Rightarrow \quad x^2(1+x)\frac{dy}{dx} + x(3x+2)\,y + C_1 = 0$$

$$\Rightarrow \quad \frac{dy}{dx} + \frac{3x+2}{x(1+x)}\,y + \frac{C_1}{x^2(1+x)} = 0 \qquad \text{... (1)}$$

This is a linear differential equation of first order.

Now, $\quad \text{I.F.} = e^{\int \frac{3x+2}{x(1+x)}dx} = e^{\int\left(\frac{2}{x}+\frac{1}{1+x}\right)dx} = e^{\left[\log x^2 + \log(1+x)\right]} = x^2(1+x).$

Thus, solution of (1) and hence that of the given equation is

$$y\,.\,x^2(1+x) = C_1 \int x^2(1+x)\,.\,\frac{1}{x^2(1+x)}\,dx + C_2$$

$$\Rightarrow \quad \boldsymbol{y\,.\,x^2(1+x) = C_1 x + C_2}.$$

Example 2. Show that the equation,

$$\boldsymbol{(2x^2+3x)\frac{d^2y}{dx^2} + (6x+3)\frac{dy}{dx} + 2y = (x+1)\,e^x}$$

by verifying it for exactness.

Solution : In the given equation,

$$a_0 = 2x^2 + 3x; \quad a_1 = 6x+3 \quad \text{and} \quad a_2 = 2; \quad f(x) = (x+1)\,e^x$$

Consider,

$$\frac{d^2a_0}{dx^2} - \frac{da_1}{dx} + a_2 = 4 - 6 + 2 = 0$$

Thus, the given equation is exact. The first integral is given by

$$a_0\frac{dy}{dx} + \left(a_1 + \frac{d\,a_0}{dx}\right)y - \int f(x)\,dx + C = 0$$

$$\Rightarrow \quad (2x^2+3x)\frac{dy}{dx} + (6x+3-4x-3)\,y = \int (x+1)\,e^x dx + C_1 \qquad \text{(where } C = -C_1\text{)}$$

$$\Rightarrow \quad (2x^2+3x)\frac{dy}{dx} + 2xy = x\,e^x + C_1$$

$$\Rightarrow \quad \frac{dy}{dx} + \left(\frac{2}{2x+3}\right)y = \frac{e^x}{2x+3} + \frac{C_1}{2x^2+3x}$$

This is a linear differential equation of first order. Now,

$$\text{I.F.} = e^{\int \frac{2}{2x+3}dx} = e^{\log(2x+3)} = 2x+3$$

Hence, solution of (1) and hence that of the given equation is

$$y(2x+3) = \int (2x+3)\left[\frac{e^x}{2x+3} + \frac{C_1}{2x^2+3x}\right]dx + C_2$$

$$\Rightarrow \quad y(2x+3) = \int\left[e^x + \frac{C_1}{x}\right]dx + C_2 \;\Rightarrow\; \boldsymbol{y(2x+3) = e^x + C_1 \log x + C_2}.$$

EXERCISES

Show that the following equations are exact and hence solve.

1. $(1-x^2)\dfrac{d^2y}{dx^2}-3x\dfrac{dy}{dx}-y=0,\ x\neq 1,$ given that $y=1$ and $\dfrac{dy}{dx}=0,$ when $x=0.$

2. $x^2\dfrac{d^2y}{dx^2}+\dfrac{dy}{dx}-2y=2x,$ given that $y = 2$, when $x = -1$ and $y = 0$, when $x = 1$.

3. $(1+x^2)\dfrac{d^2y}{dx^2}+4x\dfrac{dy}{dx}+2y=\sec^2 x.$

4. $x\dfrac{d^2y}{dx^2}+3\dfrac{dy}{dx}=\cos x,$

5. $x^2\dfrac{d^2y}{dx^2}+3\dfrac{dy}{dx}+y=0,$

6. $(1-x^2)\dfrac{d^2y}{dx^2}-2y=2x.$

Answers

1. $\sqrt{1-x^2}\ y=1$

2. $y=x^2+C_1(1+2x+x^2)+C_2x^2e^{1/x}$, where $C_1=\dfrac{e^2-1}{e^2+5}, C_2=\dfrac{6e}{e^2+5}$

3. $y(1+x^2)=\log\sec x+C_1x+C_2$

4. $yx^2=\sin x-x\cos x+\dfrac{1}{2}C_1x^2+C_2$

5. $xy=C_1\log x+C_2$

6. $2y=(C_1+1)(1+x^2)\tan^{-1}x+(C_2-1)x+2C_2(1+x^2)$